Human and Artificial Intelligence

i

Human and Artificial Intelligence

Editors

**Alessandro Micarelli
Giuseppe Sansonetti
Giuseppe D'Aniello**

Basel • Beijing • Wuhan • Barcelona • Belgrade • Novi Sad • Cluj • Manchester

Editors
Alessandro Micarelli
Roma Tre University
Roma, Italy

Giuseppe Sansonetti
Roma Tre University
Roma, Italy

Giuseppe D'Aniello
University of Salerno
Fisciano, Italy

Editorial Office
MDPI
St. Alban-Anlage 66
4052 Basel, Switzerland

This is a reprint of articles from the Special Issue published online in the open access journal *Applied Sciences* (ISSN 2076-3417) (available at: https://www.mdpi.com/journal/applsci/special_issues/Human_Artificial_Intelligence).

For citation purposes, cite each article independently as indicated on the article page online and as indicated below:

Lastname, A.A.; Lastname, B.B. Article Title. *Journal Name* **Year**, *Volume Number*, Page Range.

ISBN 978-3-0365-9873-4 (Hbk)
ISBN 978-3-0365-9874-1 (PDF)
doi.org/10.3390/books978-3-0365-9874-1

© 2024 by the authors. Articles in this book are Open Access and distributed under the Creative Commons Attribution (CC BY) license. The book as a whole is distributed by MDPI under the terms and conditions of the Creative Commons Attribution-NonCommercial-NoDerivs (CC BY-NC-ND) license.

Contents

Giuseppe Sansonetti, Giuseppe D'Aniello and Alessandro Micarelli
Special Issue on Human and Artificial Intelligence
Reprinted from: *Appl. Sci.* **2023**, *13*, 5255, doi:10.3390/app13095255 1

Antoine Falconnet, Constantinos K. Coursaris, Joerg Beringer, Wietske Van Osch, Sylvain Sénécal and Pierre-Majorique Léger
Improving User Experience with Recommender Systems by Informing the Design of Recommendation Messages
Reprinted from: *Appl. Sci.* **2023**, *13*, 2706, doi:10.3390/app13042706 5

Jiaxin Lu, Hengnian Qi, Xiaoping Wu, Chu Zhang and Qizhe Tang
Research on Authentic Signature Identification Method Integrating Dynamic and Static Features
Reprinted from: *Appl. Sci.* **2022**, *12*, 9904, doi:10.3390/app12199904 25

Elham Motamedi, Francesco Barile and Marko Tkalčič
Prediction of Eudaimonic and Hedonic Orientation of Movie Watchers
Reprinted from: *Appl. Sci.* **2022**, *12*, 9500, doi:10.3390/app12199500 43

Dejian Guan and Wentao Zhao
Adversarial Detection Based on Inner-Class Adjusted Cosine Similarity
Reprinted from: *Appl. Sci.* **2022**, *12*, 9406, doi:10.3390/app12199406 61

Hongwei Zhang, Xiaojie Wang, Si Jiang and Xuefeng Li
Multi-Granularity Semantic Collaborative Reasoning Network for Visual Dialog
Reprinted from: *Appl. Sci.* **2022**, *12*, 8947, doi:10.3390/app12188947 77

Davide Brunetti, Cristina Gena and Fabiana Vernero
Smart Interactive Technologies in the Human-Centric Factory 5.0: A Survey
Reprinted from: *Appl. Sci.* **2022**, *12*, 7965, doi:10.3390/app12167965 95

Jungpil Shin, Md. Maniruzzaman, Yuta Uchida, Md. Al Mehedi Hasan, Akiko Megumi, Akiko Suzuki and Akira Yasumura
Important Features Selection and Classification of Adult and Child from Handwriting Using Machine Learning Methods
Reprinted from: *Appl. Sci.* **2022**, *12*, 5256, doi:10.3390/app12105256 125

Evangelos Karakolis, Panagiotis Kokkinakos and Dimitrios Askounis
Provider Fairness for Diversity and Coverage in Multi-Stakeholder Recommender Systems
Reprinted from: *Appl. Sci.* **2022**, *12*, 4984, doi:10.3390/app12104984 139

Lorenzo Spagnoli, Maria Francesca Morrone, Enrico Giampieri, Giulia Paolani, Miriam Santoro, Nico Curti, et al.
Outcome Prediction for SARS-CoV-2 Patients Using Machine Learning Modeling of Clinical, Radiological, and Radiomic Features Derived from Chest CT Images
Reprinted from: *Appl. Sci.* **2022**, *12*, 4493, doi:10.3390/app12094493 159

Jesús Bobadilla, Jorge Dueñas, Abraham Gutiérrez and Fernando Ortega
Deep Variational Embedding Representation on Neural Collaborative Filtering Recommender Systems
Reprinted from: *Appl. Sci.* **2022**, *12*, 4168, doi:10.3390/app12094168 173

Alessio Ferrato, Carla Limongelli, Mauro Mezzini and Giuseppe Sansonetti
Using Deep Learning for Collecting Data about Museum Visitor Behavior
Reprinted from: *Appl. Sci.* **2022**, *12*, 533, doi:10.3390/app12020533 . . . . **185**

Amani Braham, Maha Khemaja, Félix Buendía and Faiez Gargouri
A Hybrid Recommender System for HCI Design Pattern Recommendations
Reprinted from: *Appl. Sci.* **2021**, *11*, 10776, doi:10.3390/app112210776 . . . . **207**

Giorgio Maria Di Nunzio and Guglielmo Faggioli
A Study of a Gain Based Approach for Query Aspects in Recall Oriented Tasks
Reprinted from: *Appl. Sci.* **2021**, *11*, 9075, doi:10.3390/app11199075 . . . . **233**

Laurie Carmichael, Sara-Maude Poirier, Constantinos K. Coursaris, Pierre-Majorique Léger and Sylvain Sénécal
Users' Information Disclosure Behaviors during Interactions with Chatbots: The Effect of Information Disclosure Nudges
Reprinted from: *Appl. Sci.* **2022**, *12*, 12660, doi:10.3390/app122412660 . . . . **249**

Editorial

Special Issue on Human and Artificial Intelligence

Giuseppe Sansonetti [1,*], Giuseppe D'Aniello [2] and Alessandro Micarelli [1]

1 Department of Engineering, Roma Tre University, 00146 Rome, Italy; micarel@dia.uniroma3.it
2 Department of Information and Electrical Engineering and Applied Mathematics, University of Salerno, 84084 Fisciano, Italy; gidaniello@unisa.it
* Correspondence: gsansone@dia.uniroma3.it; Tel.: +39-06-5733-3220

Although tremendous advances have been made in recent years, many real-world problems still cannot be solved by machines alone. Hence, the integration between Human Intelligence and Artificial Intelligence (AI) is needed. However, several challenges make this integration complex. The aim of this Special Issue (SI) was to provide a large and varied collection of high-level contributions presenting novel approaches and solutions to address the above issues.

This Special Issue contains 14 papers (13 research papers and 1 review paper) that deal with various topics related to human–machine interactions and cooperation. Most of these works concern different aspects of recommender systems (RSs), which are among the most widespread decision support systems. The domains covered range from healthcare to movies and from biometrics to cultural heritage. However, there are also contributions on vocal assistants and smart interactive technologies. In detail, this Special Issue includes the following papers:

- Falconnet et al. [1] analyze an aspect relating to RSs not significantly explored to date; namely, the impact of the recommendation message design generated by the system on the user's beliefs and behavior about the system and its advice. Specifically, the authors propose a model to deeply analyze the effects of different presentation choices and discuss their possible implications.
- Bobadilla et al. [2] deal with another relevant aspect of RSs. They introduce a neural model to visually represent the relationships between users and items. This can be beneficial both to the company behind the RS to increase profit and to end users to receive explanations on a particular suggestion. The latter represents one of the main objectives of our Special Issue; namely, to facilitate cooperation and communication between humans and machines.
- Karakolis et al. [3] address a relevant problem in human–computer interaction; that is, how to obtain provider fairness in terms of user coverage and diversity in RSs considering not only the target user but all the stakeholders involved in the recommendation process. The solution proposed in the literature for this problem is in the form of an optimization problem under constraints, which in this case becomes an NP-Hard problem. The authors come up with a heuristic approach for its solution and review the formulation of the problem as proposed in the literature.
- Motamedi et al. [4] explore a personality aspect related to the users' motivations underlying their consumption of multimedia resources. They advance a machine learning-based model for predicting the eudaimonic or hedonic orientation of the target user. In the movie domain, this translates into predicting whether the user is more interested in meaningful topic content or entertainment content. This can provide a significant contribution to the realization of RSs capable of increasingly satisfying the interests and preferences of the active user.
- Ferrato et al. [5] propose an approach to gather information on the behavior of museum visitors. Data are collected using low-cost instrumentation and analyzed using convolutional neural networks. This information can be exploited by the museum

Citation: Sansonetti, G.; D'Aniello, G.; Micarelli, A. Special Issue on Human and Artificial Intelligence. *Appl. Sci.* **2023**, *13*, 5255. https://doi.org/10.3390/app13095255

Received: 18 April 2023
Accepted: 20 April 2023
Published: 23 April 2023

Copyright: © 2023 by the authors. Licensee MDPI, Basel, Switzerland. This article is an open access article distributed under the terms and conditions of the Creative Commons Attribution (CC BY) license (https://creativecommons.org/licenses/by/4.0/).

curators and staff to optimize the arrangement of the artworks and by the visitors themselves to receive suggestions of personalized itineraries based on their preferences and interests.
- Unlike some SI papers describing solutions to specific problems affecting RSs in general, Braham et al. [6] themselves propose a recommender system. Specifically, their RS exploits the combination of text-based and ontology-based methods to support developers in finding the most appropriate user interface design patterns for a given design problem. The authors also report the results of a user study showing how the testers appreciated the suggested design patterns. This study is also relevant to our Special Issue for the choice of the proposed RS domain, as the selection of the most appropriate user interface is fundamental for successful cooperation between humans and machines.
- Brunetti et al. [7] present a comprehensive and in-depth survey on the advances and potential advances in the field of smart interactive technologies. In particular, they analyze the aspects that characterize the Industry 4.0 and 5.0 visions. With their contribution, the authors highlight once again the importance of considering the human factor in the design and realization of intelligent systems, which is also the main objective of this Special Issue.
- Lu et al. [8] propose a handwriting identification method that exploits both the static features of traditional pen-and-paper writing and the dynamic features of digital writing. This method allows for classification and recognition through classic machine learning models and deep neural networks.
- Guan and Zhao [9] introduce a similarity metric to detect adversarial attacks that can undermine deep neural networks, thus limiting their application in security-critical fields. Such attacks are made by adversaries simply by adding imperceptible human perturbations to normal examples, which may prevent their correct classification and recognition. The experimental results show how adopting this metric can determine a higher ability to recognize adversarial attacks with respect to similar state-of-the-art approaches.
- Zhang et al. [10] deal with a human–computer interaction task known in the literature as visual dialog, in which an agent is trained to engage in a structured conversation on an image. They present an approach to collaboratively extract information related to questions by analyzing the dialog history through coding at different granularities. The experimental results of this study on public online datasets allow us to develop increasingly effective AI-based assistants capable, for example, of helping visually impaired people to understand the content of digital images.
- Carmichael et al. [11] explore another compelling aspect of human–computer interaction systems. They analyze the impact of information disclosure nudges on the behavior of chatbot users. Based on the results of a user study, they also propose ways to make users more aware of their disclosure behavior while interacting with chatbots.
- Shin et al. [12] introduce a machine learning-based model for classifying individuals as adults or children based on handwritten text and patterns collected via a pen tablet. This model first identifies the most predictive features through a sequential forward floating selection algorithm and then performs the classification process through random forest and support vector machines. The experimental results reported in the article demonstrate the reliability of the proposed model.
- Spagnoli et al. [13] propose an approach based on semi-automatic segmentation tools and machine learning algorithms to predict the severity of SARS-CoV-2 infection from the analysis of chest computed tomography images. Such a system can support physicians in identifying patients most at risk, demonstrating how the integration of Human Intelligence and Artificial Intelligence can improve the prognostic evaluation and treatment of patients affected by COVID-19 and similar diseases.
- In the context of decision support systems supporting healthcare, it is essential to retrieve all the relevant information to best assist users with medical information needs.

Di Nunzio and Faggioli [14] address this problem by proposing an intent-aware gain metric to be used to identify the most promising query reformulation during a search session in a Consumer Health Search system. These systems represent one of the most representative examples in which Human Intelligence and Artificial Intelligence must combine and complement each other to achieve the final goal.

In summary, each paper included in this Special Issue represents a step towards a future with human–machine interactions and cooperation. We hope the readers enjoy reading these articles and may find inspiration for their research activities.

Acknowledgments: We would like to sincerely thank all the authors and peer reviewers for their valuable contributions to this Special Issue "Human and Artificial Intelligence". We would also like to express our gratitude to all the staff and people involved in this Special Issue.

Conflicts of Interest: The authors declare no conflict of interest.

References

1. Falconnet, A.; Coursaris, C.K.; Beringer, J.; Van Osch, W.; Sénécal, S.; Léger, P.M. Improving User Experience with Recommender Systems by Informing the Design of Recommendation Messages. *Appl. Sci.* **2023**, *13*, 2706. [CrossRef]
2. Bobadilla, J.; Dueãs, J.; Gutiérrez, A.; Ortega, F. Deep Variational Embedding Representation on Neural Collaborative Filtering Recommender Systems. *Appl. Sci.* **2022**, *12*, 4168. [CrossRef]
3. Karakolis, E.; Kokkinakos, P.; Askounis, D. Provider Fairness for Diversity and Coverage in Multi-Stakeholder Recommender Systems. *Appl. Sci.* **2022**, *12*, 4984. [CrossRef]
4. Motamedi, E.; Barile, F.; Tkalčič, M. Prediction of Eudaimonic and Hedonic Orientation of Movie Watchers. *Appl. Sci.* **2022**, *12*, 9500. [CrossRef]
5. Ferrato, A.; Limongelli, C.; Mezzini, M.; Sansonetti, G. Using Deep Learning for Collecting Data about Museum Visitor Behavior. *Appl. Sci.* **2022**, *12*, 533. [CrossRef]
6. Braham, A.; Khemaja, M.; Buendía, F.; Gargouri, F. A Hybrid Recommender System for HCI Design Pattern Recommendations. *Appl. Sci.* **2021**, *11*, 776. [CrossRef]
7. Brunetti, D.; Gena, C.; Vernero, F. Smart Interactive Technologies in the Human-Centric Factory 5.0: A Survey. *Appl. Sci.* **2022**, *12*, 7965. [CrossRef]
8. Lu, J.; Qi, H.; Wu, X.; Zhang, C.; Tang, Q. Research on Authentic Signature Identification Method Integrating Dynamic and Static Features. *Appl. Sci.* **2022**, *12*, 9904. [CrossRef]
9. Guan, D.; Zhao, W. Adversarial Detection Based on Inner-Class Adjusted Cosine Similarity. *Appl. Sci.* **2022**, *12*, 9406. [CrossRef]
10. Zhang, H.; Wang, X.; Jiang, S.; Li, X. Multi-Granularity Semantic Collaborative Reasoning Network for Visual Dialog. *Appl. Sci.* **2022**, *12*, 8947. [CrossRef]
11. Carmichael, L.; Poirier, S.M.; Coursaris, C.K.; Léger, P.M.; Sénécal, S. Users' Information Disclosure Behaviors during Interactions with Chatbots: The Effect of Information Disclosure Nudges. *Appl. Sci.* **2022**, *12*, 2660. [CrossRef]
12. Shin, J.; Maniruzzaman, M.; Uchida, Y.; Hasan, M.A.M.; Megumi, A.; Suzuki, A.; Yasumura, A. Important Features Selection and Classification of Adult and Child from Handwriting Using Machine Learning Methods. *Appl. Sci.* **2022**, *12*, 5256. [CrossRef]
13. Spagnoli, L.; Morrone, M.F.; Giampieri, E.; Paolani, G.; Santoro, M.; Curti, N.; Coppola, F.; Ciccarese, F.; Vara, G.; Brandi, N.; et al. Outcome Prediction for SARS-CoV-2 Patients Using Machine Learning Modeling of Clinical, Radiological, and Radiomic Features Derived from Chest CT Images. *Appl. Sci.* **2022**, *12*, 4493. [CrossRef]
14. Di Nunzio, G.M.; Faggioli, G. A Study of a Gain Based Approach for Query Aspects in Recall Oriented Tasks. *Appl. Sci.* **2021**, *11*, 9075. [CrossRef]

Disclaimer/Publisher's Note: The statements, opinions and data contained in all publications are solely those of the individual author(s) and contributor(s) and not of MDPI and/or the editor(s). MDPI and/or the editor(s) disclaim responsibility for any injury to people or property resulting from any ideas, methods, instructions or products referred to in the content.

Article

Improving User Experience with Recommender Systems by Informing the Design of Recommendation Messages

Antoine Falconnet [1], Constantinos K. Coursaris [1,*], Joerg Beringer [2], Wietske Van Osch [1], Sylvain Sénécal [3] and Pierre-Majorique Léger [1]

[1] Department of Information Technologies, HEC Montréal, Montréal, QC H3T 2A7, Canada
[2] Blue Yonder, 76149 Karlsruhe, Germany
[3] Department of Marketing, HEC Montréal, Montréal, QC H3T 2A7, Canada
* Correspondence: constantinos.coursaris@hec.ca

Abstract: Advice-giving systems such as decision support systems and recommender systems (RS) utilize algorithms to provide users with decision support by generating 'advice' ranging from tailored alerts for situational exception events to product recommendations based on preferences. Related extant research of user perceptions and behaviors has predominantly taken a system-level view, whereas limited attention has been given to the impact of message design on recommendation acceptance and system use intentions. Here, a comprehensive model was developed and tested to explore the presentation choices (i.e., recommendation message characteristics) that influenced users' confidence in—and likely acceptance of—recommendations generated by the RS. Our findings indicate that the problem and solution-related information specificity of the recommendation increase both user intention and the actual acceptance of recommendations while decreasing the decision-making time; a shorter decision-making time was also observed when the recommendation was structured in a problem-to-solution sequence. Finally, information specificity was correlated with information sufficiency and transparency, confirming prior research with support for the links between user beliefs, user attitudes, and behavioral intentions. Implications for theory and practice are also discussed.

Keywords: human–computer Interaction; user experience; decision making; recommender systems; use intention; recommendation acceptance; message design; information specificity; information sufficiency; decision-making time

1. Introduction

With the massive increase in available data in recent years, many techniques and technologies have emerged to help businesses, workers, and customers process such data more efficiently. Various recommender systems (RSs) have been developed since the mid-1990s, and many of these systems have been applied to a variety of tasks [1–3]. RSs are interactive systems that must be designed according to human-centered principles. Human–computer interaction (HCI) is the science domain that informs and validates design with respect to user interaction with such a system. In the particular case of an RS, a key design objective is to assure users understand and accept its recommendations. A large corpus of design- and development-focused RS papers exists (c.f. *Recommender Systems Handbook* by Ricci, Rokach, and Shapira, 2015 [4]).

Among RSs, we can distinguish two popular types of systems: content-based recommendation and collaborative recommendation. There is also an increasing number of hybrid recommender systems that combine different types of RSs. Other systems have appeared in previous years, such as demographic recommendation, utility-based recommendation, and knowledge-based recommendation, but they are less frequent. Differences between these RSs include the elicitation techniques (i.e., how to collect data and user

preferences), the recommendation generation algorithm, and the presentation of the recommendation (i.e., text message, image, video, sound, or a combination of these four items). In this study, we focused on an RS that produced recommendations in the form of a text message.

While RSs have received significant attention in recent years, it is important to observe that they are just one of several 'advice-giving systems'. "These include expert systems, knowledge-based systems, decision support systems, and recommender systems." [5] (p. 2). However, given the relative obscurity of the term 'advice-giving systems' and the much more frequently used term of 'recommender systems'—even for systems that support user decisions where user preferences are not the core element in the enabling machine-learning algorithm—we use the latter term of recommender systems hereafter to represent systems that generate and present recommendation messages to users.

Due to the considerable opportunities and challenges in many domains (e.g., business, government, education, and healthcare), numerous studies have been conducted on RSs [1–3], especially on the comprehension of their performance [6], their design implications [7–9], and recommendation techniques [3,10]. Thus, prior research has significantly addressed design implications at the system level [8,9]; however, researchers and scientists have mostly disregarded the design of the interface [11]. In the rare instances where the extant literature has focused on RS interface design, recommendations are produced and put forth following the collection of a user's interaction data and subsequently juxtaposed against the attributes of artifacts stored in large repositories. Yet, to date, the impact of recommendation message presentation on the user's perception, attitude, and behavior towards the recommendation has been significantly understudied.

A system's design alone will not shape its users' perceptions of trust and the likelihood of them accepting a recommendation. The nature of the message content is also likely to play a significant role in affecting users' beliefs, attitudes, and behaviors, particularly in the context of managerial decision making [9]. Therefore, as content elements and interaction elements (e.g., buttons) jointly comprise the processing of recommendations by the user, a simultaneous and granular analysis of the effects of both design elements is required. Hence, this study aims to advance the contemporary understanding of RS designs by exploring different ways to optimize the information presentation and/or interaction layers of the user–RS interactions. Related factors influence the adoption of AI in real-life contexts, which is an important aspect of successfully deploying AI. In this study, we investigated the understanding and acceptance of system-generated recommendations, which is a very prominent usage pattern of AI. Specifically, we explored presentation approaches for recommendation messages to increase the likelihood that RS users would trust and accept system-generated recommendations with minimal effort required. Specifically, our work aimed to answer the following research questions:

1. What is the effect of message design (characteristics) on a user's beliefs about system-generated recommendations?
2. What is the effect of message design (characteristics) on a user's beliefs regarding the ease of use and usefulness of the RS?
3. What is the effect of message design (characteristics) on a user's attitudes and behavioral intentions toward the RS and its recommendations?
4. What is the effect of message design (characteristics) on a user's behaviors with respect to decision-making time and the likelihood of accepting system-generated recommendations?

2. Theoretical Background

2.1. Recommender Systems

Extensive research has been performed evaluating RSs in their entirety [8,9,12,13]. RSs have progressed technologically to include machine learning and multi-modal interaction elements (e.g., Apple's Siri, Amazon's Alexa, etc.). Despite this technological progress and extensive research on RS user experience as a whole, a fundamental investigation into the optimal construction of recommendation messages has not yet been comprehensively conducted, as summarized in the following conceptual piece on the state of RS-related research:

"Explanations can vary, for example, with respect to (i) their length; (ii) the adopted vocabulary if natural language is used; (iii) the presentation format, and so on. When explanation forms are compared in user studies that are entirely different in these respects, it is impossible to understand how these details impact the results. Therefore, more studies are required to investigate the impact of these variables" [5] (p. 425). Hence, to create a more stable foundation for RS researchers and designers, additional studies on the fine-grained presentation details of recommendation messages are required [5].

2.2. Message Design in Recommender Systems

A typical recommendation message contains two core components, i.e., a described problem and a suggested solution, which is a frequent rhetorical pattern used in technical academic writing [14]. For example, in the message "I noticed that you are running out of soft drinks. Shall I order more?", the first sentence is the problem while the second is the solution. Within the problem and solution construct of recommendation messages, several elements can vary in form, including information specificity—which can relate to either the problem and/or the solution—information sequence, message styling, and situational complexity; these elements are defined below.

Problem specificity and solution specificity are motivated by the functional principle of conveying information in a clear manner [15] and the notion that people have a preference for descriptions with a higher level of detail [16]. Moreover, the accuracy of the recommendation positively affects the decision-making process preceding the uptake of the recommendation [17], while the diversity of recommendations influences user trust, leading to an increase in the adoption rate of recommendations [18]. Information sequence—i.e., presenting the problem then the solution or the solution then the problem—is motivated by extant healthcare literature that indicates merit for both sequences in health communication messages [19].

Situational complexity consists of "simple, technically complicated, socially complicated, and complex situations" [20] and is inversely related to the amount of information available [21]; that is, situational complexity arises when there is uncertainty about the available options in the specified context and how the available options intermingle with cognitive demand due to tensions between contradictory elements [22].

Lastly, the styling of text in recommendation messages (e.g., font-weight properties, such as boldface) may also affect users' perceptions, according to an empirical study on perceived professionalism in scientific question-and-answer forums [23]. Although text styling was not observed to have an impact in [23], the authors urged for the continued examination of typographical cues (i.e., boldface, italics, and underline) in other applications. For RSs, where decision-making time is critical, styling cues, such as bolding text, could help users focus on the most pertinent information at hand. This study extends prior research [9] by taking a mixed-methods approach to explore the effects of message design (characteristics) on user experience with both the presented information and the RS.

2.3. Hypotheses Development

Our hypotheses built on the ResQue (recommender systems' quality of user experience) model [9] and were partitioned into the following sets of endogenous variables: user beliefs regarding the recommendation message (information sufficiency and transparency) and the system (perceived usefulness and ease of use) as well as user attitudes toward both the recommendation message and system (see Figure 1) and behavioral outcomes (see Figure 2). The reason for the distinction between the self-reported model and the behavioral model is that the former explored the user's experience with each RS message design explicitly (via the self-reported measures) while the latter did so implicitly (via the behavioral measures). Therefore, the two models provided a complementary and more holistic view of user experience, offering a stronger foundation for emergent implications and recommendations on how to design RS messages. In the following sections, we will present our hypotheses for each set of dependent variables.

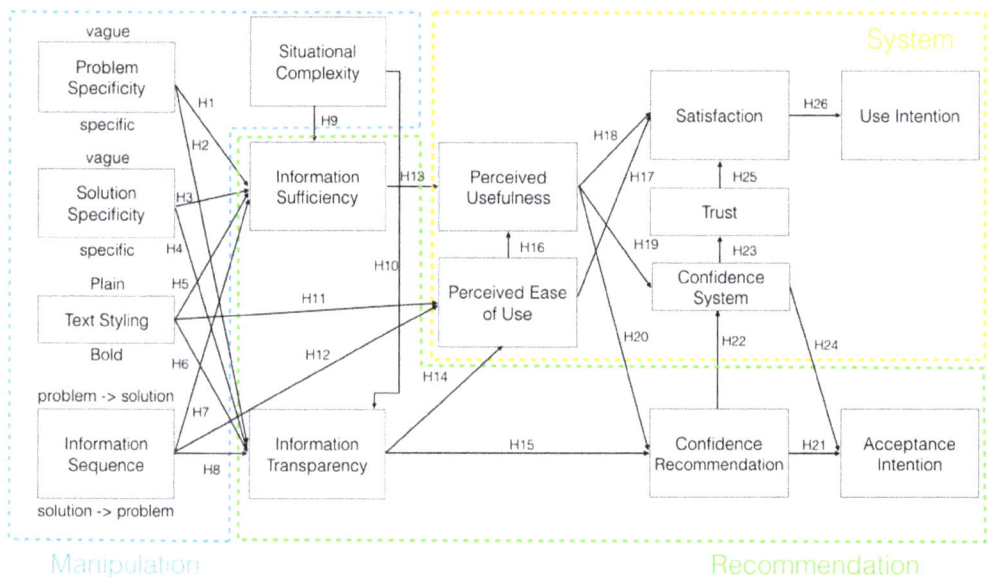

Figure 1. Proposed self-reported research model.

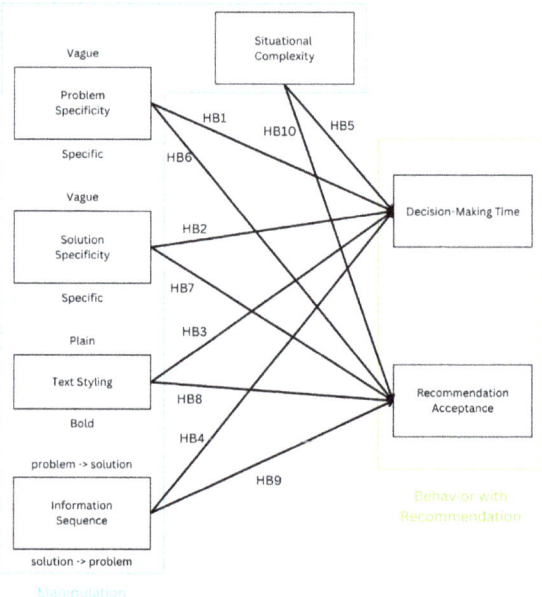

Figure 2. Proposed behavioral research model.

2.3.1. Information Transparency and Information Sufficiency

Information transparency is an aggregate user assessment of three dimensions: clarity, disclosure, and accuracy [24]. For a recommendation message consisting of two parts of information, a problem and solution, changes in problem and solution specificity are likely to yield changes in user perceptions of the clarity, disclosure, and accuracy of the information in the recommendation message, and by extension perceived information transparency as a whole. Moreover, researchers recommend that RS designers should simplify the reading of the message by the user by using navigational efficacy, design familiarity, and attractiveness [25], which will lead to clarity and transparency [9,25,26]. Indeed, the content and format of the recommendations have significant and varied impacts on users' evaluations of RS [27]. Linking the latter to message characteristics, it is plausible that user perceptions of information transparency will be positively affected by (i) changing the message (text) styling by bolding key parts of the message (i.e., bolding the object being discussed); (ii) sequencing the information presentation as problem-then-solution (rather than the reverse); and (iii) communicating simple rather than complex situations.

Thus, the following effects of a message's characteristics were hypothesized:

Hypothesis 2 (H2). *Problem specificity positively impacts information transparency.*

Hypothesis 4 (H4). *Solution specificity positively impacts information transparency.*

Hypothesis 6 (H6). *Text styling positively impacts information transparency, such that styled (bold) text is associated with greater perceived information transparency than plain text.*

Hypothesis 8 (H8). *Information sequence affects information sufficiency, such that a problem-to-solution sequence positively impacts information transparency.*

Hypothesis 10 (H10). *Situational complexity negatively impacts information transparency.*

Information sufficiency refers to whether the amount of content presented to the user is enough for the user to understand the information, and in some cases to act on it [28]. By changing the degree of information specificity (i.e., problem and/or solution) and situational complexity, the amount of information available, and the way the information is conveyed to the user, user perceptions of information are likely to be augmented [20–22]. To ensure that information is clear and easy to access, which will improve information sufficiency, the information should be structured and adapted to the needs of the receivers [26,29]. Explanations in a recommendation have an important impact on the user's behavior. Indeed, the type of explanation has a direct effect on RS use and should be different according to the desired effect [30]. For example, persuasive explanation supports the competence facet of the RS [30], while negative arguments increase the user's perceived honesty of the system regarding the recommendation [31]. Moreover, fact-based explanations (i.e., only facts with keywords) and argumentative explanations are preferred by users over full-sentence explanations [32]. The literature recommends proposing recommendations with only pertinent and cogent information [33] and excluding all information and knowledge that are not relevant for answering the request [33]. Long and strongly confident explanations can also be used to increase user acceptance [34]. Thus, choices regarding message (text) styling and information sequence may make it easier for the user to understand the presented information, and a change in information sufficiency may also be observed [25].

Thus, the following effects of a message's characteristics were hypothesized:

Hypothesis 1 (H1). *Problem specificity positively impacts information sufficiency.*

Hypothesis 3 (H3). *Solution specificity positively impacts information sufficiency.*

Hypothesis 5 (H5). *Text styling positively impacts information sufficiency, such that styled (bold) text is associated with greater perceived information sufficiency than plain text.*

Hypothesis 7 (H7). *Information sequence affects information sufficiency, such that a problem-to-solution sequence positively impacts information sufficiency.*

Hypothesis 9 (H9). *Situational complexity negatively impacts information sufficiency.*

2.3.2. Perceived Usefulness and Ease of Use

Users' attitudes are affected by their beliefs regarding a message's information properties [9]. For example, information sufficiency has been shown to impact the perceived usefulness of RSs [9,35]. In addition, the sufficiency of the information may depend on its quality [36–39], where the greater the quality of the information presented, the more useful it is found to be [40]. Furthermore, knowledgeable explanations significantly increase the perceived usefulness of an RS [41]. Thus, the following effects of a recommendation message's characteristics were hypothesized:

Hypothesis 11 (H11). *Text styling positively affects perceived usefulness, such that recommendation messages with styled (bold) text are associated with a greater perceived usefulness of the RS than plain text.*

Hypothesis 12 (H12). *Information sequence affects the perceived ease of use, such that a problem-to-solution sequence positively impacts the perceived ease of use of the RS.*

Hypothesis 13 (H13). *Information sufficiency positively impacts the perceived usefulness of the RS.*

Hypothesis 14 (H14). *Information transparency positively impacts the perceived ease of use of the RS.*

2.3.3. System and Recommendation Outcomes

For an RS to be successful with respect to its adoption, users should have confidence in the system-generated recommendations and trust the system [5]. Transparency plays an important role in users' confidence in recommendations as it may encourage or deter users' trust in a system [42,43], where recommendations perceived as transparent by users increase their confidence [44]. User perceptions of ease of use and usefulness are positively related to each other [12,45–47] and to system attitudes, such as those toward the system's use and system satisfaction [12,40,46–48]. Moreover, explanations contribute to user trust in RSs [30,49,50]. Confidence in the system positively influences trust in the system [51] and the user's behavioral intentions with the system, including their intention to accept a system-generated recommendation [40]; in addition, the degree of trust users put in the system plays an important role in the acceptance of a recommendation [52]. In the context of I.S. use, trust has been shown to positively affect satisfaction [53], which in turn has been shown to encourage use of the system [54,55]. Hence, the following hypotheses were proposed:

Hypothesis 15 (H15). *Information transparency positively impacts recommendation confidence.*

Hypothesis 16 (H16). *RS ease of use positively impacts its usefulness.*

Hypothesis 17 (H17). *RS ease of use positively impacts user satisfaction.*

Hypothesis 18 (H18). *RS usefulness positively impacts user satisfaction.*

Hypothesis 19 (H19). *RS usefulness positively impacts RS trust.*

Hypothesis 20 (H20). *RS usefulness positively impacts RS confidence.*

Hypothesis 21 (H21). *Recommendation confidence positively impacts recommendation acceptance intentions.*

Hypothesis 22 (H22). *Recommendation confidence positively impacts RS confidence.*

Hypothesis 23 (H23). *RS confidence positively impacts RS trust.*

Hypothesis 24 (H24). *RS confidence positively impacts recommendation acceptance intentions.*

Hypothesis 25 (H25). *RS trust positively impacts RS satisfaction.*

Hypothesis 26 (H26). *RS satisfaction positively impacts RS use intentions.*

2.3.4. Behavioral Outcomes

Both the content and format of a recommendation may influence users' beliefs and attitudes [8,9], and in turn affect behavioral intentions [5,8,9,12] and actual behaviors. Providing explanations for recommendations may lead to faster decision making by users and drive them to make better choices [56,57]. In addition, the way messages convey the problems faced and related information may impact users' perceptions and decisions [58–60]. Thus, the following hypotheses were proposed:

Hypothesis B1 (HB1). *Problem specificity negatively impacts (i.e., decreases) users' decision-making time.*

Hypothesis B2 (HB2). *Solution specificity negatively impacts user's decision-making time.*

Hypothesis B3 (HB3). *Text styling affects users' decision-making time, such that styled (bolded) text is associated with a shorter decision-making time than plain text.*

Hypothesis B4 (HB4). *Information sequence affects users' decision-making time, such that problem-to-solution sequencing is associated with a greater decrease in time than solution-to-problem sequencing.*

Hypothesis B5 (HB5). *Situational complexity positively impacts (i.e., increases) users' decision-making time.*

Hypothesis B6 (HB6). *Problem specificity positively impacts users' recommendation acceptance rate.*

Hypothesis B7 (HB7). *Solution specificity positively impacts users' recommendation acceptance rate.*

Hypothesis B8 (HB8). *Text styling affects users' recommendation acceptance, such that styled (bolded) text is associated with a greater recommendation acceptance rate than plain text.*

Hypothesis B9 (HB9). *Information sequence affects users' recommendation acceptance rate, such that problem-to-solution recommendations are associated with a greater acceptance rate.*

Hypothesis B10 (HB10). *Situational complexity negatively impacts users' recommendation acceptance rate.*

3. Methodology

3.1. Pilot Study

A pilot study [61] was conducted with three aims: (i) to gauge the appropriateness of the stimuli, (ii) to collect attentional and psychophysiological data to inform the main experiment, and (iii) to offer preliminary support for the hypothesized relationships. An experiment utilizing a within-subjects research design involving four factors, each with two levels (i.e., 2 × 2 × 2 × 2), was conducted. The pilot study involved fewer factors (4 instead of 5) and by extension fewer conditions (16 vs. 32) and stimuli (48 vs. 96), as well as fewer participants ($n = 6$ vs. $n = 614$) than the study presented below, which also used a different data collection approach (lab-based pilot vs. Amazon MTurk).

3.2. Experimental Design

A multi-method experiment was conducted employing a counterbalanced mixed (between-within) subjects design involving five (5) factors (i.e., counterbalanced 2 × 2 × 2 × 2 × 2 for a total of 32 conditions), tested using three (3) stimuli per condition (i.e., 96 stimuli). The factors involved (i) information sequence (problem-to-solution vs. solution-to-problem); information specificity, comprising (ii) problem specificity (vague vs. specific) and (iii) solution specificity (vague vs. specific); (iv) text styling (plain vs. bold); and (v) situation complexity (low vs. high). For the latter, complexity was varied by manipulating the product type involved in the situation—soft drinks vs. meat patties—given the following 'complicating' considerations: cost (low-cost soft drinks vs. high-cost meat patties), durability (non-perishable soft drinks vs. perishable meat patties), handling (soft drink bottles require no special handling vs. meat patties require special—refrigerated—handling). Hence, the situations with soft drinks were overall of lower complexity than those pertaining to meat patties. Additionally, we manipulated the product's availability (available vs. unavailable with available substitute product). The factor manipulations are illustrated in the sample recommendation messages in Table 1. Moreover, in order to reduce the time needed to complete a session, the study was divided into 8 groups, each comprising 4 of the 32 conditions (corresponding to 12 stimuli per participant), which required approximately 15 min to complete.

Table 1. Experimental manipulation of independent variables.

Factor	Respective Levels: Vague/Vague/Plain/Solution-to-Problem/High Complexity	Respective Levels: Specific/Specific/Bold/Problem-to-Solution/Low Complexity
Problem Specificity	I noticed that you are running low on soft drinks. I recommend ordering 20 6-packs of Coca-Cola bottles today. Shall I proceed with the order?	I noticed that you only have 10 bottles of Coca-Cola in stock. I recommend ordering 20 6-packs of Coca-Cola bottles today. Shall I proceed with the order?
Solution Specificity	I noticed that you ordered 10 6-packs of Coca-Cola bottles, but the product is no longer available from your supplier. I recommend substituting the missing product. Shall I proceed with the substitution?	I noticed that you ordered 10 6-packs of Coca-Cola bottles, but the product is no longer available from your supplier. I recommend substituting Coca-Cola with Pepsi. Shall I proceed with the substitution?
Text Styling	I recommend ordering 20 6-packs of Coca-Cola bottles today because I noticed that you only have 10 bottles of Coca-Cola left in stock. Shall I proceed with the order?	I recommend **ordering 20 6-packs of Coca-Cola bottles** today because I noticed that you only have **10 bottles of Coca-Cola left** in stock. Shall I proceed with the order?
Information Sequence	I recommend ordering 100 ground beef patties today because I noticed that you only have 10 ground beef patties left in stock. Shall I proceed with the order?	I noticed that you only have 10 ground beef patties in stock. I recommend ordering 100 ground beef patties today. Shall I proceed with the order?
Situation Complexity	I noticed that you ordered 100 ground beef patties, but the product is no longer available from your supplier. I recommend substituting ground beef patties with ground veal patties. Shall I proceed with the recommendation?	I noticed that you only have 10 bottles of Coca-Cola in stock. I recommend ordering 20 6-packs of Coca-Cola bottles today. Shall I proceed with the order?

3.3. Participants

Participants were recruited on the Amazon Mechanical Turk (MTurk) online platform. To participate in the study, these "Turkers" were screened for a minimum HIT approval rate of 90% (a human intelligence task, or HIT, is a question that needs an answer); in addition, they had to be located in the U.S. The participants were only allowed to complete a single session. Recruiting a minimum of 100 participants for each of the eight groups (i.e., per 4 conditions) resulted in a total of 843 people being recruited to our study, of which 614 yielded valid responses that were used for subsequent analysis (with a minimum of 70 responses per group). A total of 229 responses were not used, as these participants either failed the attention check ($n = 207$) or were unable to confirm their participation ($n = 22$). Participants were compensated USD 1.40 for their time.

3.4. Experimental Procedure, Stimuli, and Measurement

The experiment involved a scenario where participants assumed the role of a restaurant manager in charge of inventory and were required to make logistics decisions regarding inventory replenishment and/or order delivery rerouting based on the recommendations proposed by the RS. The RS messages themselves varied in their presentation according to the abovementioned five factors that were manipulated. Successive text-only messages showing situations (i.e., a problem and an RS-recommended solution) were used as stimuli. Two buttons ("ACCEPT" and "DETAILS") corresponding to the two decision options available to users were shown below each message (see Figure 3). Participants had to either confirm the recommendation as-is if they felt that the recommendation was appropriate for the shown situation or request additional details if they felt otherwise. The details themselves were not shown to the participant (which was indicated to them in the instructions), as doing so would introduce additional factors to the study beyond the scope of our research questions. Participants entered their choice using a keyboard and were not able to navigate backwards.

I noticed that you only have 10 bottles of Coca-Cola left in stock. I recommend ordering additional soft drinks. Shall I proceed with the order?

Figure 3. Example of study stimulus.

This multi-method experiment used the simulated recommendation messages as stimuli to evoke a reaction from the study participants, which was quantitatively measured via self-reports (survey questions) and behavioral data. Two behavioral measures were collected automatically by the experimental platform (described further below) at this point: (i) the time taken to decide (i.e., the time from stimulus exposure to choice entry) and (ii) the choice entered. After each choice, the participants were asked three (3) questions regarding the recommendation message to measure the perceived sufficiency and transparency of the information and the users' confidence in the recommendation. Information sufficiency and information transparency are both related to the quality of the information presented by the RS. After evaluating three (3) consecutive recommendation messages, seven (7) questions were asked regarding RS-related perceptions, including ease of use, usefulness, confidence, and trust, as well as participants' satisfaction with the RS and their intentions to use the RS and/or accept RS-generated recommendations (see Figure 4 for an example). The questionnaire consisted of single-item scales adapted from previously validated scales [9,12] that were used to quantitatively measure the constructs reflected in the proposed research model. Answers were provided along a 7-point Likert scale from extremely disagree (1) to extremely agree (7). To respond, participants could either click on the scale or enter the corresponding number using the keyboard. Constructs were measured through the use of adapted (reduced) single-item constructs [9], a choice that was made given the significant duration and thus cognitive burden of the experiment, as shown in Table 2.

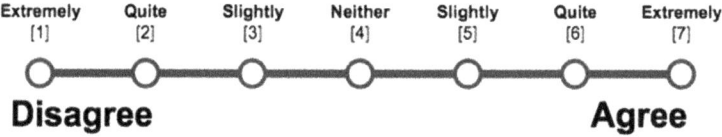

Figure 4. Example of Likert-scale question.

Table 2. Measurement items (self-reported).

	Recommendation Message		Recommender System
Sufficiency	The information provided was sufficient for me to make a decision to accept the recommendation	Ease of Use	The recommender system was easy to use
		Usefulness	The system gave me good recommendations
Transparency	I understood why this recommendation was made to me	Confidence	I am convinced of the suggestions recommended to me by the system
Confidence	I am convinced of the recommendation made to me	Trust	The recommender system can be trusted
Intention to accept recommendation	I would accept the next recommendation	Satisfaction	I am satisfied with the recommender system
		Use intention	I would use this recommender system again

3.5. Apparatus

Three web-based systems were used to conduct this study. The first was CognitionLib (BeriSoft, Inc., Redwood City, CA, USA), which is a free open-source community for ERTS Scripts that provides an online editor and is currently used by hundreds of academic institutions to create cognitive task paradigms and set up cognitive experiments. Using the ERTS language, we were able to code all the requisite elements for the experiment (as black-and-white to control for the effect of color), including the stimuli in the form of text messages, the survey questions, and the response scales. When the scripts were coded, they were imported into Cognition Lab (BeriSoft, Inc., Redwood City, CA. USA), a web-based runtime environment that hosts experiments. The third platform used was Amazon's MTurk (Amazon Inc., Bellevue, WA, USA), a crowdsourcing marketplace that connects businesses to individuals who can perform their tasks virtually, from which the participants were recruited.

4. Analysis and Results

The data were analyzed using methods appropriate for the variable types involved, as follows:

(1) For the self-reported model, we used cumulative logistic regression with random intercept for modeling the probability of having lower values. We used cumulative logistic regression because we treated the dependent variables as ordinal variables.
(2) For the behavioral model, we used an approach appropriate for the type of dependent variable as follows:
   (a) For decision-making time, we used linear regression with a random intercept because the distribution of time was roughly normal;
   (b) For recommendation acceptance, we used logistic regression with a random intercept because the behavioral decision (to accept or request details) was binary.

In all three cases, we used a random intercept model to account for the repeated-measures design; more precisely, the model allowed the intercept to vary by participant to account for the unmeasured participant-specific characteristics that were not correlated with the independent variable but had an effect on the DV (but were not measured). In the following sections, the results from the analyses corresponding to each of the study's three research questions are presented.

4.1. RQ1. What Is the Effect of Message Design (Characteristics) on a User's Beliefs about System-Generated Recommendations?

Information specificity impacted information sufficiency (problem specificity effect H1: $b = 0.3039$, $p < 0.0001$; solution specificity effect H3: $b = 0.3714$, $p < 0.0001$) and information transparency (problem specificity effect H2: $b = 0.1814$, $p < 0.0001$; solution specificity effect H4: $b = 0.1499$, $p < 0.0011$). On the other hand, the remaining hypotheses corresponding to RQ1 were not supported, i.e., those regarding the effect of text styling on information sufficiency (H5: $b = 0.0254$, $p = 0.8754$) and information transparency (H6: $b = 0.1729$, $p = 0.2697$) and the effect of information sequence on information sufficiency (H7: $b = -0.0357$, $p = 0.6756$) and information transparency (H8: $b = 0.0608$, $p = 0.4865$); in addition, situational complexity was not observed to have a significant effect on information sufficiency (H9: $b = 0.0672$, $p = 0.6804$) or information transparency (H10: $b = 0.191$, $p = 0.2258$).

4.2. RQ2. What Is the Effect of Message Design (Characteristics) on a User's Beliefs Regarding the Ease of Use and Usefulness of the RS?

The relationship between text styling (bolding) and perceived ease of use was not supported (H11: $b = 0.1327$, $p = 0.5836$), and information sequence was not found to significantly impact perceived ease of use (H12: $b = 0.0486$, $p = 0.7519$). On the other hand, the effects of information sufficiency on usefulness (H13: $b = 1.6193$, $p < 0.0001$) and information transparency on ease of use (H14: $b = 1.3461$, $p < 0.0001$) were shown to be significant.

4.3. RQ3. What Is the Effect of Message Design (Characteristics) on a User's Attitudes and Behavioral Intentions toward the RS and Its Recommendations?

Information transparency positively impacted recommendation confidence (H15: $b = 1.0743$, $p < 0.0001$), which was also affected by RS usefulness (H20: $b = 0.9111$, $p < 0.0001$) and in turn positively influenced intention to accept the recommendation (H21: $b = 1.715$, $p < 0.0001$); in addition, this intention was also affected by RS confidence (H24: $b = 1.5625$, $p < 0.0001$). The ease of use of the system positively impacted both system usefulness (H16: $b = 0.993$, $p < 0.0001$) and system satisfaction (H17: $b = 1.2254$, $p < 0.0001$). Usefulness positively impacted system satisfaction (H18: $b = 1.7252$, $p < 0.0001$) and system confidence (H19: $b = 1.5487$, $p < 0.0001$). Recommendation confidence had a positive effect on system confidence (H22: $b = 1.8567$, $p < 0.0001$), which positively impacted system trust (H23: $b = 1.402$, $p < 0.0001$) and finally system satisfaction (H25: $b = 1.7122$, $p < 0.0001$). System satisfaction positively impacted the intention to use the recommender system (H26: $b = 1.5809$, $p < 0.0001$). Lastly, all system-level mediating constructs demonstrated a good explanation of the variance in their respective DVs, including 45.2% for recommendation acceptance intention and 49.2% for system use intention, as shown in the validated model (see Figure 5a and Table 3).

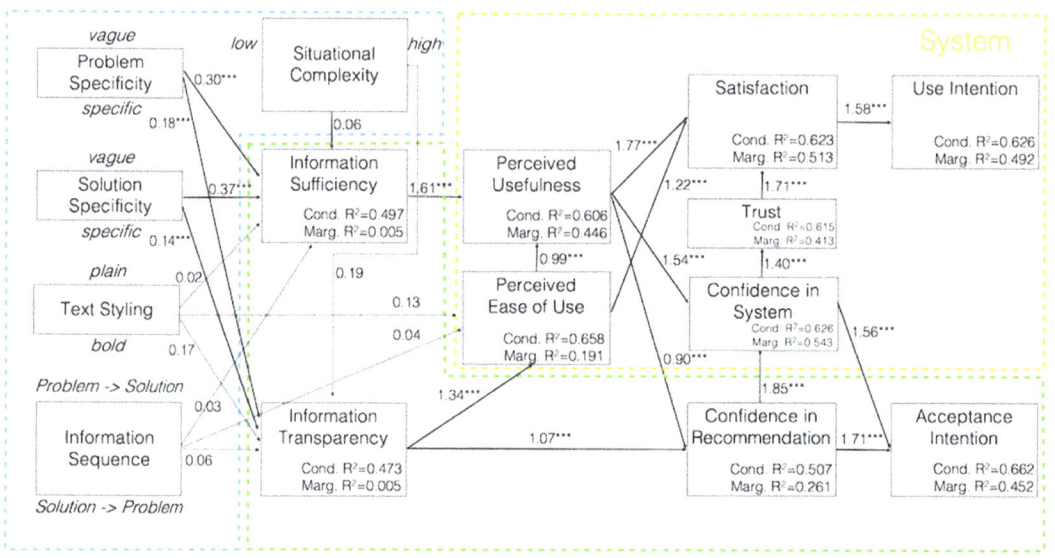

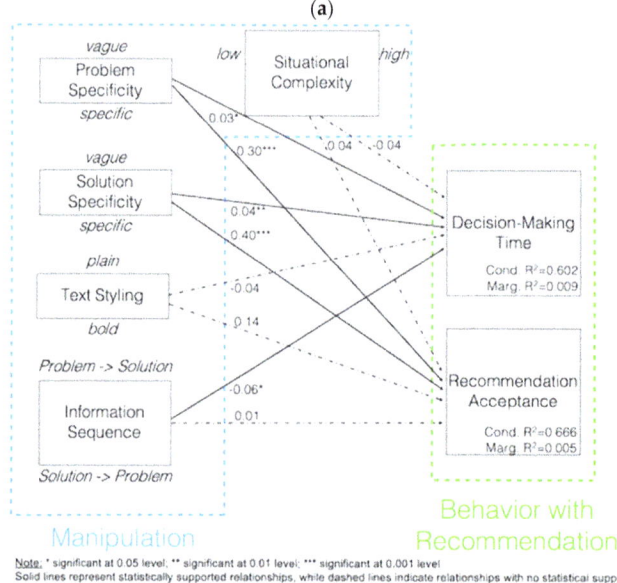

Figure 5. (a) Validated research model (self-reported data); (b) Validated research model (behavioral data).

Table 3. Summary table of all the hypotheses and their results.

Hypothesis	Dependent Variable	Independent Variable	Result	Estimate
H1	Problem specificity	Information sufficiency	Supported	0.30 ***
H2	Problem specificity	Information transparency	Supported	0.18 ***
H3	Solution specificity	Information sufficiency	Supported	0.37 ***
H4	Solution specificity	Information transparency	Supported	0.14 ***
H5	Text styling	Information sufficiency	Not supported	0.02
H6	Text styling	Information transparency	Not supported	0.17
H7	Information sequence	Information sufficiency	Not supported	0.03
H8	Information sequence	Information transparency	Not supported	0.06
H9	Situational complexity	Information sufficiency	Not supported	0.06
H10	Situational complexity	Information transparency	Not supported	0.19
H11	Text styling	Perceived ease of use	Not supported	0.13
H12	Information sequence	Perceived ease of use	Not supported	0.04
H13	Information sufficiency	Perceived usefulness	Supported	1.61 ***
H14	Information transparency	Perceived ease of use	Supported	1.34 ***
H15	Information transparency	Confidence recommendation	Supported	1.07 ***
H16	Perceived ease of use	Perceived usefulness	Supported	0.99 ***
H17	Perceived ease of use	Satisfaction	Supported	1.22 ***
H18	Perceived usefulness	Satisfaction	Supported	1.77 ***
H19	Perceived usefulness	Confidence system	Supported	1.54 ***
H20	Perceived usefulness	Confidence recommendation	Supported	0.91 ***
H21	Confidence recommendation	Acceptance intention	Supported	1.72 ***
H22	Confidence recommendation	Confidence system	Supported	1.86 ***
H23	Confidence system	Trust	Supported	1.40 ***
H24	Confidence system	Acceptance intention	Supported	1.56 ***
H25	Trust	Satisfaction	Supported	1.71 ***
H26	Satisfaction	Use intention	Supported	1.58 ***
HB1	Problem specificity	Decision-making time	Supported	0.03 *
HB2	Solution specificity	Decision-making time	Supported	0.04 **
HB3	Text styling	Decision-making time	Not supported	−0.04
HB4	Information sequence	Decision-making time	Partially supported	−0.04(All); −0.06 * (*Accept*); −0.02 (*Details*)
HB5	Situational complexity	Decision-making time	Not supported	0.02
HB6	Problem specificity	Recommendation acceptance	Supported	0.30 ***
HB7	Solution specificity	Recommendation acceptance	Supported	0.40 ***
HB8	Text styling	Recommendation acceptance	Not supported	0.14
HB9	Information sequence	Recommendation acceptance	Not Supported	0.01
HB10	Situational complexity	Recommendation acceptance	Not supported	0.04

Note: * significant at 0.05 level; ** significant at 0.01 level; *** significant at 0.001 level.

4.4. RQ4. What Is the Effect of Message DESIGN (characteristics) on a User's Behaviors with Respect to Decision-Making Time and the Likelihood of Accepting System-Generated Recommendations?

Problem specificity had a significant positive effect on decision-making time (HB1: b = 0.0337, $p < 0.05$), as did solution specificity (HB2: b = 0.03746, $p < 0.01$). In contrast, the use of a problem-to-solution information sequence did not have a significant effect on decision-making time for all messages (b = −0.03805, $p = 0.1667$); however, when looking at the time-to-decision by type of decision, it was found that information sequence significantly reduced decision-making time for accepted recommendations (HB4: b = 0.06418, $p < 0.05$) but not for those recommendations for which users requested more details (b = 0.01624, $p = 0.7185$. On the other hand, text styling and situational complexity did not impact decision-making time (HB3: b = 0.04046, $p = 0.4545$ and HB5: b = 0.02226, $p = 0.6831$).

In addition to the observed effect of information specificity on decision-making time, it was also observed that users were significantly more likely to accept the recommendation if the messages stated specific (rather than vague) problems (HB6: b = 0.3006, $p < 0.0001$) and specific (rather than vague) solutions (HB7: b = 0.4073; $p < 0.0001$). However, no effect on user behavior with respect to recommendation acceptance was observed for information sequence (HB8: b = 0.1392, $p = 0.2674$), text styling (HB8: b = 0.1392, $p = 0.2674$), and situational complexity (HB10: b = 0.03508, $p = 0.7815$).

All behavioral results are shown in Figure 5b and reported in Table 3. Lastly, a post hoc analysis further reinforced the favorable effect of information specificity: specific recommendation messages were significantly longer ($p < 0.05$ all cases) in character count, for both problem- and solution-specificity, whether counted with spaces or without spaces, yet the decision-making time on a per character basis was significantly lower ($p < 0.001$ in all cases).

5. Discussion and Conclusions

This empirical study investigated the effects of message design on user behaviors with respect to the likelihood of accepting system-generated recommendations and the time taken to decide, as well as user attitudes toward the recommendation and the RS. Our findings provide clear, evidence-based answers to the four research questions that initially motivated the study. The answers should be of interest to academic researchers, designers of RSs, and current or potential providers as well as users of RSs; however, they are also generalizable to other use contexts and domains.

5.1. Contributions to Research

Prior research on RS interface design has focused mostly on the system level. Thus, the impact of recommendation message design on the user's interaction with and behavior toward the recommendation and the RS more broadly has not been studied. This study attempted to extend the literature by identifying new factors of message design that may impact the likelihood of users accepting system-generated recommendations as well as their intention to use the recommender system. Five factors were manipulated including (i) information sequence (problem-to-solution vs. solution-to-problem); information specificity, comprising (ii) problem specificity (vague vs. specific) and (iii) solution specificity (vague vs. specific); (iv) situation complexity (simple vs. complex); and (v) text styling (plain vs. bold). Our findings revealed that the specificity of the information embedded in the recommendation message had a positive influence on users' perceptions of information sufficiency and information transparency (i.e., information sufficiency and information transparency were considered to be higher when the information regarding either the problem and/or the solution was specific rather than vague). Information specificity also impacted the users' behaviors. Higher information specificity increased the likelihood that users would accept the recommendation (from the measured acceptance intention), and it offered the additional benefit of the user needing less time to make the decision. Our findings also revealed that information sequence did not have an impact on either user perceptions of information transparency or information sufficiency and rather was found to reduce decision-making time when the information was presented in a problem-to-solution sequence. Neither text styling nor situation complexity were found to affect information transparency or information sufficiency. Focusing on the impact of user beliefs (i.e., perceived usefulness and ease of use) and attitudes (i.e., confidence in the recommendation and the RS, as well as trust and satisfaction with the RS), the study confirmed that users' beliefs influence their attitudes, which motivate their intentions to accept the recommendation and use the RS.

5.2. Contributions to Practice

The results of this study have improved our understanding of the impact of message design on users' intention to accept an algorithmically generated recommendation proposed to them and their intention to use the RS that provided the recommendation. This advances our contemporary understanding of RS designs more broadly by looking at ways to optimize the information presentation and/or interaction layers of the user–RS interactions. For example, e-tailers such as Amazon rely on algorithms to observe patterns and identify the optimal candidates to recommend for subsequent purchases to each consumer/user; however, the way that product recommendation is presented will also impact the likelihood of the consumer/user clicking through the recommended product link and potentially proceeding with the purchase of an algorithmically generated recommendation. Similarly, in a decision-support context, a supply chain management (SCM) AI-enabled system presents exception events to its operators/users; how the emergent problem and the recommended solution are presented to a user will impact their ability to process the situational information and to make a decision, thus impacting the efficacy and use of the system.

The findings of this study can inform RS designers in how to construct (design) system-generated recommendation messages in order to either minimize the time needed by the user to make a decision and/or to accept the recommendation or the system at large, for example, if the RS is used during a demo or trial period. Our results showed that information sequence helped to reduce decision-making time. Indeed, when the problem was presented before the solution, i.e., in a situational frame, users experienced less cognitive load and spent less time making a decision than when they were shown recommendations consisting of the solution followed by the problem. Hence, a best practice emerges for RS designers to structure the content of recommendation messages in a manner progressing from framing the situation (problem) to presenting the supporting information and concluding with the recommended action (solution).

However, extending from the scenario and focus of this study on new RS users, it is plausible that as user trust in the system generated recommendations increases over time (i.e., during the continued use of the RS), users may eventually prefer to quickly review the recommendation and approve it without processing the underlying problem and supporting information. In this special use case, the reverse sequence (i.e., solution-to-problem) may be preferred by the user; if so, a second recommendation might allow for the user to specify the recommendation message's construction. This would be feasible if at the system level, such recommendations are generated not as event-to-outcome rules, where recommendations are designed in full a priori for each exception event alert specifically, but are instead generated by combining message elements that are marked-up or tagged according to a library identifying each content element by its property (e.g., product name, product quantity, exception event, delivery mode, etc.) and synthesized according to the exception event. However, before formally putting forth such a recommendation, additional research is required to obtain support for this anticipated utility.

Another recommendation emerging from this study's findings is for RS designers to focus on explainability by embedding sufficient detail regarding both the problem and the solution, thereby boosting perceptions of information transparency and information sufficiency. Such perceptions contribute to a significantly more frequent acceptance of system-generated recommendations, thereby saving users time as they would not need to delve deeper (e.g., by clicking on 'details') before making a decision.

The above recommendations when implemented can serve as catalysts for both the adoption and continued use of systems that generate recommendations algorithmically for users to consider in their context. By focusing on the design of interfaces that address user needs, designers can then derive implied user requirements for interactive systems (here, RSs) and ensure that the system design is satisfactory. Consequently, users are more likely to perceive them as useful and will therefore be more likely to adopt and use them. This is especially relevant to systems that are used in corporate settings or mandated by managers, which is likely to be the case with RSs and decision-support systems.

5.3. Limitations and Opportunities for Future Research

Despite the scenario used in this study and the validation of the two levels of situational complexity through a manipulation check prior to their use, situational complexity was not found to have any effects on either user beliefs or behaviors. This was an unexpected finding as the factor and its two levels were tested through two rounds of manipulation checks involving (i) nine participants prior to the start of this study, who reported unanimously that the high-complexity situation was indeed more complex than the low-complexity situation based on the information presented in the recommendation messages, and (ii) thirty participants responding to the question "How challenging was the situation you were faced with?" (simple = 1 ... 5 = complex). It is plausible that while the high-complexity stimuli were indeed significantly higher in complexity than the low-complexity stimuli, given the brevity of the messages, the complexity was not sufficiently high to induce significantly higher levels of cognitive load (and by extension, time needed), more negative emotions (either valence or arousal), and/or worse beliefs and attitudes. Another possible explanation for the insignificance of situational complexity could be the likely lack of domain knowledge on the side of participants. Nuances in complexity of messages pertaining to product perishability and ease of product storage might require knowledge that is likely to be unique to those with significant experience in the restaurant or catering industry. Hence, we have two sets of recommendations for future research. First, future research could focus on assessing situational complexity where the role of domain-specific knowledge might be less significant. Second, future research that aims to inform RS design according to situational complexity should first explore for situational complexity 'thresholds', above which effects are observed in regard to users' cognition, emotion, or behavior, and design stimuli accordingly.

While this study undertook an investigation of an extended research model, observing the variance extracted from the mediating and dependent variables, additional explorations of the effects of additional factors on message- and system-level outcomes are needed. Other factors that can be identified in the literature and could be tested in future studies include message detail, message length, the use of subjective versus objective language, personification, affective language use, personality, and vague language [16,62–68].

Building upon results from this experiment, future work should also involve triangulated attentional and physiological measurements to gain a richer understanding of the mechanisms at play. Using eye fixation-related potential [69], future research could explore the cognitive mechanism involved in user decision making at the moment of the recommendation's consideration. Lastly, future research should explore potential interactions between message characteristics, e.g., revisiting situational complexity, such user perceptions may in fact be the outcomes of interactions with specificity and information sequence as they are inherently dimensions that add to or reduce message complexity, revealing that situational complexity may be a second-order construct comprised of various message characteristics that add to complexity. Finally, an extended study could also be performed with the three significant factors used here (i.e., problem information specificity, solution information specificity, and information sequence) to find the best combination for the optimal presentation of recommendation messages.

Author Contributions: Conceptualization and methodology, all authors; writing—original draft preparation, A.F. and C.K.C.; writing—review and editing, J.B., W.V.O., S.S. and P.-M.L.; supervision, C.K.C. and P.-M.L.; funding acquisition, C.K.C. and P.-M.L. All authors have read and agreed to the published version of the manuscript.

Funding: This research was funded by Blue Yonder (IRCPJ/514835-16) and UX Chair (IRC 505259-16).

Institutional Review Board Statement: The study was conducted in accordance with the Declaration of Helsinki and approved by the Ethics Review Board of (Redacted) (2020-3866, 28 February 2020).

Informed Consent Statement: Informed consent was obtained from all subjects involved in the study.

Data Availability Statement: The data presented in this study are available on request from the corresponding author. The data are not publicly available due to privacy considerations and adherence to the ethics review board approved protocol.

Conflicts of Interest: J.B. is an employee of the funding organization; however, the funding board of the organization had no role in writing the manuscript the design of the study; in the collection, analyses, or interpretation of data; in the writing of the manuscript; and the decision to publish the results was governed by the general framework of the research grant. All other authors declare no conflict of interest.

References

1. Candillier, L.; Jack, K.; Fessant, F.; Meyer, F. State-of-the-art recommender systems. In *Collaborative and Social Information Retrieval and Access: Techniques for Improved User Modeling*; IGI Global: Pennsylvania, PA, USA, 2009; pp. 1–22.
2. Lops, P.; de Gemmis, M.; Semeraro, G. Content-based recommender systems: State of the art and trends. In *Recommender Systems Handbook*; Springer: Berlin/Heidelberg, Germany, 2011; pp. 73–105.
3. Lu, J.; Wu, D.; Mao, M.; Wang, W.; Zhang, G. Recommender system application developments: A survey. *Decis. Support Syst.* **2015**, *74*, 12–32. [CrossRef]
4. Ricci, F.; Rokach, L.; Shapira, B. (Eds.) *Recommender Systems Handbook*; Springer: New York, NY, USA, 2015.
5. Nunes, I.; Jannach, D. A systematic review and taxonomy of explanations in decision support and recommender systems. *User Model. User-Adapt. Interact.* **2017**, *27*, 393–444. [CrossRef]
6. Sarwar, B.; Karypis, G.; Konstan, J.; Riedl, J. Application of Dimensionality Reduction in Recommender System—A Case Study. Minnesota Univ Minneapolis Dept of Computer Science, 2000. Available online: https://apps.dtic.mil/sti/citations/ADA439541 (accessed on 1 June 2022).
7. Shani, G.; Heckerman, D.; Brafman, R.I. An MDP-based recommender system. *J. Mach. Learn. Res.* **2005**, *6*, 1265–1295.
8. Xiao, B.; Benbasat, I. E-commerce product recommendation agents: Use, characteristics, and impact. *MIS Q.* **2007**, *31*, 137–209. [CrossRef]
9. Pu, P.; Chen, L.; Hu, R. A user-centric evaluation framework for recommender systems. In Proceedings of the Fifth ACM Conference on Recommender Systems, Chicago, IL, USA, 23–27 October 2011; pp. 157–164.
10. Sharma, L.; Gera, A. A survey of recommendation system: Research challenges. *Int. J. Eng. Trends Technol.* **2013**, *4*, 1989–1992.
11. Bigras, É.; Léger, P.M.; Sénécal, S. Recommendation Agent Adoption: How Recommendation Presentation Influences Employees' Perceptions, Behaviors, and Decision Quality. *Appl. Sci.* **2019**, *9*, 4244. [CrossRef]
12. Davis, F.D. Perceived usefulness, perceived ease of use, and user acceptance of information technology. *MIS Q.* **1989**, *13*, 319–340. [CrossRef]
13. Kirakowski, J.; Corbett, M. SUMI: The Software Usability Measurement Inventory. *Br. J. Educ. Technol.* **1993**, *24*, 210–212. [CrossRef]
14. Flowerdew, L. A Combined Corpus and Systemic-Functional Analysis of the Problem-Solution Pattern in a Student and Professional Corpus of Technical Writing. *TESOL Q.* **2003**, *37*, 489. [CrossRef]
15. Pettersson, R. Introduction to Message Design. *J. Vis. Lit.* **2012**, *31*, 93–104. [CrossRef]
16. Schnabel, T.; Bennett, P.N.; Joachims, T. Improving recommender systems beyond the algorithm. *arXiv* **2018**, arXiv:1802.07578.
17. Gunawardana, A.; Shani, G. Evaluating recommender systems. In *Recommender Systems Handbook*; Springer: Boston, MA, USA, 2015; pp. 265–308.
18. Panniello, U.; Gorgoglione, M.; Tuzhilin, A. Research Note—In CARSs We Trust: How Context-Aware Recommendations Affect Customers' Trust and Other Business Performance Measures of Recommender Systems. *Inf. Syst. Res.* **2016**, *27*, 182–196. [CrossRef]
19. Keller, P.A. Regulatory Focus and Efficacy of Health Messages. *J. Consum. Res.* **2006**, *33*, 109–114. [CrossRef]
20. Chazdon, S.; Grant, S. Situational Complexity and the Perception of Credible Evidence. *J. Hum. Sci. Ext.* **2019**, *7*, 4. [CrossRef]

21. Kosnes, L.; Pothos, E.M.; Tapper, K. Increased affective influence: Situational complexity or deliberation time? *Am. J. Psychol.* **2010**, *123*, 29–38. [CrossRef]
22. Johns, G. The Essential Impact of Context on Organizational Behavior. *Acad. Manag. Rev.* **2006**, *31*, 386–408. [CrossRef]
23. Zhang, Y.; Lu, T.; Phang, C.W.; Zhang, C. Scientific Knowledge Communication in Online Q&A Communities: Linguistic Devices as a Tool to Increase the Popularity and Perceived Professionalism of Knowledge Contributions. *J. Assoc. Inf. Syst.* **2019**, *20*, 1129–1173. [CrossRef]
24. Schnackenberg, A.K.; Tomlinson, E.C. Organizational transparency: A new perspective on managing trust in organization-stakeholder relationships. *J. Manag.* **2016**, *42*, 1784–1810. [CrossRef]
25. Yoo, K.H.; Gretzel, U.; Zanker, M. *Persuasive Recommender Systems: Conceptual Background and Implications*; Springer Science & Business Media: Berlin/Heidelberg, Germany, 2012.
26. Pettersson, R. Information Design Theories. *J. Vis. Lit.* **2014**, *33*, 1–96. [CrossRef]
27. Mandl, M.; Felfernig, A.; Teppan, E.; Schubert, M. Consumer decision making in knowledge-based recommendation. *J. Intell. Inf. Syst.* **2011**, *37*, 1–22. [CrossRef]
28. Ozok, A.A.; Fan, Q.; Norcio, A.F. Design guidelines for effective recommender system interfaces based on a usability criteria conceptual model: Results from a college student population. *Behav. Inf. Technol.* **2010**, *29*, 57–83. [CrossRef]
29. Pettersson, R. Information design–principles and guidelines. *J. Vis. Lit.* **2010**, *29*, 167–182. [CrossRef]
30. Holliday, D.; Wilson, S.; Stumpf, S. User trust in intelligent systems: A journey over time. In Proceedings of the 21st International Conference on Intelligent User Interfaces, Sonoma, CA, USA, 7–10 March 2016; pp. 164–168.
31. Lamche, B.; Adıgüzel, U.; Wörndl, W. Interactive explanations in mobile shopping recommender systems. In Proceedings of the 8th ACM Conference on Recommender Systems, Foster City, CA, USA, 6–10 October 2014; Volume 14.
32. Zanker, M.; Schoberegger, M. An empirical study on the persuasiveness of fact-based explanations for recommender systems. In Proceedings of the 8th ACM Conference on Recommender Systems, Foster City, CA, USA, 6–10 October 2014; Volume 1253, pp. 33–36.
33. Kunkel, J.; Donkers, T.; Michael, L.; Barbu, C.M.; Ziegler, J. Let Me Explain: Impact of Personal and Impersonal Explanations on Trust in Recommender Systems. In Proceedings of the 2019 CHI Conference on Human Factors in Computing Systems, Glasgow, UK, 4–9 May 2019; pp. 1–12.
34. Al-Taie, M.Z.; Kadry, S. Visualization of Explanations in Recommender Systems. *J. Adv. Manag. Sci.* **2014**, *2*, 140–144. [CrossRef]
35. Al-Jabri, I.M.; Roztocki, N. Adoption of ERP systems: Does information transparency matter? *Telemat. Inform.* **2015**, *32*, 300–310. [CrossRef]
36. Coursaris, C.K.; Van Osch, W.; Albini, A. Antecedents and consequents of information usefulness in user-generated online reviews: A multi-group moderation analysis of review valence. *AIS Trans. Hum.-Comput. Interact.* **2018**, *10*, 1–25. [CrossRef]
37. Cheung, C.M.K.; Lee, M.K.O.; Rabjohn, N. The impact of electronic word-of-mouth: The adoption of online opinions in online customer communities. *Internet Res.* **2008**, *18*, 229–247. [CrossRef]
38. Petty, R.E.; Cacioppo, J.T. *Communication and Persuasion: Central and Peripheral Routes to Attitude Change*; Springer: New York, NY, USA, 1986.
39. Shu, M.; Scott, N. Influence of Social Media on Chinese Students' Choice of an Overseas Study Destination: An Information Adoption Model Perspective. *J. Travel Tour. Mark.* **2014**, *31*, 286–302. [CrossRef]
40. Coursaris, C.K.; Van Osch, W.; Nah, F.F.-H.; Tan, C.-H. Exploring the effects of source credibility on information adoption on YouTube. In *International Conference on HCI in Business, Government, and Organizations*; Springer International Publishing: Berlin/Heidelberg, Germany, 2016; pp. 16–25. [CrossRef]
41. Tintarev, N.; Masthoff, J. Designing and Evaluating Explanations for Recommender Systems. In *Recommender Systems Handbook*; Springer US: Boston, MA, USA, 2010; pp. 479–510. [CrossRef]
42. Pu, P.; Chen, L. Trust building with explanation interfaces. In Proceedings of the 11th International Conference on Intelligent user Interfaces, Sydney, Australia, 29 January–1 February 2006; pp. 93–100.
43. Kizilcec, R.F. How much information? Effects of transparency on trust in an algorithmic interface. In Proceedings of the 2016 CHI Conference on Human Factors in Computing Systems, San Jose, CA, USA, 7–12 May 2016; pp. 2390–2395.
44. Sinha, R.; Swearingen, K. The role of transparency in recommender systems. In *CHI'02 Extended abstracts on Human Factors in Computing Systems*; Association for Computing Machinery: New York, NY, USA, 2002; pp. 830–831.
45. Davis, F.D. A Technology Acceptance Model for Empirically Testing New End-User Information Systems: Theory and Results. Doctoral Dissertation, Massachusetts Institute of Technology, Cambridge, MA, USA, 1985.
46. Calisir, F.; Calisir, F. The relation of interface usability characteristics, perceived usefulness, and perceived ease of use to end-user satisfaction with enterprise resource planning (ERP) systems. *Comput. Hum. Behav.* **2004**, *20*, 505–515. [CrossRef]
47. Amin, M.; Rezaei, S.; Abolghasemi, M. User satisfaction with mobile websites: The impact of perceived usefulness (PU), perceived ease of use (PEOU) and trust. *Nankai Bus. Rev. Int.* **2014**, *5*, 258–274. [CrossRef]
48. Joo, Y.J.; Lim, K.Y.; Kim, E.K. Online university students' satisfaction and persistence: Examining perceived level of presence, usefulness and ease of use as predictors in a structural model. *Comput. Educ.* **2011**, *57*, 1654–1664. [CrossRef]
49. Berkovsky, S.; Taib, R.; Conway, D. How to recommend? User trust factors in movie recommender systems. In Proceedings of the 22nd International Conference on Intelligent User Interfaces, Limassol, Cyprus, 13–16 March 2017; pp. 287–300.

50. Sharma, A.; Cosley, D. Do social explanations work? Studying and modeling the effects of social explanations in recommender systems. In Proceedings of the 22nd International Conference on World Wide Web, Rio de Janeiro, Brazil, 13–17 May 2013; pp. 1133–1144.
51. McGuirl, J.M.; Sarter, N.B. Supporting trust calibration and the effective use of decision aids by presenting dynamic system confidence information. *Hum. Factors J. Hum. Factors Ergon. Soc.* **2006**, *48*, 656–665. [CrossRef] [PubMed]
52. Jameson, A.; Willemsen, M.C.; Felfernig, A.; de Gemmis, M.; Lops, P.; Semeraro, G.; Chen, L. Human Decision Making and Recommender Systems. In *Recommender Systems Handbook*; Springer: Boston, MA, USA, 2015; pp. 611–648. [CrossRef]
53. Jarvenpaa, S.L.; Shaw, T.R.; Staples, D.S. Toward Contextualized Theories of Trust: The Role of Trust in Global Virtual Teams. *Inf. Syst. Res.* **2004**, *15*, 250–267. [CrossRef]
54. Coursaris, C.K.; Hassanein, K.; Head, M.; Bontis, N. The impact of distractions on the usability and the adoption of mobile devices for wireless data services. In Proceedings of the European Conference on Information Systems; 2007. Available online: https://aisel.aisnet.org/ecis2007/28 (accessed on 1 June 2022).
55. Coursaris, C.K.; Hassanein, K.; Head, M.M.; Bontis, N. The impact of distractions on the usability and intention to use mobile devices for wireless data services. *Comput. Hum. Behav.* **2012**, *28*, 1439–1449. [CrossRef]
56. Tintarev, N.; Masthoff, J. A Survey of Explanations in Recommender Systems. In Proceedings of the 2007 IEEE 23rd International Conference on Data Engineering Workshop, Istanbul, Turkey, 17–20 April 2007; pp. 801–810. [CrossRef]
57. Gedikli, F.; Jannach, D.; Ge, M. How should I explain? A comparison of different explanation types for recommender systems. *Int. J. Hum.-Comput. Stud.* **2014**, *72*, 367–382. [CrossRef]
58. Chan, S.H.; Song, Q. Motivational framework: Insights into decision support system use and decision performance. In *Decision Support Systems*; InTech: Rijeka, Ceoatia, 2010; pp. 1–24.
59. Kahneman, D.; Tversky, A. Prospect Theory: An Analysis of Decision under Risk. *Econometrica* **1979**, *47*, 263. [CrossRef]
60. Roy, M.C.; Lerch, F.J. Overcoming Ineffective Mental Representations in Base-Rate Problems. *Inf. Syst. Res.* **1996**, *7*, 233–247. [CrossRef]
61. Falconnet, A.; Van Osch, W.; Chen, S.L.; Beringer, J.; Fredette, M.; Sénécal, S.; Léger, P.M.; Coursaris, C.K. Beyond System Design: The Impact of Message Design on Recommendation Acceptance. In *Information Systems and Neuroscience: NeurosIS Retreat 2020*; Springer International Publishing: Berlin/Heidelberg, Germany, 2020.
62. Reiter, E. Natural Language Generation Challenges for Explainable AI. In Proceedings of the 1st Workshop on Interactive Natural Language Technology for Explainable Artificial Intelligence (NL4XAI 2019), Tokyo, Japan, 29 October 2019.
63. Adomavicius, G.; Bockstedt, J.; Curley, S.; Zhang, J. Reducing recommender systems biases: An investigation of rating display designs. *Forthcom. MIS Q.* **2019**, *43*, 1321–1341.
64. Schreiner, M.; Fischer, T.; Riedl, R. Impact of content characteristics and emotion on behavioral engagement in social media: Literature review and research agenda. *Electron. Commer. Res.* **2019**, *21*, 329–345. [CrossRef]
65. Rzepka, C.; Berger, B. User Interaction with AI-Enabled Systems: A Systematic Review of IS Research. 2018. Available online: https://www.researchgate.net/profile/Benedikt-Berger-2/publication/329269262_User_Interaction_with_AI-enabled_Systems_A_Systematic_Review_of_IS_Research/links/5bffb55392851c63cab02730/User-Interaction-with-AI-enabled-Systems-A-Systematic-Review-of-IS-Research.pdf (accessed on 1 June 2022).
66. Li, H.; Chatterjee, S.; Turetken, O. Information Technology Enabled Persuasion: An Experimental Investigation of the Role of Communication Channel, Strategy and Affect. *AIS Trans. Hum.-Comput. Interact.* **2017**, *9*, 281–300. [CrossRef]
67. Matsui, T.; Yamada, S. The effect of subjective speech on product recommendation virtual agent. In Proceedings of the 24th International Conference on Intelligent User Interfaces: Companion, Marina del Ray, CA, USA, 16–20 March 2019; pp. 109–110.
68. Chattaraman, V.; Kwon, W.-S.; Gilbert, J.E.; Ross, K. Should AI-Based, conversational digital assistants employ social- or task-oriented interaction style? A task-competency and reciprocity perspective for older adults. *Comput. Hum. Behav.* **2018**, *90*, 315–330. [CrossRef]
69. Léger, P.M.; Sénécal, S.; Courtemanche, F.; de Guinea, A.O.; Titah, R.; Fredette, M.; Labonte-LeMoyne, É. Precision is in the eye of the beholder: Application of eye fixation-related potentials to information systems research. *Assoc. Inf. Syst.* **2014**. [CrossRef]

Disclaimer/Publisher's Note: The statements, opinions and data contained in all publications are solely those of the individual author(s) and contributor(s) and not of MDPI and/or the editor(s). MDPI and/or the editor(s) disclaim responsibility for any injury to people or property resulting from any ideas, methods, instructions or products referred to in the content.

Article

Research on Authentic Signature Identification Method Integrating Dynamic and Static Features

Jiaxin Lu [1,2], Hengnian Qi [1,2,*], Xiaoping Wu [1,2], Chu Zhang [1,2] and Qizhe Tang [1,2]

1. School of Information Engineering, Huzhou University, Huzhou 313000, China
2. Zhejiang Province Key Laboratory of Smart Management & Application of Modern Agricultural Resources, Huzhou University, Huzhou 313000, China
* Correspondence: qhn@zjhu.edu.cn

Featured Application: This study focuses on fusing the static features of traditional pen-and-paper writing with the dynamic features of digital writing, seeking more understandable features for precise signature identification.

Abstract: In many fields of social life, such as justice, finance, communication and so on, signatures are used for identity recognition. The increasingly convenient and extensive application of technology increases the opportunity for forged signatures. How to effectively identify a forged signature is still a challenge to be tackled by research. Offline static handwriting has a unique structure and strong interpretability, while online handwriting contains dynamic information, such as timing and pressure. Therefore, this paper proposes an authentic signature identification method, integrating dynamic and static features. The dynamic data and structural style of the signature are extracted by dot matrix pen technology, the global and local features, time and space features are fused and clearer and understandable features are applied to signature identification. At the same time, the classification of a forged signature is more detailed according to the characteristics of signature and a variety of machine learning models and a deep learning network structure are used for classification and recognition. When the number of classifications is 5, it is better to identify simple forgery signatures. When the classification number is 15, the accuracy rate is mostly about 96.7% and the highest accuracy reaches 100% on CNN. This paper focuses on feature extraction, incorporates the advantages of dynamic and static features and improves the classification accuracy of signature identification.

Keywords: signature identification; dynamic characteristics; static characteristics

Citation: Lu, J.; Qi, H.; Wu, X.; Zhang, C.; Tang, Q. Research on Authentic Signature Identification Method Integrating Dynamic and Static Features. *Appl. Sci.* **2022**, *12*, 9904. https://doi.org/10.3390/app12199904

Academic Editors: Alessandro Micarelli, Giuseppe Sansonetti and Giuseppe D'Aniello

Received: 22 August 2022
Accepted: 26 September 2022
Published: 1 October 2022

Publisher's Note: MDPI stays neutral with regard to jurisdictional claims in published maps and institutional affiliations.

Copyright: © 2022 by the authors. Licensee MDPI, Basel, Switzerland. This article is an open access article distributed under the terms and conditions of the Creative Commons Attribution (CC BY) license (https://creativecommons.org/licenses/by/4.0/).

1. Introduction

Biometric recognition is based on human characteristics and signatures are considered one of the most common biological features [1,2]. The active mode of handwriting is widely associated with signature identification in biometric user authentication systems [3]. In a sense, handwriting is a behavioral manifestation of human thought, especially signatures, which have unique characteristics and strong personal style color [4]. Signature identification is required for office approval in corporate units, signing in cell phone business offices and banks, corroboration in the judicial industry and identification in examination scenarios. With the further development of information technology, the increasing popularity of handwritten signature acquisition devices and the maturity of digital writing technology have led to the replacement of pen-and-paper writing in the traditional sense. In the process of signature verification, dynamic features are the trend and static structure is the basis. If more comprehensive, simple and accurate methods appear, they will have a profound impact on various industries. However, as signature verification and identification systems are often used for forgery and fraud detection [5], the emergence of forged signatures complicates simple programs and even causes huge losses.

Currently, there are two types of signature identification: offline and online. Offline handwriting identification materials use traditional writing tools to write handwriting information on paper, which is then captured as a picture by a camera or scanner [6]. The features extracted from offline images can be combined to form a variety of effective features with uniqueness that cannot be ignored. Online signatures are obtained by signing on touch screen devices, such as tablets and cell phones, and many features are obtained by using a special pen and tablet and a scanned signature image [7]. Online handwriting recognition can be performed by collecting rich information, such as writing speed, angle, strength used by writers and stroke order online [8]. The online data are very clear, captured on a digital device, consisting of a discrete number of samples [9] and contain some additional supporting information [10].

With the popularity of paperless scenarios, online signature verification is widely used in various fields [11]. Electronic signatures are influenced by writing carriers and writing tools, resulting in many handwriting feature changes [12]. Handwriting is especially important from the perspective of handwriting verification but relying on handwriting signatures alone also loses some important features.

This study uses a dot matrix pen tool to identify handwriting by combining the static features of traditional pen-and-paper writing with the dynamic features of digitized writing. The dot matrix digital pen is a writing tool that captures the pen's motion track of the pen through the high-speed camera at the front, obtains the pen tip pressure data by a pressure sensor and transmits the dynamic information of the writing process through coordination and pressure changes simultaneously. After preprocessing the written dynamic data and static image information, the easily understood structured static features and fine dynamic features are extracted and then the training and test sets are divided for each subtask and different models are used to study the fused features for classification and discrimination.

The paper is organized as follows: Section 2 describes the work related to this study. Section 3 describes materials and methods. Section 4 shows the results of the study. Section 5 is the discussion. Section 6 provides a summary.

2. Related Works

Handwriting identification is based on human handwriting to determine the identity of the writer [13]. Offline signature verification is more practical than online signature verification because it is more popular and its structural information is more intuitive to reflect the characteristics of the writer. The online signature verification mode is more robust than the offline signature verification mode because it captures the dynamic information of the signature in real time and is not easy for impersonators to copy [14].

There are three types of signature forgery: simple, random and skilled. In the case of simple forgery, the forger knows the name information of the signer, but does not know the real signature of the signer. In the case of random forgery, the forger knows the name of the signer or one of the real signatures. In the case of skilled forgery, both the signer's name information and real signature information are known to the forger and the forger often practices imitating the signature of the signer [15].

The features used in the identification method can be divided into global features, local features, statistical features and dynamic features. For Chinese offline handwriting, Qingwu Li et al. [16] generated handwriting feature sets to identify handwriting samples by extracting curvature features of the stroke skeleton in four directions: horizontal, vertical, apostrophe and down. The samples were divided into reference handwriting and query handwriting. The similarity measurement method was used to find the writer of the corresponding handwriting. The handwriting of 10 people was randomly selected for the query and the number of characters per sample was 30, with an identification rate of 86%. Ding et al. [17] proposed an offline signature identification method based on scale invariant feature transform (SIFT) for local details of signature images, which detects SIFT feature points of the signature image and extracts feature descriptors, performs matching according to the Euclidean distance, filters matching pairs through the ratio of adjacent distances and the angle difference of feature points and performs histogram statistics on the angle difference in the matched feature points to form an ODH feature vector. Finally, the identification is completed according to the number of matching pairs and the similarity of ODH feature vectors. 4NsigComp2010 Database has fake signature, including real signature, fake signature and fake signature. The real signature is a signature written by the same author as the reference signature, the imitation signature is a signature written by other authors imitating the reference signature and the fake signature is a signature written by the same author as the reference signature but deliberately concealing the writing method. Tested on the local database, the error acceptance rate (EAR) was 5.3%, the error rejection rate (ERR) was 7%, the equal error rate (EER) was 6.7% and the EER was 20% on the 4NsigComp2010 Database. GRAPHJ is a forensic tool for handwriting analysis that implements automatic detection of lines and words in handwritten documents. The main focus in feature extraction is to measure the number of parts, such as the distance between text and characters, as well as the height and width of characters. The relative position of the punctuation on the "i" character is also used as a parameter to infer the authorship [18,19].

Huang Feiteng et al. [20] conducted a study on recognition of electronic signatures based on dynamic features, using writing duration, number of strokes and average writing strength per stroke as feature classification and collected three types of signature samples: simple, general and complex for classification. The results of discriminant analysis (DA), K-nearest neighbor (KNN), random forest (RF) and support vector machine (SVM) were all above 77% or more, which, to some extent, shows the feasibility of machine learning algorithms for classification of electronic signature handwriting recognition. Bhowal P et al. [21] designed an online signature verification system to extract three different types of features from the online signature, namely, physical features, frequency-based features and statistical features. The first ensemble using the feature classifier strategy combines the results of the seven classifiers using the sum of the normalized distribution, while the second ensemble, using the majority voting strategy, uses the decision of the first ensemble to make the final prediction, which is evaluated on the SVC 2004 and MCYT-100. The dataset includes real signatures and skilled forged signatures with 98.43% accuracy on the SVC 2004 dataset and 97.87% accuracy on MCYT-100.

Yelmati et al. [22] obtained a total of 42 feature vectors containing static and dynamic features, such as average velocity, pen up/down ratio, maximum pressure, pressure range, x-velocity variance, signature width, signature height, etc. They obtained better accuracy and faster training time on the SVC2004 Dataset but used fewer static features and weak interpretability. Kunshuai Wu [23] extracted GLCM and LBP features and fused them. After extracting texture features, he proposed an extraction method for signature stroke depth features, taking depth as the dynamic feature of the signature. The rules of using the GPDS dataset are consistent with the local dataset and are divided into three parts: real signature, skilled forged signature and random pseudo-signature, collecting 10 real signatures and 10 skilled signatures for a total of 20 groups. The highest overall correct rate of 87.75% for texture feature identification and 97.378% for depth feature identification was achieved, but an attempt was not made to combine the two, fusing dynamic and static feature information. Zhou et al. [6] proposed a handwritten signature verification method based on improved combined features. Based on the acquisition of offline images and online data, texture features were extracted using GLCM and HOG and nine geometric features were extracted. In addition to the horizontal and vertical coordinates and pressure contained in the online data, four dynamic features, velocity, acceleration, angle and radius of curvature, were also extracted. Support vector machine (SVM) and dynamic time warping (DTW) were used to verify the results. The forged signature is obtained by finding 2–3 experimenters to provide real signatures and forging them after pre-training. A total of 20 authors was collected and 1200 signatures were forged. Thus, 3, 5, 8 and 10 real signatures were selected for training small samples. The remaining signatures were used as test samples. After feature fusion, the highest accuracy rate of 10 samples was 97.83% and the false accept rate (FAR) value was 1.00%, The value of false reject rate (FRR) was 3.33%, but the characteristics of Chinese signatures are not well utilized and the task of forging signature is not detailed enough.

3. Materials and Methods

3.1. Sample Collection

Handwritten handwriting identification has become a very active research direction because of its wide application fields and numerous advantages [24]. The establishment of handwriting database is the basis of the research. Although the issue of signature identification has been discussed for many years, with the continuous update of science and technology, there is no practical database for the dynamic and static combination of forged signature handwriting database. Therefore, this study design is designed to establish a practical Chinese signature forged handwriting database for research purposes.

Collect the forged signature handwriting of the writer and establish a Chinese signature forged handwriting database. The database is a Chinese signature database, including 44 signatures of different signers. The acquisition device used is a dot matrix pen, which is composed of a high-speed camera and a pressure sensor. It can not only collect the coordinate information and pressure values of the sampling points during the writing process, but also collect the offline images for writing signatures. The multi-task design covers the issues that can be involved more comprehensively from two perspectives: the complexity of the strokes and the difficulty of the imitation. The signature handwriting is divided into two types: simple forged signature and skilled forged signature, as shown in Figure 1. The signature handwriting is collected according to different degrees and each type of signature is written 10 times under natural conditions in compliance with the personal habits of the writer to collect as many signature handwriting samples as possible. The online raw data are X and Y coordinate points, pressure, timestamp and pen up–down marks. Pen up–down marks refer to when the pen is lifted and the pen is dropped. Simple forged signature is written when the writer does not know the real signature. According to the complexity of the signature strokes, the signature is divided into simple signature, general signature and complex signature. The simple forgery signature task of different writers is shown in Table 1, where P1, P2 and P3 are different writers (the same below). Skilled forgery signature is to write and practice imitation when the writer knows the real signature. The real signature is shown in Table 2. According to the imitation degree of signature imitation, the signature is divided into simple imitation, general imitation and complex imitation. The skilled forgery signature tasks of different writers are shown in Table 3. Task 1 is a simple forgery task and task 2 is a skilled forgery task, containing a total of 2640 images and the corresponding signature data.

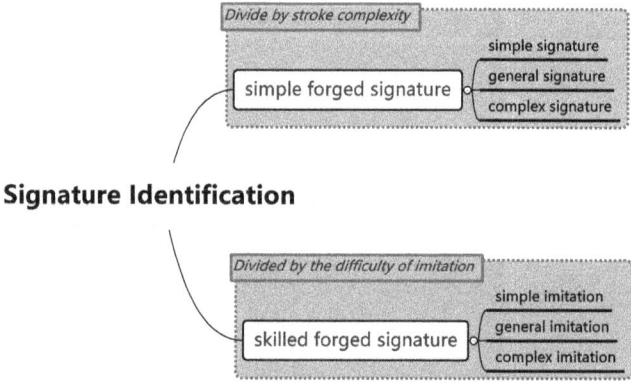

Figure 1. Sample collection tasks.

Table 1. Simple forgery task.

	Simple Signature	General Signature	Complex Signature
P1	王一	李云龙	诸葛佳晔
P2	王一	李云龙	诸葛佳晔
P3	王一	李云龙	诸葛佳晔

Table 2. Genuine signature.

Simple Genuine Signature	General Genuine Signature	Complex Genuine Signature
王宇全	李云龙	陈泉

Table 3. Skilled forgery task.

	Simple Imitation	General Imitation	Complex Imitation
P1	王宇全	李云龙	陈泉
P2	王宇全	李云龙	陈泉
P3	王宇全	李云龙	陈泉

The X and Y coordinate points, pressure, time and signature images can be obtained by writing, which are the original data collected by the sample. Figure 2 shows the X and Y points of the signature data, which are the changes in the X and Y coordinate points with time during the writing process.

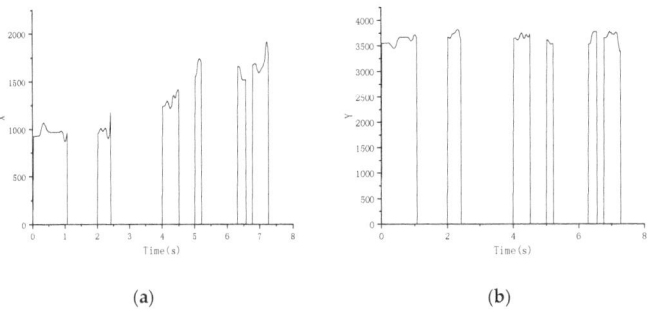

(a)   (b)

Figure 2. X and Y sites of signature data. (a) Change in X coordinate point; (b) change in Y coordinate point.

3.2. Preprocessing

In the preprocessing stage, the collected handwriting information is processed to remove irrelevant information, enhance the availability of information and facilitate feature extraction. The collected handwriting information is mainly divided into online data and offline images, which are preprocessed, respectively. A flow chart of data preprocessing is shown in Figure 3. By further processing the original data such as X and Y coordinates and pressure obtained by online writing, the dynamic information such as speed, acceleration and dead time is calculated as shown in Table 4, so as to improve the diversity of dynamic data and enhance the quality of signature data. For offline images, after selecting the required samples, each signature is trimmed with a fixed size and then de-noising, opening and closing operations and binarization operations are carried out. Finally, the binary image is refined using a fast refinement algorithm to extract the skeleton as in Figure 4, as a way to reduce the interference of the external influencing environment.

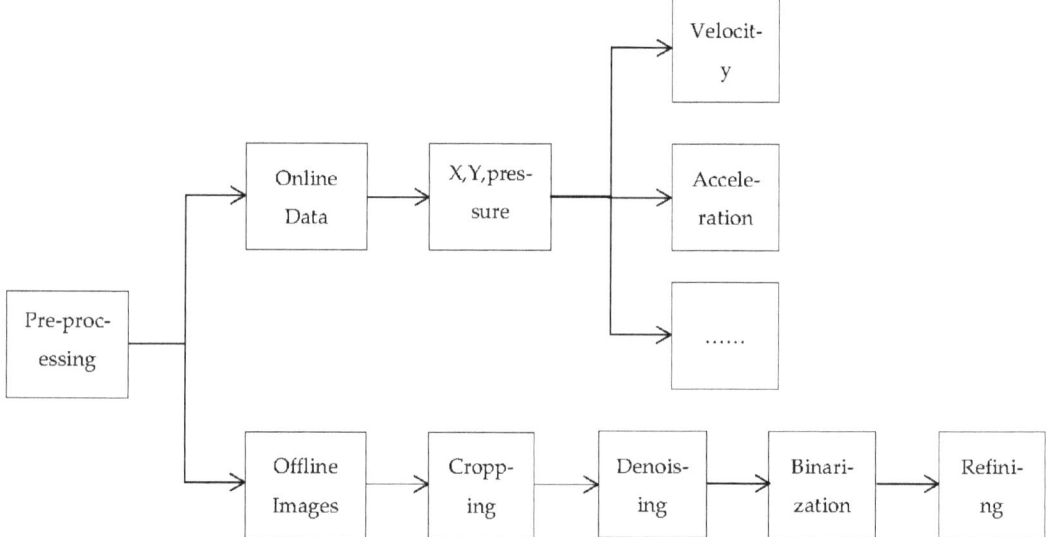

Figure 3. Data preprocessing.

Table 4. Online image preprocessing.

Raw Data	Processed Data
X	StrokeSum
Y	HangTime
Pressure	StrokeTime
State	StrokeLength
StrokeNum	Velocity
Timestamp	Acceleration
	Pressure

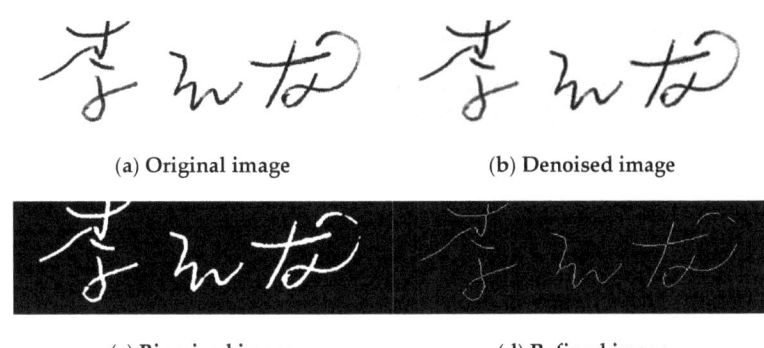

(a) Original image   (b) Denoised image

(c) Binarized image   (d) Refined image

Figure 4. Offline image preprocessing.

3.3. Classification Model

After feature selection, a variety of classification methods is used for verification. The training data are input into the classifier to learn the model and then the model predicts the label. Finally, the accuracy of model prediction is calculated to achieve the effect of signature identification.

Machine learning is a way to realize artificial intelligence, including different kinds of algorithms. It is a method of using data, training models and then using models to predict. For smaller datasets, classical machine learning algorithms are usually better than deep learning, which often require a large amount of data. The experiments are mainly conducted using four traditional algorithms: discriminant analysis, K-nearest neighbor, random forest and support vector machine.

Convolutional neural network (CNN) is a kind of feedforward neural network with convolution calculation and deep structure, which has been used to varying degrees in image processing, natural language processing, etc. [25]. The advantage is that multiple convolutional filters are used to extract high-level information from low-level information and the disadvantage is that encapsulation is not conducive to network performance improvement. The model used in the experiment has 9 layers, including 4 convolution layers, 4 maximum pooling layers and 1 fully connected layer.

Long short-term memory (LSTM) is a special recurrent neural network (RNN) model, which solves the short-term memory problem of RNN for solving the gradient disappearance and gradient explosion problems during the training of long sequences [26]. LSTM is very suitable for dealing with problems highly related to time series, such as machine translation, conversation generation, encoding and decoding, etc. The model used in the experiment has 4 layers, including 1 input layer, 2 hidden layers and 1 output layer

4. Results

4.1. Feature Extraction

Feature selection is a process of removing irrelevant features, retaining relevant features and transforming the original data that cannot be recognized by the algorithm into data features that can be recognized. The selection of features usually follows the principles of representativeness, stability and comprehensibility. Too much will increase the amount of calculation and too little will lead to a loss of information. To extract the features of the preprocessed data, first judge the possible effective features, select the time and space features, local and global features, construct features according to the style characteristics of the signature and finally conduct a further screening of all features. The dynamic and static features of feature extraction are shown in Table 5.

Table 5. Feature extraction.

Dynamic Feature	Static Feature
StrokeSum	AspectRatio
AveragePressure	Area
HangTime	Center of Gravity
StrokeTime	SpindleDirection
SpeedMax	Quadrilateral defining Signature structure
SpeedMin	ChainCode

4.1.1. Dynamic Feature Extraction

When writing a signature, a series of movement tracks will be left. Each person's stroke characteristics, writing strength and speed will be different. Dynamic features are obtained by further processing the information on attributes, such as speed, time and pressure, obtained when writing online signatures. It has higher accuracy and can reflect the writing style of the writer to a certain extent.

In the process of screening dynamic features, the obtained dynamic features are analyzed by a thermal map. The heat map shows the color shades corresponding to different correlation coefficients to explore the correlation between the identity of the writer and each feature; as in Figure 5, the selected features have larger values and lighter colors, the results indicate that the selected dynamic features are more effective.

(1) Total strokes

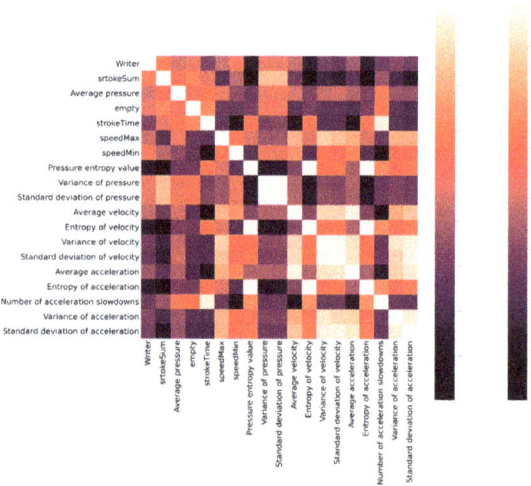

Figure 5. Heat map.

As a representative of the extremely strong embodiment of writing style, the number of strokes can show the connection of the written signature, as shown in Figure 6, which provides favorable conditions for identification. Especially in the imitation of complex signatures, the writing habits and psychological states of different signers will have a certain impact on the number of strokes.

(2) Average pressure

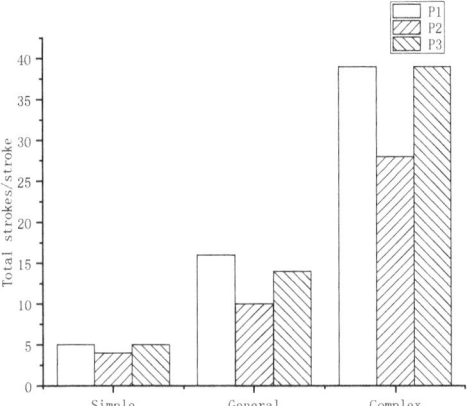

Figure 6. Total strokes.

Pressure is the force exerted on the paper by the individual through the nib when writing, as in Figure 7. The behavior of writing a signature is a dynamic process. As a continuous and hard-to-copy feature, pressure is difficult for forgers to accurately reproduce. Although there are different degrees of pressure values in continuous strokes, considering the number of strokes, the average pressure value of the signature is better.

(3) Total hang time

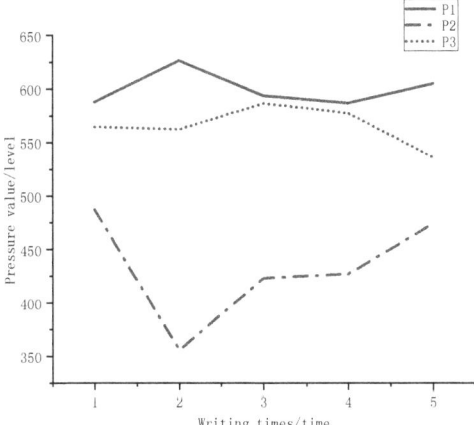

Figure 7. Pressure value.

This refers to the total pause time between each stroke during writing, as shown in Figure 8. Different authors have different proficiency in signature and personal writing habits and the pause time during writing is also different.

(4) Total time

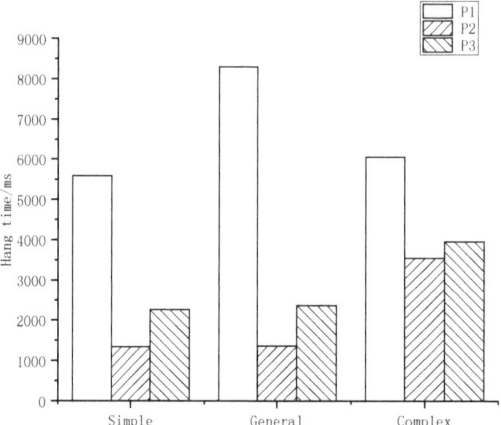

Figure 8. Hang time.

In the case of forged signatures, the time spent writing varies from author to author, as shown in Figure 9, which is directly related to the author's original writing speed, the difficulty of signing and the author's proficiency in forging signatures.

(5)   Maximum velocity

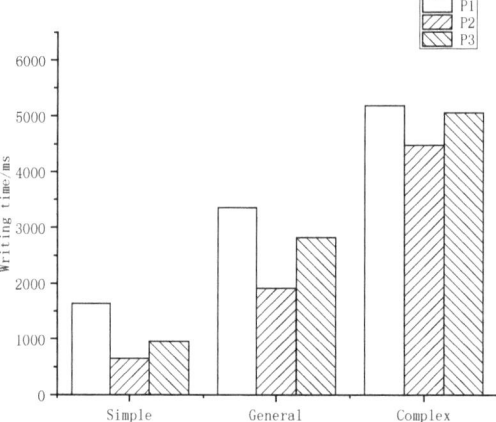

Figure 9. Writing time.

The maximum velocity of the author's writing is shown in Figure 10. Velocity is a key feature that cannot be ignored when writing and it can express the natural degree and accuracy of a signature.

(6)   Minimum velocity

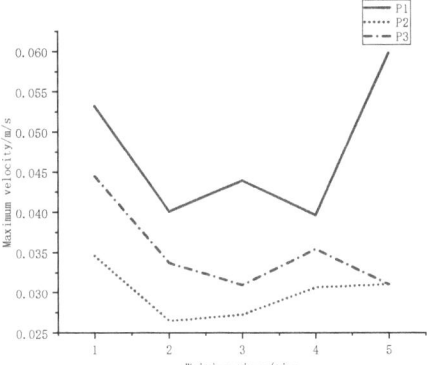

Figure 10. Maximum velocity.

It shows the minimum value of the writer's speed when writing, as in Figure 11.

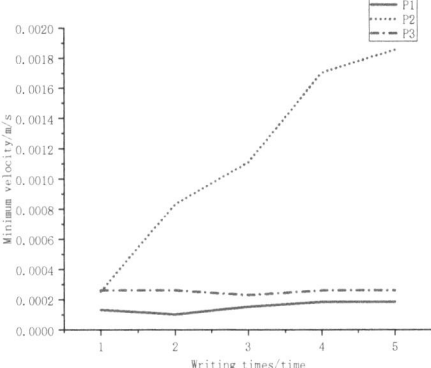

Figure 11. Minimum velocity.

4.1.2. Static Feature Extraction

Static features are the features extracted from offline signature images, which are similar to the results of visual analysis. They mainly distinguish different writers by analyzing the image structure, including the shape, position and writing style of the signature. The characteristics of the combined signatures are different through the different treatment methods of different strokes by the writers and the extracted static characteristics have a certain relationship with the public's cognitive judgment, which is easier for the public to understand.

(1) Aspect ratio

It is the horizontal and vertical span of the signature image, such as Figure 12, which reflects the habits of the writer when writing the signature, such as flat or rounded. Although this is not a unique key feature, the writing characteristics reflected by the aspect ratio are less likely to change for different writers.

(2) Area

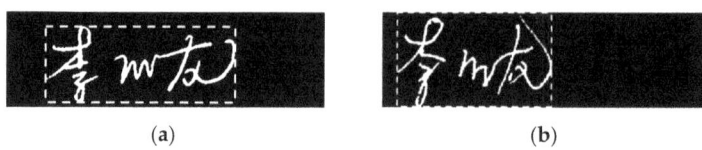

Figure 12. Aspect ratio. (**a**) A signature with a certain aspect ratio A; (**b**) a signature with a certain aspect ratio B.

The area is the most basic feature to describe the size of the block pattern. The pattern area in the image can be represented by the number of pixels in the same marked area, as shown in Figure 13. This refers to the total number of pixels after the binarization image and the sample area reflect the size of signatures of different people to a certain extent. It is informative to supplement with the aspect ratio feature.

(3) Center of gravity

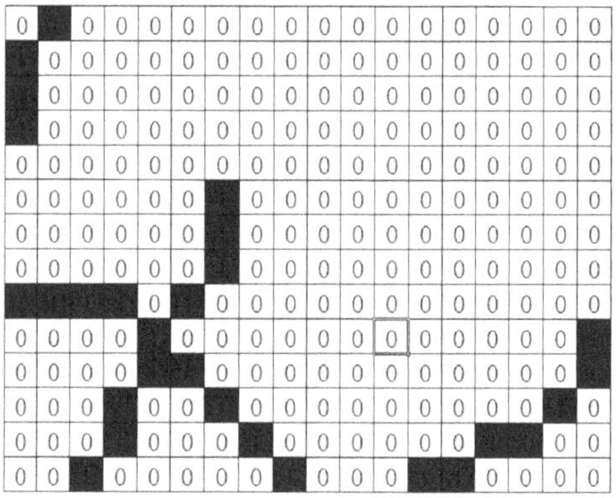

Figure 13. Area.

The center of gravity of the signature is the center point of the weight for the whole signature, as shown in Figure 14. The center of gravity is the foundation of writing and a good grasp of the center of gravity will result in flatter words, which will vary from writer to writer.

(4) Spindle direction

Figure 14. Graphic center of gravity. (**a**) A signature with a center of gravity A; (**b**) a signature with a center of gravity B.

Among the axes that pass through the center of gravity of the graph, the longest axis is called the principal axis of the graph. The angle between the principal axis and the i-axis is called the principal axis direction angle θ, as in Figure 15, which can be used to represent the position of the signature graph.

(5) Quadrilateral defining signature structure

(a)　　　　　　　　　　　　　　　　(b)

Figure 15. Spindle direction. (**a**) Directional angle; (**b**) signature corresponds to spindle direction.

For the signature, which has obvious personalized features and is not easy to change, it is possible to find some representative features that make a significant contribution to distinguishing different signatures. The strokes, such as horizontal, vertical, apostrophe, down, dot and hook, are all handled differently by different signers and the shape of the structural quadrilateral formed is also completely different. According to this feature, the general writing characteristics of the writer can be deduced. We processed the refined sample image to find the most edge points of the image in the four directions of up, down, left and right and connected the four edge points, in turn, to get a quadrilateral, as in Figure 16. When reflecting personal characteristics, we extract the four internal angles of the quadrilateral of the edge points.

(6) Chain code for signature quadrilateral

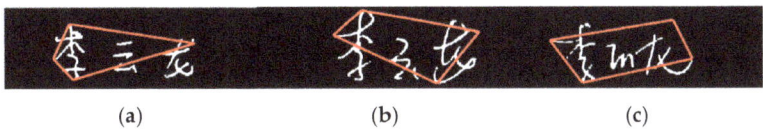

(a)　　　　　　　　(b)　　　　　　　　(c)

Figure 16. Quadrilateral defining signature structure. (**a**) signature written by P1; (**b**) signature written by P2; (**c**) signature written by P3.

Starting from the construction of a vertex of a quadrilateral, mark the edges in anti-clockwise order and classify the boundary between each edge and the horizontal direction into chain codes of 0–7 different numbers, as shown in Figure 17. The chain code is adjusted on the basis of the direction of the meter character grid. According to the floating offset, when people write horizontally and vertically, the original direction of the meter character grid is expanded by 45 degrees to both sides; that is, the angle of each direction has a limit of 45 degrees, which can well eliminate the error caused by different handwriting. The chain code complements the internal angle feature and distinguishes signatures with the same angle but different directions, as shown in Figure 18.

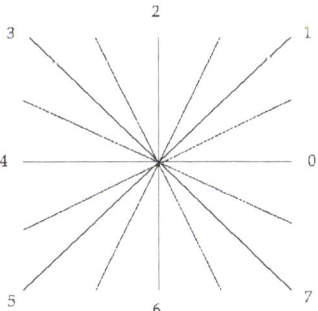

Figure 17. Chain code.

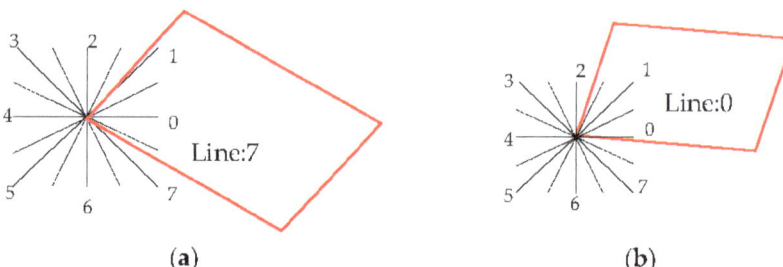

Figure 18. Signature quadrilateral with different chain codes at the same angle. (**a**) Signature quadrilateral with chain code 7; (**b**) signature quadrilateral with chain code 0.

4.2. Classification Results

To verify the feasibility of this method, experiments were conducted on a local dataset. The extracted dynamic features are fused with static features to obtain the signature feature set, which consists of 12 features, including 6 dynamic features and 6 static features. The number of words contained in the signature is small and identification is difficult. The local dataset contains the signatures of 44 writers. The experiments were conducted with a sample of 15 writers' signatures for two major categories of simple forgery and skilled forgery and the samples were divided into training and test sets, according to 7:3, which were randomly selected in the original dataset. A variety of typical methods of machine learning was used for classification and, in addition, RNN and CNN were used for experiments with this feature, which showed that the method of fusing dynamic and static features was better in terms of classification accuracy. We found that any classifier does not work well when trained and tested on dynamic features or static feature sets alone. This is because we selected features that incorporate a representative part unique to dynamic and static, both of which complement each other, discarding relatively redundant parts and highlighting the uniqueness of dynamic and static features, while increasing interpretability and making it easier for the public to understand.

For the multi-classification experiment of forged signatures, in order to verify the effectiveness of the selected features in the multi classification, the machine learning algorithm is used to conduct separate experiments for a different number of writers. The results are shown in Table 6. The effect is better when the number of writers is 5 or less and basically 100% can be classified correctly, the effect is above 90% when the number of writers is 10 and the effect is stable at about 90% when the number of writers is 15. Table 7 shows the experimental results of forged signature identification when the number of writers is 15. Overall, simple forged signatures are better than skilled forged signatures in terms of identification, which is consistent with the characteristics of forged signatures. Among simple forgeries, complex signatures are best identified with a DA classification accuracy of 100% and, among skilled forgeries, simple imitations are best identified with a DA classification accuracy of 93.3%.

When deep learning is used for classification, CNN and LSTM network structures are mainly used and attention modules are added to the network. Table 8 shows the experimental results of forged signatures when the number of writers is 15. On the whole, the classification results of simple forgery are better than those of skilled forgery. The classification accuracy of complex signatures is better in simple forgery and the classification accuracy is 96.7%. The classification of complex forgery is better in skilled forgery, the average classification accuracy was 96.7% and the highest was 100%.

Table 6. Result of multi-classification experiment.

Number of Writers	Simple Forged Signature			Skilled Forged Signature		
	Simple Signature	General Signature	Complex Signature	Simple Imitation	General Imitation	Complex Imitation
2	100.0	100.0	100.0	100.0	100.0	100.0
3	100.0	100.0	100.0	100.0	100.0	100.0
4	100.0	100.0	100.0	100.0	100.0	100.0
5	94.3	100.0	100.0	97.1	97.1	100.0
6	90.5	100.0	97.6	97.6	97.6	97.6
7	93.9	98.0	100.0	98.0	98.0	91.8
8	91.1	96.4	98.2	98.2	94.8	91.1
9	92.1	96.8	93.7	96.8	95.2	93.7
10	92.9	94.3	91.4	97.1	95.7	90.0
11	84.4	96.1	94.8	98.7	96.1	92.2
12	84.5	98.8	92.9	95.2	95.2	94.0
13	90.1	96.7	94.5	90.1	94.5	91.2
14	83.7	92.9	91.8	89.8	93.9	85.7
15	86.7	97.1	90.5	88.6	93.3	86.7

Table 7. Result on multiple classifiers.

	Simple Forged Signature			Skilled Forged Signature		
	Simple Signature	General Signature	Complex Signature	Simple Imitation	General Imitation	Complex Imitation
KNN	73.3	86.7	95.6	80.0	75.6	82.2
DA	75.6	91.1	100.0	93.3	77.8	93.3
RF	80.0	88.9	95.6	88.9	75.6	84.4
SVM	75.6	77.8	95.6	73.3	75.6	75.6

Table 8. Deep learning network classification results.

	Simple Forged Signature			Skilled Forged Signature		
	Simple Signature	General Signature	Complex Signature	Simple Imitation	General Imitation	Complex Imitation
CNN	90.0	90.0	96.7	96.7	83.3	93.0
CNN + Att	96.7	96.7	96.7	93.3	90.0	100.0
LSTM	95.7	96.7	96.7	90.0	90.0	96.7
LSTM + Att	95.7	96.7	96.7	93.3	80.0	96.7
CNN-LSTM	83.3	90.0	86.7	90.0	83.3	93.3
CNN-LSTM + Att	83.3	93.3	93.3	100.0	83.3	96.7

5. Discussion

In this study, the method of combining dynamic and static feature extraction is used to achieve better results. For the whole signature, dynamic features pay more attention to fine and clear information and use the value of each sampling point to obtain other data, while offline images pay more attention to the overall and structural information and use static features to complete the macro supplement. From the two dimensions, we can integrate more comprehensive features and complement each other. In the aspect of dynamic features, the more prominent and special features in the writing process, such as writing speed and pressure, are selected. In the aspect of static features, the public's impression of the signature, such as the aspect ratio (the signature is flat or square) and the angle of the signature quadrangle (whether the signature is inclined to the left or right as a whole), is referred to. The features that can best represent the dynamic and static features are screened out, which can also achieve better results and be more easily recognized by the public.

At present, there are few studies on dynamic and static signature identification and it is difficult to find a database for research. At the same time, we made a comparative analysis of the results of existing studies. Zhou et al. gradually improved the accuracy when training with real samples of 3, 5, 8 and 10. The highest classification result was 97.83% when 10 samples were used [6]. In addition to fewer selected features, the others are basically consistent with the design of the dichotomous experiment in this study. When the number of writers was two, the accuracy of all tasks was 100%. When Huang et al. studied the multi-classification recognition of electronic signatures, the multi-classification results of the machine learning algorithm for 3000 samples from 30 authors were more than 90% [20]. In this study, when the number of writers was 15, the machine learning algorithm was about 90%, but the effect was good and stable in the deep learning network. The selection of features by Yelmti et al. is largely consistent with the features we used in the extraction of dynamic features [22] and, through correlation analysis, we know that the standard deviation and variance in velocity and other features have relatively low correlation. From the above discussion, we can see that this study is relatively comprehensive in feature extraction. When combining dynamic and static features, structural static features are added. When the amount of multi-classification identification is small, it can reach 100% and when the number of authors is 15, most of them can reach 96.7%. Under the condition of high accuracy, it is easier to understand, but the identification of individual tasks still needs to be improved.

The scope of application of signature verification has spread throughout people's daily life and it is essential for the general public to identify reasonably and effectively. Alice J. et al., for identity recognition, relied on human facial and body expressions from static and dynamic situations, incorporating different conditions. Experiments have shown that a fusion of static and dynamic features, which focus on different directions, works better and achieves perfect performance [27]. Such feature fusion is not only for identity recognition; for example, the line interruption caused by typhoon can be predicted through the coordination of static and dynamic data [28] and multi-scale features and hierarchical features can be extracted for super-resolution image detection [29]. The effectiveness of feature fusion is fully demonstrated in various fields. A better result can be achieved by starting from multiple dimensions, looking at things in a comprehensive way, learning from each other's strong points to complement each other's weak points and explaining in simple terms.

6. Conclusions

This paper proposes a handwriting identification method that incorporates dynamic and static features and establishes a Chinese signature forgery handwriting database. By combining the static features of traditional paper-and-pen writing and the dynamic features of digital writing, the feasibility of the used features for multi-classification forgery handwriting identification is verified to some extent by experimenting and comparing different classification number cases using multiple classifiers.

The fusion of dynamic and static features makes the handwriting identification more interpretable and the effective features obtained are more comprehensive. It can better identify the forged signature handwriting and obtain better accuracy. Multi-classification experiments on forged signatures are a new way of thinking for handwriting identification.

Author Contributions: Conceptualization, J.L. and H.Q.; methodology, J.L. and H.Q.; software, J.L.; validation, J.L.; formal analysis, J.L.; investigation, J.L.; resources, H.Q.; data curation, J.L.; writing—original draft preparation, J.L.; writing—review and editing, J.L., H.Q., C.Z. and Q.T.; supervision, H.Q., X.W., C.Z. and Q.T.; project administration, H.Q., X.W., C.Z. and Q.T.; All authors have read and agreed to the published version of the manuscript.

Funding: This work was supported by Zhejiang Key R&D Plan (Grant number: 2017C03047).

Institutional Review Board Statement: Not applicable.

Informed Consent Statement: Informed consent was obtained from all subjects involved in the study.

Data Availability Statement: The data presented in this study are available on request from the corresponding author. Correspondence: qhn@zjhu.edu.cn (H.Q.)

Conflicts of Interest: The authors declare no conflict of interest.

References

1. Liu, L. The Application of Weighted DTW on Handwritten Signature Verification. Ph.D. Thesis, Shandong Normal University, Jinan, China, 2011.
2. Plamondon, R.; Srihari, S.N. Online and off-line handwriting recognition: A comprehensive survey. *IEEE Trans. Pattern Anal. Mach. Intell.* **2000**, *22*, 63–84. [CrossRef]
3. Dhieb, T.; Boubaker, H.; Njah, S.; Ben Ayed, M.; Alimi, A.M. A novel biometric system for signature verification based on score level fusion approach. *Multimed. Tools Appl.* **2022**, *81*, 7817–7845. [CrossRef]
4. Ye, J. Online Signature Verification Based on SVM and One-class SVM. Ph.D. Thesis, South China University of Technology, Guangzhou, China, 2016.
5. Naz, S.; Bibi, K.; Ahmad, R. DeepSignature: Fine-tuned transfer learning based signature verification system. *Multimed. Tools Appl.* **2022**, *81*, 38113–38122. [CrossRef]
6. Zhou, Y.; Zheng, J.; Hu, H.; Wang, Y. Handwritten Signature Verification Method Basedon Improved Combined Features. *Appl. Sci.* **2021**, *11*, 5687. [CrossRef]
7. Saleem, M.; Kovari, B. Online signature verification using signature down-sampling and signer-dependent sampling frequency. *Neural Comput. Appl.* **2021**, 1–13. [CrossRef]
8. Chen, S.; Wang, Y. A Robust Off-line Writer Identifi-cation Method. *Acta Autom. Sin.* **2020**, *46*, 108–116.
9. Yapıcı, M.M.; Tekerek, A.; Topaloğlu, N. Deep learning-based data augmentation method and signature verification system for offline handwritten signature. *Pattern Anal. Appl.* **2021**, *24*, 165–179. [CrossRef]
10. Jain, A.; Singh, S.K.; Singh, K.P. Handwritten signature verification using shallow convolutional neural net-work. *Multimed. Tools Appl.* **2020**, *79*, 19993–20018. [CrossRef]
11. Okawa, M. Online Signature Verification Using LocallyWeighted Dynamic Time Warping via Multiple Fusion Strategies. *IEEE Access* **2022**, *10*, 40806–40817. [CrossRef]
12. Wang, S.; Wang, J. Analysis on the Changes of Hand-writing and Electronic Signature. *J. Railw. Police Collge* **2019**, *29*, 7.
13. Zhang, C.; Tong, X.; Wang, J. Review of Handwritten Signature Identification Based on Machine Learning. *J. Jiangsu Police Inst.* **2021**, *36*, 6.
14. Rohilla, S.; Sharma, A.; Singla, R. Role of sub-trajectories in online signature verification. *Array* **2020**, *6*, 100028. [CrossRef]
15. Hameed, M.M.; Ahmad, R.; Kiah, M.L.M.; Murtaza, G. Machine learning-based offlinesignature verification systems: A systematic review. *Signal Process. Image Commun.* **2021**, *93*, 116139. [CrossRef]
16. Li, Q.; Ma, Y.; Zhou, Y.; Zhou, L. Method of Writer Identification Based on Curvature of Strokes. *J. Chin. Inf. Processing* **2016**, *30*, 6.
17. Ding, Y.; Zhan, E.; Zheng, J.; Wang, Y. Offline signature identification based on improved SIFT. *Appl. Esearch Comput.* **2017**, *34*, 5.
18. Guarnera, L.; Farinella, G.M.; Furnari, A.; Salici, A.; Ciampini, C.; Matranga, V.; Battiato, S. GRAPHJ: A forensics tool for hand-writing analysis. In Proceedings of the International Conference on Image Analysis and Processing; Springer: Berlin/Heidelberg, Germany, 2017; pp. 591–601.
19. Guarnera, L.; Farinella, G.M.; Furnari, A.; Salici, A.; Ciampini, C.; Matranga, V.; Battiato, S. Forensic analysis of handwritten documents with GRAPHJ. *J. Electron. Imaging* **2018**, *27*, 051230. [CrossRef]
20. Huang, F.; Hao, H.; Chen, W.; Sun, J.; Shi, W.; Zhang, L.; Wnag, Z. Research on Electronic Signature Handwriting Recognition Based on Dynamic Features. *Mod. Comput.* **2020**, *5*, 84–88. [CrossRef]
21. Bhowal, P.; Banerjee, D.; Malakar, S.; Sarkar, R. A two-tier ensemble approachfor writer dependent online signature verification. *J. Ambient. Intell. Humaniz. Comput.* **2022**, *13*, 21–40. [CrossRef]
22. Yelmati, S.R.; Rao, J.H. Online Signature Verification Using Fully Connected Deep Neural Networks. *Int. J. Eng. Manuf.* **2021**, *11*, 41–47.
23. Wu, K. Signatures Verification Based on Texture Feature and Depth Feature. Ph.D. Thesis, Chinese Academy of Sciences, Beijing, China, 2020.
24. Batool, F.E.; Attique, M.; Sharif, M.; Javed, K.; Nazir, M.; Abbasi, A.A.; Iqbal, Z.; Riaz, N. Offline signature verification system: A novel techniqueof fusion of GLCM and geometric features using SVM. *Multimed. Toolsand Appl.* **2020**, 1–20. [CrossRef]
25. Zhang, B. Off-line signature verification and identification by pyramid histogram of oriented gradients. *Int. J. Intell. Comput. Cybern.* **2010**, *3*, 611–630. [CrossRef]
26. Hochreiter, S.; Schmidhuber, J. Long short-term memory. *Neural Comput.* **1997**, *9*, 1735–1780. [CrossRef] [PubMed]
27. O'Toole, A.J.; Phillips, P.J.; Weimer, S.; Roark, D.A.; Ayyad, J.; Barwick, R.; Dunlop, J. Recognizing people from dynamic and static faces and bodies: Dissecting identity with a fusion approach. *Vis. Res.* **2011**, *51*, 74–83. [CrossRef] [PubMed]
28. Tang, L.; Xie, H.; Wang, Y.; Zhu, H.; Bie, Z. Predicting typhoon-induced transmission line outages with coordination of static and dynamic data. *Int. J. Electr. Power Energy Syst.* **2022**, *142*, 108296. [CrossRef]
29. Luo, J.; Liu, L.; Xu, W.; Yin, Q.; Lin, C.; Liu, H.; Lu, W. Stereo super-resolution images detection based on multi-scale feature extraction and hierarchical feature fusion. *Gene Expr. Patterns* **2022**, *45*, 119266. [CrossRef]

Article

Prediction of Eudaimonic and Hedonic Orientation of Movie Watchers

Elham Motamedi [1], Francesco Barile [2] and Marko Tkalčič [1,*]

[1] Faculty of Mathematics, Natural Sciences and Information Technologies, University of Primorska, Glagoljaška 8, SI-6000 Koper, Slovenia
[2] Faculty of Science and Engineering, Department of Advanced Computing Sciences, Maastricht University, 6200 MD Maastricht, The Netherlands
* Correspondence: marko.tkalcic@famnit.upr.si

Abstract: Personality accounts for how individuals differ in their enduring emotional, interpersonal, experiential, attitudinal and motivational styles. Personality, especially in the form of the Five Factor Model, has shown usefulness in personalized systems, such as recommender systems. In this work, we focus on a personality model that is targeted at motivations for multimedia consumption. The model is composed of two dimensions: the (i) eudaimonic orientation of users (EO) and (ii) hedonic orientation of users (HO). While the former accounts for how much a user is interested in content that deals with meaningful topics, the latter accounts for how much a user is interested in the entertaining quality of the content. Our research goal is to devise a model that predicts the EH and HO of users from interaction data with movies, such as ratings. We collected a dataset of 350 users, 703 movies and 3499 ratings. We performed a comparison of various predictive algorithms, as both regression and classification problems. Finally, we demonstrate that our proposed approach is able to predict the EO and HO of users from traces of interactions with movies substantially better than the baseline approaches. The outcomes of this work have implications for exploitation in recommender systems.

Keywords: personality prediction; eudaimonic orientation; hedonic orientation; user modeling

1. Introduction

Personality traits have been defined as the *most important ways in which individuals differ in their enduring emotional, interpersonal, experiential, attitudinal and motivational styles* [1]. These factors have shown their utility in many online applications, such as recommender systems [2] and targeted advertising [3]. However, the usage of lengthy questionnaires, which is the typical method of assessing a user's personality, is not practical for online applications because they are intrusive and time-consuming. For this reason, personality prediction from the digital traces of users emerged about a decade ago using mostly social media sources to infer users' personality traits [4–7]. As of today, advanced algorithms [8] and even off-the-shelf tools (https://applymagicsauce.com/demo (accessed on 15 September 2022), https://www.ibm.com/no-en/cloud/watson-personality-insights (accessed on 15 September 2022)) exist for personality prediction.

The majority of personality prediction approaches focuses on the prediction of personality in the form of the Five Factor Model (FFM), with the five factors being Openness, Conscientiousness, Extraversion, Agreeableness and Neuroticism [9]. This is a generic personality model that covers all areas of human activity. For more specific domains, other personality models have been devised, such as the Bartle model for gaming [10], the vocational RIASEC model [11], and the conflict-coping Thomas–Kilman model [12].

In this paper, we focus on a personality model that describes the motivations for multimedia consumption. The model is composed of two factors—namely, the *eudaimonic* and *hedonic orientation of users*. The first factor, the eudaimonic orientation of users, accounts for how much a user is looking for meaning in multimedia content, while the the second factor, hedonic orientation, accounts for how much a user is looking for entertaining quality [13]. This model has potential for user modeling and recommender systems in the entertainment domain, as demonstrated in [14].

In particular, in scenarios, when items are described with their eudaimonic and hedonic qualities (e.g., the movie *Manchester by the Sea* has high eudaimonic but low hedonic quality), a content-based filtering approach can be useful for recommendations [15]. Thus far, there has been no work on predicting the eudaimonic and hedonic orientation (EHO) personality model of users from digital traces.

In this paper, we propose a model that takes features generated from traces of interactions with movies to predict EHO. We demonstrate the quality of the predictions on a dataset of 350 users, 703 movies and 3499 ratings. The results of our analysis are examined based on the information users provide, the interactions they have with the system, or both. According to our analysis, our proposed approach predicts the EO and HO of users substantially better than baseline approaches.

2. Related Work

This work builds upon two bodies of related work: (i) personality research from psychology and (ii) social media-based predictions using machine-learning models.

2.1. Personality Research from Psychology

As mentioned earlier, psychology research defines personality traits as the *most important ways in which individuals differ in their enduring emotional, interpersonal, experiential, attitudinal and motivational styles* [1]. In order to account for the diversity of users, several personality models have been devised. Over the past few decades, factor analysis has been used to reduce personality traits to their underlying factors [16–18].

Several independent researchers have contributed to this research line by introducing various factors. The most known is the Five Factor Model (FFM), which is composed of the factors Openness, Conscientiousness, Extraversion, Agreeableness and Neuroticism. The model is based on the lexical hypothesis, i.e., the things that are important eventually end up in words we use. Studying the usage of language, the FFM was constructed [1].

The FFM is not the only personality model that has been used by researchers but it is the most extensively used. There are other alternative models with different numbers of factors that were introduced to argue for the comprehensibility of FFM. For example, the HEXACO model is an alternative personality model that includes six factors. The sixth factor in this model is called Honesty–Humility [19].

Although the FFM is an established model, it does not necessarily fit into specific domains. For this reason, several domain-specific models have been developed. In the domain of human resources, the RIASEC model, composed of the types realistic, investigative, artistic, social, enterprising and conventional is used [11]. These types are correlated with the FFM factors but account for more fine-grained variance in the job domain and also offer more explanational power.

For explaining the behavior of users in groups, the Thomas–Kilman model has been proposed [12]. This model has two factors, assertiveness and cooperativeness, that account for how people cope in conflict situations. The usage of the Thomas–Kilman model has been investigated in the domain of group recommender systems [20,21]. As an example, a recent work done by Abolghasemi et al. [21] proposed a group recommendation approach that takes the users' personality into account [21].

They modeled the influence of users on other group members using influence graphs in which nodes represent users and edges represent users' influence on other group members. Their proposed model suggested that a user's influence on another member would be stronger if they are more assertive and less cooperative when the other members are less assertive and more cooperative.

Further domain-specific models include the Bartle gaming types, which distinguishes killers, achievers, explorers and socializers [10], and in particular, the eudaimonic and hedonic orientation (EHO) model, which is the topic of prediction of this paper.

The EHO model has roots in positive psychology [22]. Human experiences have been modeled using the concepts of eudaimonia and hedonia [23]. The eudaimonic experience is composed of two factors: (i) deeper reflection, which encompasses relatedness, central values and personal growth; and (ii) life evaluation, which encompasses the purpose of life, self-acceptance and autonomy.

The hedonic experience, on the other hand, is about pure pleasure. The two factors have a low correlation, which indicates their almost-orthogonality. The work of Oliver and Raney showed that users have different propensities for eudaimonic and hedonic experiences, which leads to the concept of traits in the EHO model [13]. Furthermore, they devised an instrument for measuring the EHO factors of users in the domain of movie consumption [13]. This questionnaire has six questions for the eudaimonic orientation (EO) and six questions for the hedonic orientation (HO).

2.2. Social-Media-Based Predictions Using Machine-Learning Models

Instruments in the form of questionnaires are time-consuming and obtrusive for end users in online applications. For example, the widely used Big Five Inventory (BFI) for measuring the FFM personality factors is composed of 44 items [24]. As personality accounts for variance in user behavior, including online behavior, unobtrusive models that use user digital traces have been devised for inferring personality traits. However, the majority of the personality prediction work has concentrated on predicting the FFM values.

Through meta data analysis, Azucar et al. [8] examined the predictive value of digital traces on each of the personality factors in FFM. The Pearson correlation values for all factors ranged from 0.30 to 0.40, indicating that the digital traces of users are a reliable source of predicting their personality traits. Early work from Quercia et al. [4] used micro blogs for generating features and achieved a solidly low error. In their seminal work, Kosinski et al. [6] used Facebook likes and applied singular value decomposition to obtain a matrix of latent features. The authors predicted FFM factors of users and several other personal characteristics with high accuracy.

Several other types of digital traces have been used to achieve good FFM predictions. For example, Ferwerda et al. [25] used low-level features from Instagram photos. In another work, they used drug profiles for predicting FFM factors [26]. A successful approach was also the usage of eye gaze data for predicting FFM factors, as demonstrated by Berkovsky et al. [27].

High accuracy was achieved with a hybrid model that uses two social media sources as feature generators: Instagram and Twitter [7]. The authors used linguistic features as well as low-level image features to predict FFM factors. In a meta-review of FFM prediction models, their work demonstrated the highest accuracy [8].

The aforementioned related work shows that personality information is embedded in our online behavior and the digital traces we leave behind. Given the lack of predictive models of EHO and the potential of the EHO model for personalized applications, we fill this gap in knowledge by devising a predictive model of EHO-based on digital traces of user-movie interactions. More specifically, we address the following research questions (RQs):

- RQ1: How are users clustered based on their EHO values?
- RQ2: How do different machine-learning algorithms perform in predicting the EHO of users?
- RQ3: How do prediction algorithms perform with different groups of features?

The rest of this paper is organized as follows. In Section 3.1, the data and data collection method are introduced. In Section 3.2, the machine-learning pipeline is described. In Section 4, user clustering based on their eudaimonic and hedonic orientation (EHO) is discussed. Moreover, the results of the EHO prediction as regression and classification problems followed by discussions are provided. Our conclusions are drawn in Section 6.

3. Methods and Materials

In order to show predict the EHO of users, we (i) first collected a dataset; then (ii) trained a predictive model, both as a regression and a classification problem; and finally (iii) performed the evaluation.

3.1. Data Acquisition

For our study, we decided to collect the following data about the participants:

- EHO of users.
- Personality.
- Genre preferences.
- Film sophistication.

Furthermore, for each participant, we wanted to collect assessments about movies.

We conducted a user study to collect the data required. In total, we had 350 users providing 3499 assessments of 703 movies. We generated a pool of 1000 popular movies from the Movielens 25M dataset, from which 55 movies were randomly selected to be shown to each study participant.

In the first step, we measured the demographics, genre preferences, personality, EHO and the film sophistication of each participant. The features extracted from these questions are summarized in Table 1 as U–F. The answers to the demographics questions are referred to as DEMQ. DEMQ includes questions about gender, education and age. In order to specify the gender, the user could choose among male, female, other and prefer not to say. For this study, we considered six educational categories: primary school or lower, secondary school, university bachelor's degree, university master's degree, university PhD and other professional education degrees. The user was required to input their age, which was verified to be over 18.

GPREFQ refers to the genre preference answers. Each user was asked to rate different genres of movies including action, adventure, comedy, drama, fantasy, history, romance, science fiction and thriller with a score ranging from 1 to 5.

For measuring personality traits, we used the Big Five 44 Item Inventory measure proposed by John and Srivastava [24] and Ten-Item Personality Inventory (TIPI) measure proposed by Gosling et al. [28]. Due to the higher correlation value of *Extraversion* and *Openness* traits with EO and HO, we used questions from the Big Five 44 Item Inventory. For *Agreeableness*, *Conscientiousness* and *Neuroticism*, we used TIPI questions. The answers to the questions are referred to as BFIQ and range from 1 to 7. Based on these answers, we calculated the value associated with each factor in the FFM, which we refer to as BFT. Using the list of FFM questionnaires in the same order provided by John and Srivastava [24], the *Extraversion* and *Openness* traits are calculated as follows:

$$Ex = \frac{BFIQ_1 + (8 - BFIQ_6) + (BFIQ_{26}) + (BFIQ_{36})}{4} \quad (1)$$

$$Op = \frac{BFIQ_5 + BFIQ_{10} + (8 - BFIQ_{35}) + (8 - BFIQ_{41})}{4} \quad (2)$$

where Ex and Op stand for the personality traits *Extraversion* and *Openness*, respectively. $BFIQ_n$ is the n-th question in the BFI questionnaire proposed by John and Srivastava [24]. Using the list of TIPI questionnaires in the same order provided by Gosling et al. [28], the *Agreeableness*, *Conscientiousness* and *Neuroticism* traits are calculated as follows:

$$Ag = \frac{(8 - BFIQ_2) + BFIQ_7}{2} \quad (3)$$

$$Co = \frac{BFIQ_3 + (8 - BFIQ_8)}{2} \quad (4)$$

$$Ne = \frac{(8 - BFIQ_4) + BFIQ_9}{2} \quad (5)$$

where *Ag*, *Co* and *Ne* stand for the personality traits *Agreeableness*, *Conscientiousness* and *Neuroticism*, respectively. $BFIQ_n$ is the n-th question in the TIPI questionnaire proposed by Gosling et al. [28].

Oliver and Raney [13] included six statements related to *EO* and six statements related to HO. In accordance with the correlation between the questions proposed by Oliver and Raney [13] and *EO/HO* values, we selected three statements for measuring each. We asked users to tell us to which degree they agree with the statements on a scale from 1 to 7. Assuming the same order of questions as in Oliver and Raney [13], $EHOQ_n$ refers to the n-th question. *EO* and *HO* are calculated as follows:

$$EO = \frac{EHOQ_1 + EHOQ_3 + EHOQ_4}{3} \qquad (6)$$

$$HO = \frac{EHOQ_7 + EHOQ_9 + EHOQ_{12}}{3} \qquad (7)$$

Müllensiefen et al. [29] proposed a factor structure of a reduced self-report inventory for measuring the music sophistication index. Based on this work, the questionnaire of the *Goldsmiths Musical Sophistication Index (Gold MSI)* (https://shiny.gold-msi.org/gmsiconfigurator/ (accessed on 15 September 2022)) was designed for the music domain. In order to measure film sophistication, we adapted the music sophistication index questionnaire to fit the movie domain. The answers to the film sophistication questionnaire are referred to as SFIQ, which are on a scale from 1 to 7. From these answers, the following two film sophistication factors referred to as SFI are calculated: (i) *Active Engagement* and (ii) *Emotions*. Assuming the same order of questions in the Gold MSI questionnaire, $SFIQ_n$ refers to the n-th question for each sophistication factor. *Active Engagement* and *Emotions* are calculated as follows:

$$AE = \frac{SFIQ_7 + SFIQ_9}{2} \qquad (8)$$

$$EM = \frac{SFIQ_1 + SFIQ_2}{2} \qquad (9)$$

where *AE* and *EM* stand for the film sophistication factors of *Active Engagement* and *Emotions*, respectively.

In the second step, among 55 movies presented to the participants, they were asked to select ten. They were instructed to choose movies they have watched or they were familiar enough to judge in terms of their preferences and feelings while watching them.

FPREFQ is the rating of users on a scale from 1 to 5. The eudaimonic and hedonic perceptions (EHP) of users from movies were measured with the questionnaire, adapted from the one proposed by Oliver and Raney [13] (EHPQ). According to the correlation between the questions proposed by Oliver and Raney [13] and the eudaimonic perception (EP)/hedonic perception (HP), we selected two statements for measuring each. We asked users to tell us to which degree they agree with the statements on a scale from 1 to 7. Assuming the same order of questions as in Oliver and Raney [13], $EHPQ_n$ refers to the n-th question. *EP* and *HP* are calculated as follows:

$$EP = \frac{EHPQ_1 + EHPQ_6}{2} \qquad (10)$$

$$HP = \frac{EHPQ_7 + EHOQ_9}{2} \qquad (11)$$

Table 1. Feature list. Answers to the question sets provided to users are represented by feature groups ending in Q. BFT, EHO and SFI were calculated based on user responses. The features that describe users and interactions are U-F and I-F, respectively.

Feature Groups	Feature Subgroups	Description	Range of Values
U-F	DEMQ	Demographic questions (ie: gender, education and age)	Age > 18 Others: categorical features
	GPREFQ	Genre preference questions (including: action, adventure, comedy, drama, fantasy, history, romance, science fiction, thriller)	5 likert scale
	BFIQ	Big five inventory questions	7 likert scale
	EHOQ	Eudaimonic and hedonic orientation questions	7 likert scale
	SFIQ	Sophisticaiton index questions	7 likert scale
	BFT	Big five traits (calculated from BFIQ)	$r \in R : \{1 \leq r \leq 7\}$
	EHO	Eudaimonic and hedonic orientation of users (calculated from EHOQ)	$r \in R : \{1 \leq r \leq 7\}$
	SFI	sophistication indexes including: active engagement, emotion (calculated from SFIQ)	$r \in R : \{1 \leq r \leq 7\}$
I-F	FPREFQ	Film preference questions	5 likert scale
	EHPQ	Questions related to eudaimonic and hedonic perceptions of users from films	7 likert scale
	EHP	Eudaimonic and hedonic perceptions of users from films (calculated from EHPQ)	$r \in R : \{1 \leq r \leq 7\}$

3.2. Machine-Learning Workflow

The machine-learning pipeline was implemented using the *Scikit-learn* library (https://scikit-learn.org/stable/ (accessed on 15 September 2022)) in Python and is depicted in Figure 1.

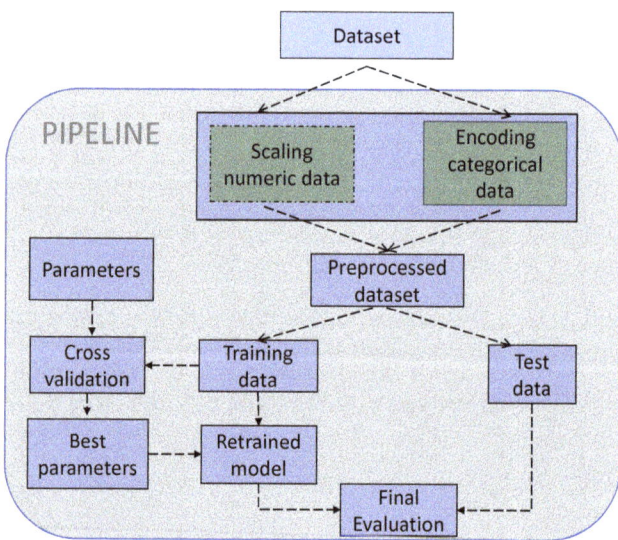

Figure 1. The machine-learning pipeline (symbols with dashed line edges illustrate that that step is not always performed).

The goal of the machine-learning algorithm was to predict two user characteristics, the eudaimonic orientation and the hedonic orientation values, from features collected in the user study. We approached this prediction in two ways: (i) as a regression problem and (ii) as a classification problem, where we used median splitting to label users with high- and low-eudaimonic orientation and high- and low-hedonic orientation.

As can be seen in Figure 1, the collected data is fed into the pipeline. In our dataset, the features are either numerical, including integer and float types, or categorical, including nominal or ordinal features. There are two categorical features in the dataset: i.e, gender and education. We assumed gender as a nominal categorical feature and therefore used *OneHotEncoder* class from the *Scikit-learn* library for encoding it, which encodes categorical features as a one-hot numeric array.

We assumed that education is an ordinal categorical feature, and therefore we encoded it with *OrdinalEncoder* from the *Scikit-learn* library. All the other features in the dataset are numerical data. Since feature scaling is required only for machine-learning estimators that consider the distance between observations and not every estimator, this step is not always performed. The list of machine-learning algorithms that use the scaling step can be seen in Tables 2 and 3. In the case of performing feature scaling, we used the *StandardScale* class from the *Scikit-learn*.

Table 2. ML models and hyperparameters (regression problem). Except for XGBRegressor, Scikit-learn was used to implement all models (https://scikit-learn.org/stable/ accessed on 15 September 2022). Python's xgboost library was used to implement XGBRegressor. (https://xgboost.readthedocs.io/en/stable/parameter.html accessed on 15 September 2022).

Model	Scaled	Parameters	Tested Values
Ridge	✓	alpha (regularization parameter)	0, 0.0001, 0.001, 0.01, 0.1, 1, 5, 10
Lasso	✓	alpha (regularization parameter)	0, 0.0001, 0.001, 0.01, 0.1, 1, 5, 10
SVR	✓	Kernel	linear, rbf, poly
		C	0.001, 0.01, 0.1, 1.0, 10, 100
		gamma	scale, auto
		epsilon	0.001, 0.01, 0.1, 0.2, 0.5, 0.3, 1.0, 2.0, 4.0
KNeighborsRegressor	✓	n_neighbors	3, 5, 7, 9
		weights	uniform, distance
		algorithm	ball_tree, kd_tree, brute
		p	1, 2
DecisionTreeRegressor	X	criterion	mae, mse
		splitter	best, random
		max_depth	1, 3, 5, 7, 9, 11, 12
		min_weight_fraction_leaf	0.1, 0.2, 0.3, 0.4, 0.5
		max_features	auto, log2, sqrt, None
		max_leaf_nodes	None, 10, 20, 30, 40, 50, 60, 70, 80, 90
RandomForestRegressor	X	n_estimators	1, 3, 5, 7, 9, 11, 13, 15, 17, 20
		max_features	auto, sqrt
		max_depth	1, 3, 5, 7, 9, 11, 13, 15, 17, 20
		min_samples_split	2, 3, 4, 5, 10
		min_samples_leaf	1, 2, 4
		bootstrap	True, False
XGBRegressor	X	max_depth	3, 4, 5, 6, 7, 8, 9, 10, 11
		min_child_weight	1, 2, 3, 4, 5, 6, 7
		eta	0.001, 0.01, 0.1, 0.2, 0.5, 0.3, 1
		subsample	0.7, 0.8, 0.9, 1.0
		colsample_bytree	0.7, 0.8, 0.9, 1.0
		objective	reg:squarederror

Table 3. ML models and hyperparameters (classification problem). Except for XGBRegressor, Scikit-learn was used to implement all models (https://scikit-learn.org/stable/, accessed on 15 September 2022). Python's xgboost library was used to implement XGBRegressor. (https://xgboost.readthedocs.io/en/stable/parameter.html, accessed on 15 September 2022).

Model	Scaled	Parameters	Tested Values
Ridge	✓	alpha (regularization parameter)	0, 0.0001, 0.001, 0.01, 0.1, 1, 5, 10
SVC	✓	Kernel	linear, rbf, poly
		C	0.001, 0.01, 0.1, 1.0, 10, 100
		gamma	0.001, 0.01, 0.1, 1
KNeighborsClassifier	✓	n_neighbors	3, 5, 7, 9
		weights	uniform, distance
		algorithm	ball_tree, kd_tree, brute
		p	1, 2
DecisionTreeClassifier	X	criterion	gini, entropy
		splitter	best, random
		max_depth	1, 3, 5, 7, 9, 11, 12
		min_weight_fraction_leaf	0.1, 0.2, 0.3, 0.4, 0.5
		max_features	auto, log2, sqrt, None
		max_leaf_nodes	None, 10, 20, 30, 40, 50, 60, 70, 80, 90
RandomForestClassifier	X	n_estimators	1, 5, 9, 13, 17, 20
		max_depth	1, 5, 9, 13, 17, 20
		min_samples_split	2, 3, 4, 5, 10
		min_samples_leaf	1, 2, 4
		bootstrap	True, False
XGBClassifier	X	max_depth	3, 4, 5, 6, 7, 8, 9, 10, 11
		min_child_weight	1, 2, 3, 4, 5, 6, 7
		eta	0.001, 0.01, 0.1, 0.2, 0.5, 0.3, 1
		subsample	0.7, 0.8, 0.9, 1.0
		colsample_bytree	0.7, 0.8, 0.9, 1.0
		objective	reg:squarederror

For training the model, we used a nested K fold cross-validation approach in which we could optimize the hyperparameters of the model. Different numbers of folds were used for outer cross-validation (where we evaluated the dataset) and inner cross-validation (where we tuned parameters on the evaluation sets). Feature selection was made by both manual and automatic methods. Manual feature selection was performed in the initial steps by limiting the features to the list of desired features.

We also performed the automated feature selection by feeding the varied number of features (k parameter of SelectKBest class) as a hyperparameter in the pipeline (referred to as automated feature selection in Figure 2). Automatic feature selection was performed using SelectKBest class from scikit-learn Library, in which mutual information between individual features and the target variable was used to decide on the final set with k features. For parameter k of SelectKBest, we used all integer numbers (n) in the range of:

$$3 \leq n \leq max(features)$$

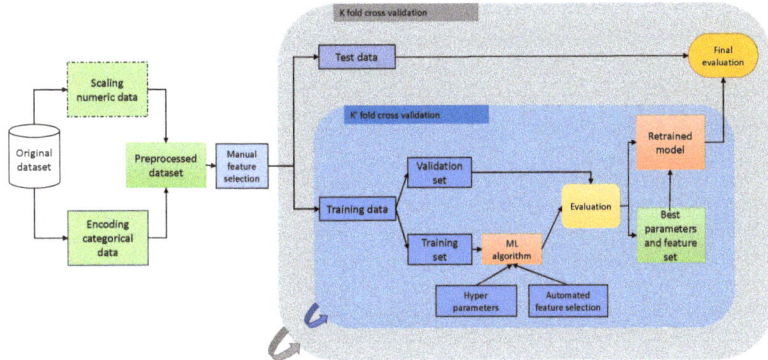

Figure 2. The machine-learning pipeline in more details (symbols with dashed line edges illustrate that that step is not always performed).

We trained seven machine-learning algorithms to predict the EHO values of users: Lasso, Ridge, SVR, K-nearest neighbors, decision tree, random forest and gradient boosted trees (XGBoost). We selected these models to investigate a range of models, including linear and non-linear models. Given the fact that the choice of different hyper parameters may change the results considerably, we chose a varied range of values for different hyper parameters [30].

The list of machine-learning algorithms and the corresponding hyper parameters is provided in Table 2. In this paper, we also define two classification problems. One for predicting users' classes based on their eudaimonic orientation: (i) high eudaimonic oriented, (ii) low eudaimonic oriented; the other for predicting users' classes based on their hedonic orientation: (i) high hedonic oriented, (ii) low hedonic oriented. The list of machine-learning algorithms and the hyper parameters of the classification problem is provided in Table 3.

4. Results and Discussion

4.1. User Clustering along Eudaimonic and Hedonic Orientation

Figure 3 shows the distribution of users along their eudaimonic and hedonic orientations. We performed k-means clustering over all the hedonic and eudaimonic variables. From a visual inspection of Figure 3b, we can see that users can not be easily placed in a specific number of clusters. We applied two approaches to determine the optimal number of clusters k: (i) the elbow method and (ii) the silhouette method.

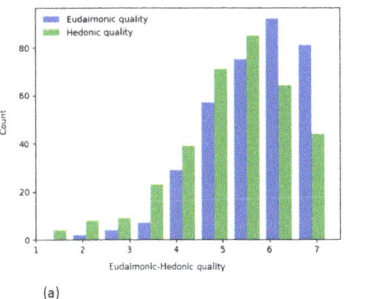
(a)

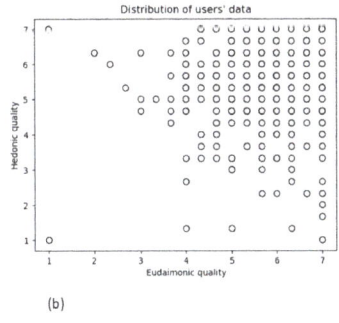
(b)

Figure 3. The distribution of users: (**a**) histogram of hedonic (blue) and eudaimonic (green) values and (**b**) eudaimonic vs. hedonic quality.

We used the elbow method [31] to determine the optimal number of clusters k, using the *KMeans* clustering algorithm. Based on the elbow diagram showed in Figure 4, we decided to investigate k values from 2 to 5. The clustering outcome is depicted in Figure 5.

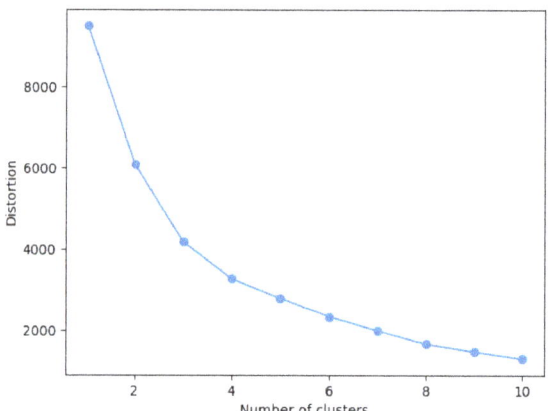

Figure 4. Elbow diagram. x-axis = Number of clusters; y-axis = Distortion (sum of square errors) of data points in the clusters. The clustering was performed with *KMeans* over eudaimonic and hedonic variables.

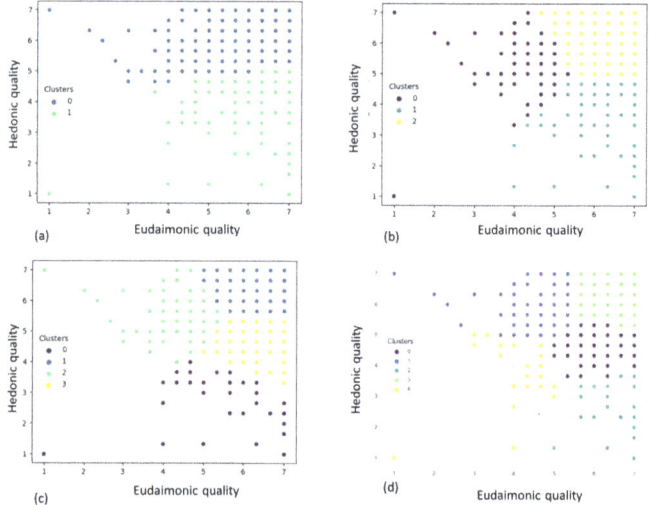

Figure 5. Different clusters formed by *KMeans* over eudaimonic and hedonic variables. (**a**): k = 2, (**b**): k = 3, (**c**): k = 4 and (**d**): k = 5. The parameter k of *KMeans* in *scikit-learn* library determines the number of clusters.

For choosing the best number of clusters we also employed the silhouette method. In this method, the silhouette coefficients are calculated for each data point, which measures the degree of similarity of a data point to its cluster compared to other clusters. The silhouette coefficients can be between -1 to 1. The higher the value of the silhouette coefficient, the more similar the data point is to its cluster than others. A Silhouette coefficient of 1 indicates that the sample is far from other clusters. If the silhouette coefficient is 0, then the sample is close to the decision boundary of two neighboring clusters. If the silhouette coefficient is −1, the sample most likely belongs to another cluster. Silhouette plots are shown in Figure 6.

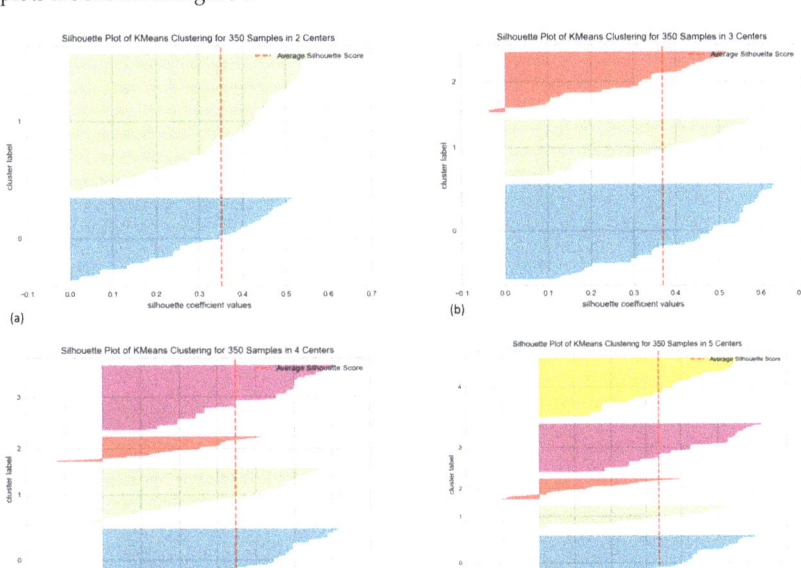

Figure 6. The different number of clusters using the *silhouette method* over eudaimonic and hedonic variables. (**a**): k = 2, (**b**): k = 3, (**c**): k = 4 and (**d**): k = 5. Parameter *k* determines the number of clusters. The average value of the Silhouette coefficients is shown with a red dashed line.

Based on Figure 6, the average Silhouette coefficient is between 0.3 and 0.4, regardless of k's value. k = 5 and k = 4 are not good choices, as all samples in one of the clusters are below average Silhouette coefficient. For k = 2, one of the clusters is larger in size than the other one, which gives this intuition that it can be divided into two subclusters. However, outliers emerge for k = 3. For k = 2 and k = 3, we can see silhouette plots and cluster diagrams showing the distribution of clusters in Figure 7. Figure 7 indicates that clusters are more distinguished based on hedonic orientation if we have two clusters (diagram (a)). Eudaimonic orientation impacts the final cluster more if there are three clusters compared to two clusters.

4.2. Eudaimonic and Hedonic Orientation Prediction

The results of EHO prediction are presented as a regression and as a classification problem.

4.2.1. Regression

We evaluated seven machine-learning-regression models. As the baseline, we used the predictor of the mean value of the target variable. We considered two feature sets in the manual feature selection step: (i) U-F features, which contain user-related features (except for EHOQ and EHO variables, which are the target variables) and (ii) I-F features, which describe the users' interactions. The detailed list of features is reported in Table 1.

To evaluate the prediction of the various algorithms we used the mean absolute error (MAE) and the root mean squared error (RMSE), which penalizes large errors more than MAE [32]. The results are reported in Tables 4–6, for feature sets U-F, I-F and both, respectively.

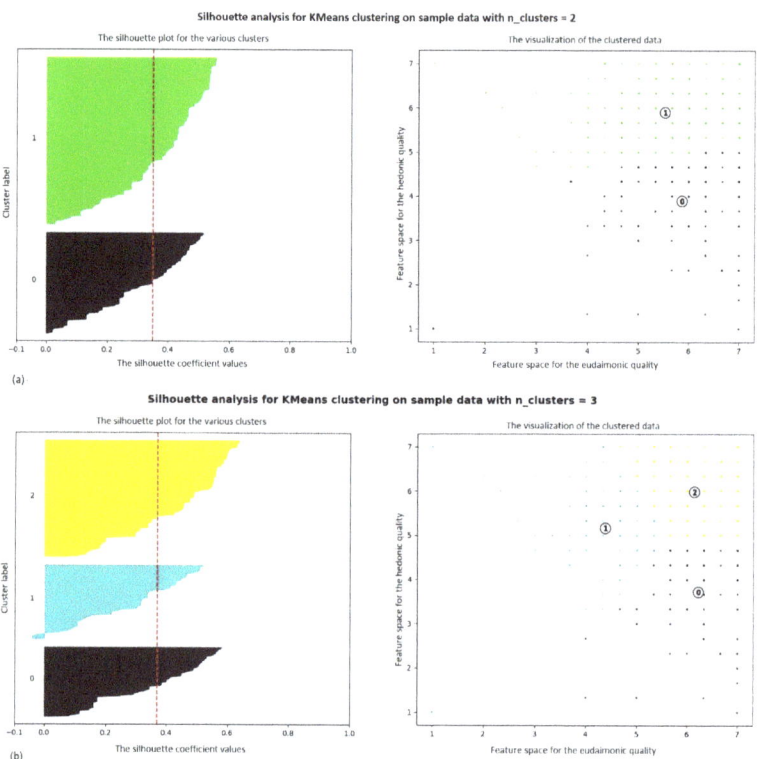

Figure 7. The different number of clusters using *silhouette method* over eudaimonic and hedonic variables. (**a**): number of clusters = 2 and (**b**): number of clusters = 3. Parameter $n\_clusters$ determines the number of clusters. The average value of the Silhouette coefficients is shown with a red dashed line.

Table 4. EHO prediction results (regression problem) using U-F features. Nested cross-validation with 10 outer and five inner splits. EO (Eudaimonic orientation) and HO (Hedonic orientation). Numbers are the mean value of 10 outer splits. Numbers in parenthesis indicate the standard deviation values.

ML Algorithm	RMSE (EO)	MAE (EO)	RMSE (HO)	MAE (HO)
Base	1.09 (0.00)	0.85 (0.00)	1.24 (0.00)	0.97 (0.00)
Ridge	0.88 (0.07)	0.48 (0.07)	1.02 (0.07)	0.86 (0.10)
Lasso	0.88 (0.06)	0.48 (0.07)	1.02 (0.07)	0.86 (0.11)
SVR	0.96 (0.10)	0.66 (0.11)	1.10 (0.09)	1.17 (0.17)
KNN	1.07 (0.09)	1.01 (0.15)	1.11 (0.08)	1.14 (0.16)
Decision Tree	1.09 (0.10)	1.06 (0.15)	1.15 (0.08)	1.37 (0.21)
Random Forest	1.07 (0.09)	1.00 (0.17)	1.09 (0.11)	1.09 (0.19)
XGBoost	1.06 (0.15)	0.98 (0.26)	1.08 (0.07)	1.05 (0.13)

In the regression problem, in the case of using U-F features (Table 4), we can see that Ridge and Lasso performed better compared to the baseline. The SVR also performed better than the baseline based on the MAE and RMSE value in predicting the eudaimonic orientation of users. In terms of hedonic orientation prediction by SVR, only the RMSE value is lower than the baseline. Random Forest and XGBoost provided slightly lower values of RMSE than the baseline but higher MAE values. The decision tree algorithm was only better at predicting hedonic orientation based on the RMSE metric than other methods.

In the regression problem, in the case of using I-F features (Table 5), we can see that the results are more or less close to the baseline, whereas using both U-F and I-F features (Table 6) results are better, except for the MAE value for the KNN algorithm, which is not better but still close to the baseline.

Table 5. EHO prediction results (regression problem) using I-F features. Nested cross-validation with 10 outer and five inner splits. EO (Eudaimonic orientation) and HO (Hedonic orientation). Numbers are the mean value of 10 outer splits. Numbers in parenthesis indicate the standard deviation values.

ML Algorithm	RMSE (EO)	MAE (EO)	RMSE (HO)	MAE (HO)
Base	1.09 (0.00)	0.85 (0.00)	1.24 (0.00)	0.97 (0.00)
Ridge	1.02 (0.09)	0.87 (0.12)	1.04 (0.08)	0.94 (0.12)
Lasso	1.02 (0.09)	0.86 (0.12)	1.05 (0.07)	0.94 (0.11)
SVR	1.04 (0.09)	0.92 (0.13)	1.07 (0.09)	1.01 (0.15)
KNN	1.05 (0.10)	0.95 (0.14)	1.10 (0.10)	1.16 (0.18)
Decision Tree	1.08 (0.11)	1.06 (0.18)	1.10 (0.08)	1.15 (0.15)
Random Forest	1.06 (0.11)	0.99 (0.15)	1.10 (0.09)	1.15 (0.17)
XGBoost	1.07 (0.12)	1.03 (0.19)	1.09 (0.09)	1.09 (0.16)

Table 6. EHO prediction results (regression problem) using U-F and I-F features. Nested cross-validation with 10 outer and five inner splits. EO (Eudaimonic orientation) and HO (Hedonic orientation). Numbers are the mean value of 10 outer splits. Numbers in parenthesis indicate the standard deviation values.

ML Algorithm	RMSE (EO)	MAE (EO)	RMSE (HO)	MAE (HO)
Base	1.09 (0.00)	0.85 (0.00)	1.24 (0.00)	0.97 (0.00)
Ridge	0.88 (0.04)	0.48 (0.05)	1.03 (0.04)	0.87 (0.07)
Lasso	0.87 (0.05)	0.47 (0.05)	1.02 (0.05)	0.86 (0.07)
SVR	0.97 (0.07)	0.69 (0.06)	1.02 (0.05)	0.86 (0.08)
KNN	1.04 (0.08)	0.90 (0.10)	1.05 (0.04)	0.95 (0.08)
Decision Tree	1.01 (0.07)	0.82 (0.08)	1.05 (0.06)	0.93 (0.09)
Random Forest	1.01 (0.06)	0.82 (0.08)	1.04 (0.07)	0.90 (0.11)
XGBoost	0.98 (0.04)	0.74 (0.06)	1.04 (0.07)	0.89 (0.10)

4.2.2. Class Prediction

For the classification problem, the target variables were converted to classes using median splitting. In order to evaluate the classifiers, we used the accuracy, precision, recall, F1 score and the area under the curve (AUC) in the receiver operating characteristic (ROC). Again, we compared U-F and I-F features. As the baseline, we used the majority classifier. Since we used median splitting, the metrics values should be 0.5; however, due to slight inequalities caused by data splitting, they are slightly off the 0.5 value.

When using the U-F features (Tables 7 and 8), the Ridge, SVC and Random Forest algorithms performed better than the baseline. The decision tree and KNN performed close but still better than the baseline.

Table 7. EO prediction results (classification problem) using U-F features. Nested cross-validation with 10 outer and five inner splits. Classification problems on two classes: (a): High_EO (High Eudaimonic Orientated) and (b): Low_EO (Low Eudaimonic Oriented). Numbers are the mean value of 10 outer splits. Numbers in parenthesis indicate the standard deviation values.

ML Algorithm	Accuracy	Precision	Recall	F1 Score	ROC AUC
Base	0.51 (0.00)	0.00 (0.00)	0.00 (0.00)	0.00 (0.00)	0.50 (0.00)
Ridge	0.78 (0.08)	0.75 (0.12)	0.86 (0.12)	0.79 (0.08)	0.89 (0.07)
SVC	0.65 (0.07)	0.66 (0.09)	0.61 (0.12)	0.63 (0.08)	0.71 (0.08)
KNN	0.57 (0.11)	0.57 (0.15)	0.55 (0.15)	0.55 (0.13)	0.60 (0.10)
Decision Tree	0.53 (0.11)	0.53 (0.13)	0.53 (0.13)	0.52 (0.12)	0.52 (0.11)
Random Forest	0.62 (0.08)	0.62 (0.09)	0.58 (0.09)	0.60 (0.07)	0.68 (0.08)

Table 8. HO prediction results (classification problem) using U-F features. Nested cross-validation with 10 outer and five inner splits. Classification problems on two classes: (a): High_HO (High Hedonic Orientated) and (b): Low_HO (Low Hedonic Oriented). Numbers are the mean value of 10 outer splits. Numbers in parenthesis indicate the standard deviation values.

ML Algorithm	Accuracy	Precision	Recall	F1 Score	ROC AUC
Base	0.56 (0.00)	0.00 (0.00)	0.00 (0.00)	0.00 (0.00)	0.50 (0.00)
Ridge	0.66 (0.07)	0.63 (0.10)	0.63 (0.11)	0.62 (0.05)	0.72 (0.07)
SVC	0.65 (0.07)	0.60 (0.13)	0.61 (0.15)	0.60 (0.12)	0.69 (0.10)
KNN	0.59 (0.08)	0.55 (0.15)	0.48 (0.08)	0.51 (0.10)	0.60 (0.09)
Decision Tree	0.53 (0.11)	0.53 (0.13)	0.53 (0.13)	0.52 (0.12)	0.52 (0.11)
Random Forest	0.63 (0.06)	0.63 (0.15)	0.52 (0.12)	0.54 (0.10)	0.65 (0.06)

In the classification problem of predicting EO and HO classes using I-F features (Tables 9 and 10), the Ridge, SVC and Random Forest algorithms performed better than the baseline. The KNN performed close but still better than the baseline. We can see that the decision tree performed better in predicting HO classes than in predicting EO classes.

Table 9. EO prediction results (classification problem) using I-F features. Nested cross-validation with 10 outer and five inner splits. Classification problems on two classes: (a): High_EO (High Eudaimonic Orientated) and (b): Low_EO (Low Eudaimonic Oriented). Numbers are the mean value of 10 outer splits. Numbers in parenthesis indicate the standard deviation values.

ML Algorithm	Accuracy	Precision	Recall	F1 Score	ROC AUC
Base	0.51 (0.00)	0.00 (0.00)	0.00 (0.00)	0.00 (0.00)	0.50 (0.00)
Ridge	0.56 (0.02)	0.56 (0.04)	0.54 (0.04)	0.55 (0.01)	0.59 (0.02)
SVC	0.60 (0.03)	0.60 (0.03)	0.58 (0.05)	0.59 (0.02)	0.63 (0.03)
KNN	0.57 (0.04)	0.57 (0.04)	0.55 (0.05)	0.56 (0.04)	0.60 (0.04)
Decision Tree	0.62 (0.04)	0.61 (0.05)	0.61 (0.06)	0.61 (0.05)	0.64 (0.06)
Random Forest	0.61 (0.06)	0.60 (0.06)	0.61 (0.06)	0.60 (0.05)	0.66 (0.07)

Table 10. HO prediction results (classification problem) using I-F features. Nested cross-validation with 10 outer and five inner splits. Classification problems on two classes: (a): High_HO (High Hedonic Orientated) and (b): Low_HO (Low Hedonic Oriented). Numbers are the mean value of 10 outer splits. Numbers in parenthesis indicate the standard deviation values.

ML Algorithm	Accuracy	Precision	Recall	F1 Score	ROC AUC
Base	0.56 (0.00)	0.00 (0.00)	0.00 (0.00)	0.00 (0.00)	0.50 (0.00)
Ridge	0.61 (0.03)	0.56 (0.04)	0.54 (0.05)	0.55 (0.03)	0.65 (0.03)
SVC	0.63 (0.02)	0.60 (0.04)	0.53 (0.04)	0.56 (0.03)	0.69 (0.02)
KNN	0.63 (0.03)	0.59 (0.07)	0.53 (0.04)	0.56 (0.04)	0.67 (0.04)
Decision Tree	0.66 (0.04)	0.63 (0.06)	0.59 (0.10)	0.60 (0.06)	0.72 (0.06)
Random Forest	0.65 (0.06)	0.62 (0.07)	0.56 (0.07)	0.59 (0.07)	0.70 (0.06)

For most machine-learning algorithms, the results of classification problems using I-F features or U-F features are better than those using the baseline in both cases. When I-F features are used, the decision tree provides better results. With other machine-learning algorithms, the results are close or better than the baseline, regardless of whether I-F or U-F features are used. The baseline performed better in terms of accuracy only when using the decision tree model to predict HO classes using U-F features. For both EO and HO class prediction (Tables 11 and 12), the Ridge as well as Random Forest algorithms outperformed the baseline significantly.

Table 11. EO prediction results (classification problem) using I-F and U-F features. Nested cross-validation with 10 outer and five inner splits. Classification problems on two classes: (a): High_EO (High Eudaimonic Orientated) and (b): Low_EO (Low Eudaimonic Oriented). Numbers are the mean value of 10 outer splits. Numbers in parenthesis indicate the standard deviation values.

ML Algorithm	Accuracy	Precision	Recall	F1 Score	ROC AUC
Base	0.51 (0.00)	0.00 (0.00)	0.00 (0.00)	0.00 (0.00)	0.50 (0.00)
Ridge	0.77 (0.08)	0.73 (0.08)	0.83 (0.13)	0.77 (0.09)	0.85 (0.08)
SVC	0.55 (0.09)	0.54 (0.16)	0.46 (0.17)	0.49 (0.15)	0.58 (0.12)
KNN	0.50 (0.07)	0.50 (0.09)	0.47 (0.09)	0.48 (0.08)	0.51 (0.07)
Decision Tree	0.55 (0.09)	0.54 (0.14)	0.58 (0.11)	0.55 (0.11)	0.55 (0.10)
Random Forest	0.64 (0.06)	0.63 (0.10)	0.64 (0.10)	0.63 (0.08)	0.70 (0.08)

Table 12. HO prediction results (classification problem) using I-F and U-F features. Nested cross-validation with 10 outer and five inner splits. Classification problems on two classes: (a): High_HO (High Hedonic Orientated) and (b): Low_HO (Low Hedonic Oriented). Numbers are the mean value of 10 outer splits. Numbers in parenthesis indicate the standard deviation values.

ML Algorithm	Accuracy	Precision	Recall	F1 Score	ROC AUC
Base	0.56 (0.00)	0.00 (0.00)	0.00 (0.00)	0.00 (0.00)	0.50 (0.00)
Ridge	0.65 (0.06)	0.62 (0.14)	0.60 (0.09)	0.59 (0.07)	0.70 (0.07)
SVC	0.55 (0.07)	0.50 (0.15)	0.46 (0.08)	0.47 (0.09)	0.56 (0.08)
KNN	0.54 (0.09)	0.49 (0.16)	0.49 (0.14)	0.48 (0.13)	0.54 (0.10)
Decision Tree	0.61 (0.06)	0.58 (0.16)	0.59 (0.11)	0.56 (0.07)	0.62 (0.07)
Random Forest	0.64 (0.07)	0.61 (0.18)	0.53 (0.11)	0.56 (0.12)	0.68 (0.06)

5. Discussion

A visual inspection of Figure 3 indicates that the users who participated in our user study were mostly users with a high value of eudaimonic characteristics. We are missing norms of the distributions of the EO and HO for the general population to compare our sample with. Such norms do exist for instruments that have been available for a longer time, such as the BFI [33]. We speculate that the participants who decided to take part in our data collection were more film-savvy and hence had high EO. In future work, we would need to collect data with a more representative sample of participants.

We clustered the participants in the space of EO and HO. Based on the elbow method, three to four clusters appear to be a reasonable choice. However, upon inspecting the clusters in the EHO space, three clusters seems better as the users are nicely separated according to their orientations. In fact, looking at Figure 5b, we can see that cluster 0 is formed of people with low eudaimonic orientation but high hedonic orientation (pleasure seekers). Cluster 1 is characterized by low HO and high EU (meaning seekers). Finally, the users in cluster 2 exhibit both high EO and high HO, which is a novel cluster compared to our previous work [14].

In terms of which ML algorithm is the best for the task at hand, the ridge and lasso regressors and the ridge classifier appears to yield the best results across different setups. There are differences in the performance due to various feature sets but these algorithms constantly outperform others independently on the feature sets, the target variable and the metric.

Generally, the U-F set yields better prediction performance than the I-F set of features. However, the results show that the metric used accounts for how much the features improve over the baseline. When using the U-F features only, there is a substantial improvement both in the RMSE and MAE in both target variables compared to the baseline. When using the I-F features, however, the difference is smaller. Furthermore, in terms of the MAE, the difference is almost negligible, while in terms of the RMSE there is a stronger improvement of prediction over the baseline.

In the classification problem, the different metrics used (accuracy, precision, recall, F-score and area under curve) show different aspects of the classifier performance. However, in our case, accuracy and precision are the most important. Although, here, the U-F features again perform better than the I-F features, the difference is not as pronounced as in the regression case. It is clear that both sets of features account for a substantial amount of variance in HO and EO of users.

6. Conclusions

In this paper, we presented the experimental results of a machine-learning model that predicts the eudaimonic and hedonic orientation of users in the domain of movies. We proposed an approach where features are extracted from user preferences for movies and demographic characteristics, which is a typical set of information present in movie recommender systems. We evaluated the proposed approach on a moderate-sized dataset of 350 users and 703 movies and showed that understanding the hedonic and eudaimonic characteristics of movies and user preferences are good indicators of the eudaimonic and hedonic orientations of users.

Based on the research questions, we showed that (i) There were no distinct clusters of users in the EO-HO space. However, the elbow method indicated that three clusters appears to be a reasonable interpretation: (a) users with high HO and EO, (b) users with high HO and low EO and (c) users with low HO and high HO. Furthermore, we showed that, on our dataset, (ii) the ridge and lasso regressors and the ridge classifier performed the best across a range of different feature sets and metrics. Finally, we showed that (iii) I-F features improved the prediction over the baseline, both in regression and classification; however, the major boost was given by the U-F features.

Future work should address the issue of the potentially non-representative sample. Furthermore, other domains than movies should be explored—for example, music and books.

We also plan to examine the performance of predicting the eudaimonic/hedonic qualities of movies from their subtitles using state-of-the-art algorithms, such as support vector machines, convolution neural networks, recurrent neural networks and Bert [34]. Finally, the proposed approach should be integrated into a recommender system pipeline to evaluate how much the proposed approach helps in improving the quality of recommendations.

Author Contributions: Conceptualization, E.M. and M.T.; methodology, E.M., F.B. and M.T.; software, E.M.; validation, E.M., F.B. and M.T.; formal analysis, E.M.; investigation, E.M., F.B. and M.T.; resources, E.M.; data curation, E.M.; writing—original draft preparation, E.M.; writing—review and editing, E.M., F.B. and M.T.; visualization, E.M.; supervision, M.T.; project administration, M.T.; funding acquisition, M.T. All authors have read and agreed to the published version of the manuscript.

Funding: This research received no external funding.

Informed Consent Statement: Informed consent was obtained from all subjects involved in the study.

Data Availability Statement: Data are available at the corresponding author upon written request.

Conflicts of Interest: The authors declare no conflict of interest.

Abbreviations

The following abbreviations are used in this manuscript:

BFI	Big Five Inventory
EHO	eudaimonic and hedonic orientation of users
EO	eudaimonic orientation of users
HO	hedonic orientation of users
EHP	eudaimonic and hoedonic perception of users
EP	eudaimonic perception of users
HP	hedonic perception of users
FFM	Five Factor Model
MSI	Musical Sophistication Index
TIPI	Ten-Item Personality Inventory

References

1. McCrae, R.R.; John, O.P. An Introduction to the Five-Factor Model and its Applications. *J. Personal.* **1992**, *60*, 175–215. [CrossRef] [PubMed]
2. Tkalčič, M.; Chen, L. Personality and Recommender Systems. In *Recommender Systems Handbook*; Ricci, F., Rokach, L., Shapira, B., Eds.; Springer: New York, NY, USA, 2022; pp. 757–787. [CrossRef]
3. Matz, S.C.; Kosinski, M.; Nave, G.; Stillwell, D.J. Psychological targeting as an effective approach to digital mass persuasion. *Proc. Natl. Acad. Sci. USA* **2017**, *114*, 12714–12719. [CrossRef] [PubMed]
4. Quercia, D.; Kosinski, M.; Stillwell, D.; Crowcroft, J. Our twitter profiles, our selves: Predicting personality with twitter. In Proceedings of the 2011 IEEE International Conference on Privacy, Security, Risk and Trust and IEEE International Conference on Social Computing, PASSAT/SocialCom 2011, Boston, MA, USA, 9–11 October 2011; pp. 180–185. [CrossRef]
5. Golbeck, J.; Robles, C.; Edmondson, M.; Turner, K. Predicting Personality from Twitter. In Proceedings of the 2011 IEEE Third Int'l Conference on Privacy, Security, Risk and Trust and 2011 IEEE Third Int'l Conference on Social Computing, Boston, MA, USA, 9–11 October 2011; pp. 149–156. [CrossRef]
6. Kosinski, M.; Stillwell, D.; Graepel, T. Private traits and attributes are predictable from digital records of human behavior. *Proc. Natl. Acad. Sci. USA* **2013**, *110*, 5802–5805. [CrossRef] [PubMed]
7. Skowron, M.; Tkalčič, M.; Ferwerda, B.; Schedl, M. Fusing Social Media Cues. In Proceedings of the 25th International Conference Companion on World Wide Web-WWW '16 Companion, Montréal, QC, Canada, 11–15 May 2016; ACM Press: New York, NY, USA, 2016; pp. 107–108. [CrossRef]
8. Azucar, D.; Marengo, D.; Settanni, M. Predicting the Big 5 personality traits from digital footprints on social media: A meta-analysis. *Personal. Individ. Differ.* **2018**, *124*, 150–159. [CrossRef]
9. Matz, S.; Chan, Y.W.F.; Kosinski, M. Models of Personality. In *Emotions and Personality in Personalized Services: Models, Evaluation and Applications*; Tkalčič, M., De Carolis, B., de Gemmis, M., Odić, A., Košir, A., Eds.; Springer International Publishing: Cham, Switzerland, 2016; pp. 35–54. [CrossRef]
10. Stewart, B. *Personality and Play Styles: A Unified Model*. 2011. pp. 1–11. Available online: https://mud.co.uk/richard/hcds.htm (accessed on 15 September 2022).

11. Holland, J.L. *Making Vocational Choices: A Theory of Vocational Personalities and Work Environments*; Psychological Assessment Resources; PrenticeHall: Englewood Cliffs, NJ, USA, 1997; Volume 3, pp. XIV, 303 S.
12. Thomas, K.W.; Kilmann, R.H.; Trainer, J. *Thomas–Kilmann Conflict Mode Instrument COMPETING ASSERTIVENESS*; US, Jane Trainer Acme: Belmont, CA, USA, 2010.
13. Oliver, M.B.; Raney, A.A. Entertainment as Pleasurable and Meaningful: Identifying Hedonic and Eudaimonic Motivations for Entertainment Consumption. *J. Commun.* **2011**, *61*, 984–1004. [CrossRef]
14. Tkalčič, M.; Ferwerda, B. Eudaimonic Modeling of Moviegoers. In Proceedings of the 26th Conference on User Modeling, Adaptation and Personalization, Singapore, 8–11 July 2018; ACM: New York, NY, USA, 2018; pp. 163–167. [CrossRef]
15. Tkalcic, M.; Motamedi, E.; Barile, F.; Puc, E.; Mars Bitenc, U. Prediction of Hedonic and Eudaimonic Characteristics from User Interactions. In Proceedings of the Adjunct Proceedings of the 30th ACM Conference on User Modeling, Adaptation and Personalization, Barcelona, Spain, 4–7 July 2022; ACM: Barcelona, Spain, 2022; pp. 366–370. [CrossRef]
16. Allport, G.W. Concepts of trait and personality. *Psychol. Bull.* **1927**, *24*, 284. [CrossRef]
17. Goldberg, L.R. An alternative "description of personality": The big-five factor structure. *J. Personal. Soc. Psychol.* **1990**, *59*, 1216. [CrossRef]
18. Fiske, D.W. Consistency of the factorial structures of personality ratings from different sources. *J. Abnorm. Soc. Psychol.* **1949**, *44*, 329. [CrossRef] [PubMed]
19. Feher, A.; Vernon, P.A. Looking beyond the Big Five: A selective review of alternatives to the Big Five model of personality. *Personal. Individ. Differ.* **2021**, *169*, 110002. [CrossRef]
20. Delic, A.; Neidhardt, J.; Nguyen, T.N.; Ricci, F. An observational user study for group recommender systems in the tourism domain. *Inf. Technol. Tour.* **2018**, *19*, 87–116. [CrossRef]
21. Abolghasemi, R.; Engelstad, P.; Herrera-Viedma, E.; Yazidi, A. A personality-aware group recommendation system based on pairwise preferences. *Inf. Sci.* **2022**, *595*, 1–17. [CrossRef]
22. Botella, C.; Riva, G.; Gaggioli, A.; Wiederhold, B.K.; Alcaniz, M.; Baños, R.M.; Erino, S.I.S.; Ipresso, P.I.C.; Aggioli, A.N.G.; Allavicini, F.E.P.; et al. The Present and Future of Positive Technologies. *Cyberpsychol. Behav. Soc. Netw.* **2012**, *15*, 78–84. [CrossRef] [PubMed]
23. Wirth, W.; Hofer, M.; Schramm, H. Beyond Pleasure: Exploring the Eudaimonic Entertainment Experience. *Hum. Commun. Res.* **2012**, *38*, 406–428. [CrossRef]
24. John, O.; Srivastava, S. The Big Five trait taxonomy: History, measurement, and theoretical perspectives. In *Handbook of Personality: Theory and Research*, 2nd ed.; Pervin, L.A., John, O.P., Eds.; Guilford Press: New York, NY, USA, 1999; Volume 2, pp. 102–138.
25. Ferwerda, B.; Tkalčič, M. You Are What You Post: What the Content of Instagram Pictures Tells About Users' Personality. In Proceedings of the ACM IUI 2018 Workshops, Tokyo, Japan, 7–11 March 2018; Volume 2068.
26. Ferwerda, B.; Tkalčič, M. Exploring the Prediction of Personality Traits from Drug Consumption Profiles. In Proceedings of the Adjunct Publication of the 28th ACM Conference on User Modeling, Adaptation and Personalization, Genoa, Italy, 12–18 July 2020; ACM: Genoa, Italy, 2020; pp. 2–5. [CrossRef]
27. Berkovsky, S.; Taib, R.; Koprinska, I.; Wang, E.; Zeng, Y.; Li, J.; Kleitman, S. Detecting Personality Traits Using Eye-Tracking Data. In Proceedings of the 2019 CHI Conference on Human Factors in Computing Systems-CHI '19, Glasgow, UK, 4–9 May 2019; pp. 1–12. [CrossRef]
28. Gosling, S.D.; Rentfrow, P.J.; Swann, W.B., Jr. A very brief measure of the Big-Five personality domains. *J. Res. Personal.* **2003**, *37*, 504–528. [CrossRef]
29. Müllensiefen, D.; Gingras, B.; Musil, J.; Stewart, L. The musicality of non-musicians: An index for assessing musical sophistication in the general population. *PLoS ONE* **2014**, *9*, e89642. [CrossRef] [PubMed]
30. Van Rijn, J.N.; Hutter, F. Hyperparameter importance across datasets. In Proceedings of the 24th ACM SIGKDD International Conference on Knowledge Discovery & Data Mining, London, UK, 19–23 August 2018; pp. 2367–2376.
31. Nainggolan, R.; Perangin-angin, R.; Simarmata, E.; Tarigan, A.F. Improved the performance of the K-means cluster using the sum of squared error (SSE) optimized by using the Elbow method. *J. Phys. Conf. Ser.* **2019**, *1361*, 012015. [CrossRef]
32. Dou, Z.; Sun, Y.; Zhang, Y.; Wang, T.; Wu, C.; Fan, S. Regional Manufacturing Industry Demand Forecasting: A Deep Learning Approach. *Appl. Sci.* **2021**, *11*, 6199. [CrossRef]
33. Srivastava, S. Norms for the Big Five Inventory and other personality measures. *Hardest Sci.* **2012**, *17*. Available online: https://thehardestscience.com/2012/10/17/norms-for-the-big-five-inventory-and-other-personality-measures/ (accessed on 15 September 2022).
34. Hu, Y.; Ding, J.; Dou, Z.; Chang, H. Short-text classification detector: A bert-based mental approach. *Comput. Intell. Neurosci.* **2022**, *2022*, 8660828. [CrossRef] [PubMed]

Article

Adversarial Detection Based on Inner-Class Adjusted Cosine Similarity †

Dejian Guan and Wentao Zhao *

College of Computer, National University of Defense Technology, Changsha 410000, China
* Correspondence: wtzhao@nudt.edu.cn
† This paper is an extended version of our paper published in IEEE, Dejian Guan, Dan Liu, Wentao Zhao. Adversarial Detection based on Local Cosine Similarity. 2022 IEEE International Conference on Artificial Intelligence and Computer Applications (ICAICA), Dalian, China, 24–26 June 2022.

Abstract: Deep neural networks (DNNs) have attracted extensive attention because of their excellent performance in many areas; however, DNNs are vulnerable to adversarial examples. In this paper, we propose a similarity metric called inner-class adjusted cosine similarity (IACS) and apply it to detect adversarial examples. Motivated by the fast gradient sign method (FGSM), we propose to utilize an adjusted cosine similarity which takes both the feature angle and scale information into consideration and therefore is able to effectively discriminate subtle differences. Given the predicted label, the proposed IACS is measured between the features of the test sample and those of the normal samples with the same label. Unlike other detection methods, we can extend our method to extract disentangled features with different deep network models but are not limited to the target model (the adversarial attack model). Furthermore, the proposed method is able to detect adversarial examples crossing attacks, that is, a detector learned with one type of attack can effectively detect other types. Extensive experimental results show that the proposed IACS features can well distinguish adversarial examples and normal examples and achieve state-of-the-art performance.

Keywords: adversarial detection; inner-class adjusted cosine similarity; adversarial examples; deep learning

1. Introduction

In recent years, deep neural networks (DNNs) have attracted extensive attention and provided excellent performance in many fields. However, researchers discovered that DNNs were vulnerable to adversarial examples [1,2]. Szegedy et al. [1] first demonstrated that by adding human imperceptible perturbations on normal examples, adversaries could confuse the judgment of DNNs. This property of DNNs significantly hinders their application in security-critical areas.

There are works trying to explain the reason why there are adversarial examples in DNNs. Szegedy et al. [1] offered a simple explanation that the set of adversarial examples was of extremely low probability, and never or barely appeared in the training and test set. Later, Goodfellow et al. [3] pointed out that the linearity of DNN models is enough to form adversarial examples and they argued that adversarial examples can be explained as a property of high-dimensional dot products; they also highlighted that the direction of perturbation, rather than the specific point in space, mattered most. Tanay et al. [4] argued that the existence of adversarial examples was closely related to model classification boundary and introduced the "boundary tilting" perspective that adversarial examples existed when the classification boundary lay close to the submanifold of normal examples.

The discovery of the fragility of DNNs to adversarial examples triggered a range of research interests in adversarial attacks and defenses. A growing number of methods have been proposed to generate adversarial examples including L-BFGS [1], FGSM [3], and so on. In order to defend against these attacks, researchers also introduced a range of defense methods to counter attacks by enhancing the robustness model [3,5–9], preprocessing input data [10–12], or attempting to differentiate adversarial examples from normal examples [13–17].

As an intuitive defense means, adversarial detecting has attracted a lot of attention. These methods can be divided into two categories: collecting disentangled features in the input space [18–20] or the activation space of target models [13,17,21]. Furthermore, most detection methods rely too much on target models to extract disentangled features. If we cannot get the target model, the methods may not work.

In this work, we propose a novel adversarial example detection method that is independent of whether we can get the target model or not. Our method utilizes the natural adaptation characteristics of the cosine distance to high-dimensional data and introduces predicted label information to measure the similarity between test data and normal data. In Figure 1, we outline our detection method. The extracted feature map from DNNs and the predicted label information are used to estimate the IACS values and the IACS estimates serve as features to train a linear regression classifier to classify the test data. The contribution of this paper is mainly threefold:

- We propose a similarity metric called the inner-class adjusted cosine similarity (IACS) and apply it to detect adversarial examples.
- Our detection method is independent of whether we can get the target model or not, and the extracted IACS values are stable enough to detect adversarial examples crossing attacks.
- Extensive experiments have been conducted and confirm that our method has excellent advantages in detecting adversarial examples compared with other detection methods. Moreover, our method further confirms that the direction of the adversarial perturbation matters most.

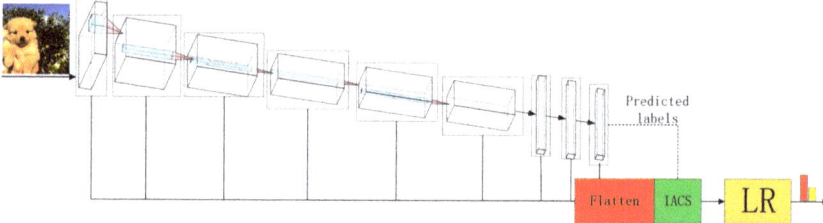

Figure 1. An overview of our detection method based on inner-class adjusted cosine similarity (IACS): We first extract the features of each layer and flatten them into one dimension. Then, the extracted features and predicted label information are used to calculate the IACS and further train the linear regression classifier to discriminate the IACS values of adversarial examples from those of normal examples.

2. Related Works

In this section, we discuss related works which include two parts: adversarial attack and adversarial defense.

2.1. Adversarial Attack

Adversarial attacks try to force deep neural networks(DNNs) to make mistakes by crafting adversarial examples with human imperceptible perturbations. We denote x as the input of DNN, C_x as the label of input x, and $f(\cdot)$ as the well-trained DNNs. Given x and network $f(\cdot)$, we can obtain the label of input x through forward propagation; in general, we can call x an adversarial example if $f(x) \neq C_x$. Here, we introduce five mainstream attack methods including FGSM, PGD, DeepFool, JSMA, and CW. They are all typical attack methods ranging from L_0, L_2, to $L\infty$ norms.

- **FGSM:** The fast gradient sign method(FGSM) was proposed by Goodfellow et al. [3] and is a single-step attack method. The elements of the imperceptibly small perturbation are equal to the sign of the elements of the gradient of the loss function with respect to the input; therefore, it is a typical l_∞-norm attack method. The discovery of the FGSM also proved that the direction of the perturbation, rather than the specific point in space, mattered most.
- **PGD:** The projected gradient descent (PGD) was proposed by Madry et al. [7] and is a multistep attack method. As in the FGSM [3], it also utilizes the gradient of the loss function with regard to the input to guide the generation of adversarial examples. However, the method introduces random perturbations and replaces one big step with several small steps; therefore, it can generate more accurate adversarial examples but it also requires a higher computation complexity.
- **JSMA:** The Jacobian based saliency map attack(JSMA) [22] was proposed by Papernot et al. and is a typical l_0-norm method. It aims to change as few pixels as possible by perturbing the most significant pixels to mislead the model. In this process, the approach updates a saliency map to guide the choice of the most significant pixel at each iteration. The saliency map can be calculated by:

$$S(X,t)[i] = \begin{cases} 0, \; if \; \frac{\partial F_t(X)}{\partial X_i} > 0 \; or \; \sum_{j \neq t} \frac{\partial F_j(X)}{\partial X_i} < 0, \\ (\frac{\partial F_t(X)}{\partial X_i})|\sum_{j \neq t} \frac{\partial F_j(X)}{\partial X_i}|, otherwise \end{cases} \qquad (1)$$

where i is a pixel index of the input.
- **DeepFool:** This algorithm was proposed by Dezfooli et al. [23] and is a nontarget attack method. It aims to find minimal perturbations. The method views the model as a linear function around the original sample and adopts an iterative procedure to estimate the minimal perturbation from the sample to its nearest decision boundary. By moving vertically to the nearest decision boundary at each iteration, it reaches the other side of the classification boundary. Since the DeepFool algorithm can calculate the minimal perturbations, therefore, it can reliably quantify the robustness of DNNs.
- **CW:** This refers to a series of attack methods for the L_0, L_2, and L_∞ distance metrics proposed by Carlini and Wagner [24]. In order to generate strong attacks, they introduced confidence to strengthen the attack performance, and to ensure the modification yielded a valid image, they introduced a *change of variables* to deal with the "box constraint" problem. As a typical optimization-based method, the overall optimization function can be defined as follows:

$$minimize D(x, x+\delta) + c * f(x+\delta), \qquad (2)$$

where c is the confidence, D is the distance function, and $f(\cdot)$ is the cost function. We adopted the l_2-norm attack in the following experiments.

Furthermore, there are black-box adversarial attack methods. Compared with white-box adversarial attacks, they are harder to work or need more perturbations, therefore are easier to be detected. In this paper, we focus on white-box attacks to test detectors.

2.2. Adversarial Defense

In general, adversarial defense can be roughly categorized into three classes: (i) improving the robustness of the network, (ii) input modification, and (iii) detecting-only and then rejecting adversarial examples.

The methods aimed to build robust models try to classify the adversarial example as the right label. As an intuitive method, adversarial training has been extended to many versions from its original version [3] to fitting on large-scale datasets [25] and to ensemble adversarial training [6]. Currently it is still a strong defense method. Although adversarial training is useful, it is computationally expensive. Papernot et al. [8] proposed a defensive distillation to conceal the information of the gradient to defend against adversarial examples. Later, Ross et al. [26] refuted that the defensive distillation could make the models more vulnerable to attacks than an undefended model under certain conditions, and proposed to enhance the model with an input gradient regularization.

The second line of research is input modification, which modifies the input data to filter or counteract the adversarial perturbations. Data compression as a defense method has attracted a lot of attention. Dziugaite et al. [11] studied the effects of JPG compression and observed that JPG compression could actually reverse the drop in classification accuracy of adversarial images to a large extent. Das et al. [12] proposed an ensemble JPEG compression method to counteract the perturbations. Although data compression methods achieve a resistance effect to a certain extent, compression also results in a loss of the original information. In the article [10], the authors proposed a thermometer encoding to defend against adversarial attacks which could ensure no loss of the original information.

Detection-only defense is the other way to defend against adversarial attacks. We divided these methods into two categories: (i) detecting adversarial examples in the input space with raw data and (ii) using latent features of the models to extract disentangled features. For the first category of methods, Kheerchouche et al. [18] proposed to collect natural scene statistics (NSS) from input space to detect adversarial examples. Grosse et al. [19] proposed to train a new $N + 1$ class for adversarial examples classification. Gong et al. [20] constructed a similar method to train a new binary classifier with normal examples and adversarial examples.

The second category of adversarial detection methods uses the target model to extract disentangled features to discriminate adversarial examples. Yang et al.[17] observed that the feature attribution map of an adversarial example near the decision boundary was always different from the corresponding original example. They proposed to calculate the feature attributions from the target model and use the leave-one-out method to measure the differences in feature attributions between adversarial examples and normal examples and further detect adversarial examples. feinman et al. [21] proposed to detect the adversarial examples by kernel density estimates in the hidden layer of a DNN. They trained kernel density estimates (KD) on normal examples according to different classes, and the probability density values of adversarial examples should be less than that of those normal examples, by which they formed an adversarial detector. Schwinn et al. [27] analyzed the geometry of the loss landscape of neural networks based on the saliency maps of the input and proposed a geometric gradient analysis (GGA) to identify the out-of-distribution (OOD) and adversarial examples.

Most related to our work, Ma et al. [13] proposed to use the local intrinsic dimensionality (LID) to detect adversarial examples; the estimator of the LID of x was defined as follows:

$$\hat{LID}(x) = -\left(\frac{1}{k}\sum_{i=1}^{k}log\frac{r_i(x)}{r_k(x)}\right)^{-1}, \qquad (3)$$

where $r_i(x)$ denotes the distance between x and its ith nearest neighbor in the activation space and the $r_k(x)$ is the largest distance among the k-nearest neighbors. They calculated the LID value of samples in each layer and trained a linear regression classifier to discriminate the adversarial examples from normal examples. Our method used the same intuition, that is, we compared the test data with normal data, but we introduced the concept of inner class to limit the comparison scope within the same class label and unlike the LID calculating a Euclidean distance, we used a different basic similarity metric, the cosine similarity.

3. Method

In this section, we introduce our method in detail. Our method stems from the core idea of the fast gradient sign method (FGSM) [3] where the authors pointed out that the direction of the perturbation mattered most. In other words, the adversarial perturbation was sensitive to angles or direction. As a result, we intuitively attempted to use the cosine similarity as the basic metric to discriminate the adversarial examples from normal examples. We studied the cosine similarity and its variant the adjusted cosine similarity [28], which introduces the normalization on the basis of cosine similarity. Furthermore, in order to fit the anomaly detection task, we introduced the predicted label information to extract the disentangled feature between normal examples and adversarial examples. The code is available at https://github.com/lingKok/adversarial-detection-based-on-IACS.

3.1. Basic Metric and Inner-Class Metric

On the basis of a basic metric, we introduce the idea of inner class and propose the inner-class cosine similarity (ICS) and inner-class adjusted cosine similarity (IACS). In this section, we introduce the basic metrics, the cosine similarity (CS) and adjusted cosine similarity (ACS), and the inner-class metrics, the ICS and IACS.

3.1.1. Cosine Similarity

The cosine similarity (CS) is a classical similarity measurement method that measures the similarity between two vectors. With the increase of dimensionality, similarities based on the Euclidean distance face the curse of dimensionality and their characterization ability cannot be guaranteed. Unlike the Euclidean distance, the cosine similarity can effectively measure the relationship between high-dimensional data. The cosine similarity (CS) can be formulated as follows:

$$CS(x,y) = \frac{x \cdot y}{\|x\|\|y\|}, \tag{4}$$

where $(\cdot)$ denotes the dot-product of two vectors.

3.1.2. Adjusted Cosine Similarity

The adjusted cosine similarity (ACS) is a variant of the cosine similarity. Although the cosine similarity can deal with the curse of dimensionality, it is more concerned with the relationship between the angles of vectors and is not sensitive to the absolute value of specific data such as size and length. Therefore, Sarvar et al. [28] proposed the concept of an adjusted cosine similarity. The adjusted cosine similarity offsets the shortcoming by subtracting the corresponding feature mean value. The adjusted cosine similarity of a sample x_i and sample x_j is given by:

$$ACS(x_i, x_j) = \frac{(x_i - \bar{x}) \cdot (x_j - \bar{x})}{\|x_i - \bar{x}\|\|x_j - \bar{x}\|}, \tag{5}$$

where $\bar{x}$ denotes the mean value of samples.

3.1.3. Inner-class Cosine Similarity

The inner-class cosine similarity introduces the concept of inner class on the basis of cosine similarity, which computes the cosine similarity limited to the same predicted class. Given the category of x, the ICS of x is calculated by:

$$ICS(x) = \frac{1}{|C(x)|} \sum_{x_j \in C(x)} CS(x, x_j), \tag{6}$$

where $C(x)$ denotes the set of samples with the same class as x, and $|C(x)|$ denotes the number of elements in set $C(x)$.

3.1.4. Inner-class Adjusted Cosine Similarity

Similar to ICS, the inner-class adjusted cosine similarity (IACS) computes the adjusted cosine similarity limited to the same predicted class. Given the category of x, the IACS of x is calculated by:

$$IACS(x) = \frac{1}{|C(x)|} \sum_{x_j \in C(x)} ACS(x, x_j). \tag{7}$$

3.2. Adversarial Detection Based on Inner-Class Cosine Similarity

In this section, we describe the implementation of the detection method in detail.

3.2.1. Notation and Terminology

Given a well-trained deep neural network classifier $f(\cdot)$, we denote the mixture data as $x_i \in D_{mix}$ (including normal and adversarial examples), the baseline data as $x_j \in D_{bsd}$ (only including normal data), $f_k(\cdot)$ as the output of the k_{th} layer of the classifier ($0 <= k <= n$), and $L_k(\cdot)$ as the flattened feature of the $f_k(\cdot)$.

3.2.2. Detector Training

In the training phase, we first collect the flattened features of each classifier layer. The flattening operation of sample x can be formulated as follows:

$$L_k(x) = flatten(f_k(x)), \tag{8}$$

where the $flatten(\cdot)$ denotes the flattening operation, which flattens the multidimensional data into one dimension.

Then, we calculate the adjusted cosine similarity of the mixture data $x_i \in D_{mix}$ with $x_j \in D_{bsd}$, which can be formulated as follows:

$$ACS(L_k(x_i), L_k(x_j)) = \frac{(L_k(x_i) - \overline{L_k}(x_i)) \cdot (L_k(x_j) - \overline{L_k}(x_j))}{\|(L_k(x_i) - \overline{L_k}(x_i))\| \|(L_k(x_j) - \overline{L_k}(x_j))\|}, \tag{9}$$

where $\overline{L_k}(x_i)$ denotes the average of the lth layer output features of the mixture examples, and $\overline{L_k}(x_j)$ denotes the average of the lth layer output features of the baseline examples (normal examples). This means we calculate the ACS values between the mixture data and normal data.

In order to better fit the anomaly detection task, we propose the inner-class adjusted cosine similarity (IACS) metric to detect adversarial examples. Given some label information predicted by classifier $f(\cdot)$, the adjusted cosine similarity (ACS) with the same predicted label as the x's label is selected to calculate the mean value, which is used as the IACS value of the sample x at the k_{th} layer, as shown in Equation (10).

$$IACS_k(x) = \frac{1}{|C_k(x)|} \sum_{L_k(x_j) \in C_k(x)} ACS(L_k(x), L_k(x_j)), \tag{10}$$

where $C_k(x)$ denotes the set of the k_{th} layer output features of normal samples with the same label as sample x.

We next describe how the inner-class adjusted cosine similarity (IACS) estimates can serve as features to train a detector to discriminate adversarial examples from normal examples. Just as Algorithm 1 shows, the IACS values associated with each mixture sample are estimated with the baseline samples by Equations (9) and (10). Then, we use the IACS values (one value for one layer) to train a linear regression classifier, in which the IACS values from adversarial examples are labeled as 1 and the IACS values from normal examples are labeled as 0 in the experiment.

Algorithm 1 Adversarial detection algorithm based on IACS.

Require:
  $f(\cdot)$: A target classifier trained well by normal examples.
  D_{mix}: Mixture dataset D_{mix}, $x_i \in D_{mix}$.
  D_{bsd}: Baseline dataset D_{bsd}, $x_j \in D_{bsd}$.
Ensure: Linear regression classifier LR.
1: Extract the output of $f(\cdot)$'s layer: $\{f_k(x)\}_1^n$.
2: Flatten the output and get: $\{L_k(x)\}_1^n$.
3: **for** k=1:n (number of layer) **do**
4:   Calculate the mean value of $L_k(x_i)$ and $L_k(x_j)$ in a minibatch, $x_i \in M$, $x_j \in N$.
5:   Calculate the adjusted cosine similarity by Equation (9) and get $ACS(L_k(x_i), L_k(x_j))$.
6:   Calculate the IACS by Equation (10) and get $IACS_k(x_i)$.
7: **end for**
8: Set the feature $IACS(x_i)$ as 1 if x_i is from adversarial example else 0;
  $IACS(x_i) = [IACS_1, IACS_2, ..., IACS_n]^T$
9: Train a linear regression classifier LR on $(IACS_{pos}, IACS_{neg})$.

In addition, note that there is no need to choose a very big baseline dataset (normal examples) to calculate the IACS values, provided that the baseline data is chosen relatively randomly and there are enough samples in the same category to fully maintain its inner-class characteristics. This can significantly reduce the computation load. In the experiment, we found that the detecting performance could be efficiently ensured even for a size of baseline data as small as 100, that is, 10 normal samples per class.

3.2.3. Detector Assessment

In the detecting phase, the test data can be classified by its IACS values. In fact, the trained linear regression classifier (LR) is a binary classifier, therefore, we used the AUC score to measure the performance of the LR. The AUC score denotes the area under the receive operating characteristic which can efficiently avoid the difference caused by manual selection thresholds. The closer the AUC score is to 1, the better the performance is and the closer it is to 0.5, the worse the performance of the LR is.

In experiments, we divided the mixture dataset into a training set and the test set with the ratio of 7:3. That is, we used the IACS values of the training set to train the LR and calculated the AUC score to measure the performance of the LR.

4. Experiments and Results

In this section, we evaluated the discrimination ability of IACS values between adversarial examples and normal examples and tested these features on the MNIST, SVHN, and CIFAR10 datasets. We conducted a comparison with the state-of-art methods including kernel density estimates (KD)-based method [21], local intrinsic dimensionality (LID)-based method [13] and natural scene statistics (NSS)-based method [18].

4.1. Experiment Settings

Hardware setup: All our experiments were conducted on a computer that was equipped with an Intel(R) Core(TM) i9-10920X CPU and an RTX 3080 GPU.

Model: The pretrained DNN model structure used for MNIST and SVHN was the same, that is, a Convnet with $3 \times 3 \times 16$, $3 \times 3 \times 32$, and $3 \times 3 \times 64$ convolutional layers followed by 2×2 max pooling layers and two 200-unit fully connected layers. They achieved an accuracy of 99.34% and 87.39% on MNIST and SVHN, respectively. For CIFAR10, we trained a fine-tuned Resnet20 with an additional linear layer. This model reported an accuracy of 87.09%. Refer to Table 1 for the detailed training parameters.

Table 1. Parameters set for training the classifier.

Parameter	MNIST	SVHN	CIFAR
Optimization Method	SGD	SGD	Adam
Learning Rate	0.05	0.05	0.001
Momentum	0.9	0.9	-
Batch Size	200	100	100
Epoch	20	40	200

Adversarial examples: We implemented five attacks based on an open uniform platform for security analysis [29], including: fast gradient sign method (FGSM) [3], projected gradient descent (PGD) [7], Jacobian based saliency map attack (JSMA) [22], DeepFool [23], and CW_2 [24].

- **FGSM**: We set the perturbation amplitude ϵ. For MNIST, we set the amplitude ϵ as 0.3. For SVHN and CIFAR10, we set it as 0.1.
- **PGD**: There were two parameters to set: the number of iterations it and the perturbation amplitude ϵ. In the experiment, we set it as 1000 for the three datasets and we set ϵ as 0.3 for MNIST and 0.1 for both SVHN and CIFAR10.
- **JSMA**: The perturbation coefficient θ was set to 1 and the modified pixel number was limited by the parameter γ, which was set to 0.2 for the three datasets.
- **DeepFool**: We set the number of iterations it as 50 and the overshoot coefficient as 0.02 for all datasets.
- **CW_2**: There were four parameters that could affect the adversarial examples: the number of iterations it, the confidence coefficient c, the number of search step n_s, and the learning rate lr. We set $c = 0, it = 1000, n_s = 10$, and $lr = 0.002$ for all datasets.

For each attack, we chose 1000 candidate samples from the test dataset (which were classified correctly by the target classifier) and generated the adversarial examples. We also chose an equal number of test samples as baseline data.

4.2. Evaluation of the Discrimination Ability of IACS

In this section, we evaluated the differences between adversarial examples and normal examples based on IACS values. Figure 2 shows the IACS values (from the penultimate layer of Resnet for CIFAR10) of 100 randomly selected adversarial examples (green) generated by CW_2 [24] and those of 100 random normal examples (red) from the CIFAR10 test dataset. We found that the IACS values for the normal examples were significantly larger than the IACS values for the adversarial examples. This met our expectation that the similarity between a normal example and a normal example was greater than that between an adversarial example and a normal example.

We also studied the cosine similarity as a basic metric. We evaluated the AUC score with just a single layer detector with IACS values and ICS values.

In Figure 3, we show the AUC score of each layer from the start layer to the end layer. We found that the overall performance of the IACS was better than that of the ICS. Notice that we only output one IACS or ICS value for each Resnet block for convenience.

4.3. Comparison with Other Methods

We conducted comparative experiments with other three state-of-the-art methods: kernel density estimates (KD)-based method [21], local intrinsic dimensionality (LID)-based method [13], and natural scene statistic (NSS)-based method [18], which are all supervised methods, as is our method. As Tables 2 and 3 show, we report the AUC score of different detection methods on different datasets with different attacks. We found our method achieved good results in almost all datasets and attacks. Especially on CW_2 [24], JSMA [22], DeepFool [23] attacks, our method had obvious advantages.

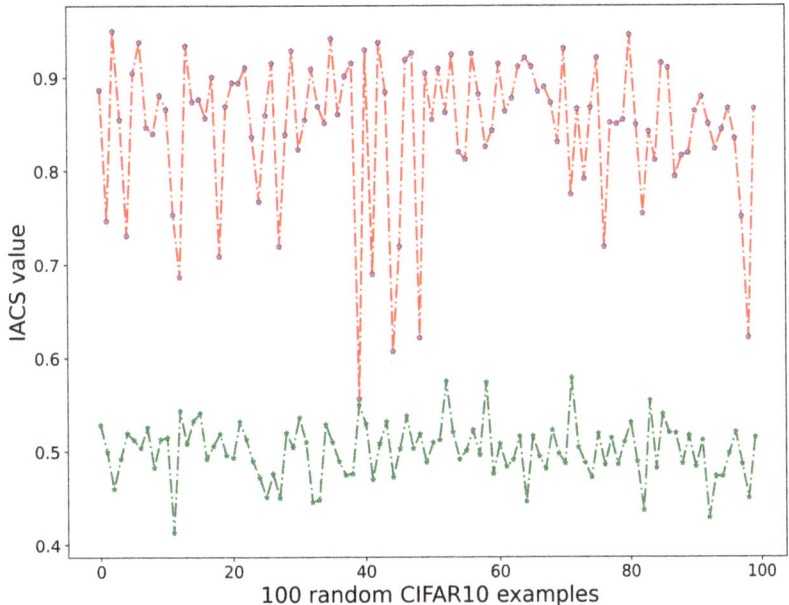

Figure 2. IACS comparison between normal and adversarial examples. The red points denote the normal examples' IACS values, and the green points denote the adversarial examples' IACS values.

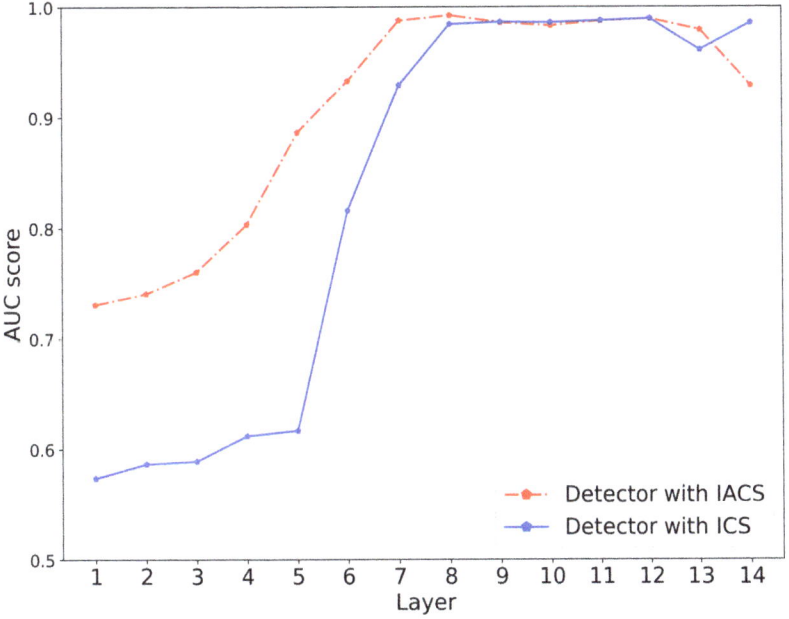

Figure 3. Single layer detector's AUC score with IACS and ICS.

Table 2. The AUC score of different detection methods including the KD-based method, the LID-based method, the NSS-based method, and the IACS-based (our method) method on MNIST and SVHN datasets. The best results are highlighted in bold.

	MNIST				SVHN			
	KD	LID	NSS	IACS	KD	LID	NSS	IACS
FGSM	0.9284	0.9907	**1.0000**	**1.0000**	0.6787	0.996	**1.0000**	0.9987
PGD	0.8938	0.8929	**1.0000**	**1.0000**	0.7926	0.9735	**1.0000**	0.9982
DeepFool	0.9597	0.9844	**1.0000**	**1.0000**	0.5494	0.8048	0.5102	**0.9996**
JSMA	0.9711	0.983	**1.0000**	**1.0000**	0.6801	0.9225	0.9961	**1.0000**
CW_2	0.9847	0.9872	**1.0000**	**1.0000**	0.5163	0.7709	0.6250	**1.0000**

Table 3. The AUC score of different detection methods including the KD-based method, the LID-based method, the NSS-based method, and the IACS-based (our method) method on the CIFAR10 dataset. The best results are highlighted in bold.

	CIFAR			
	KD	LID	NSS	IACS
FGSM	0.7355	0.9950	**0.9999**	0.9832
PGD	0.9774	0.9950	**0.9995**	0.9898
DeepFool	0.6434	0.9109	0.5214	**0.9837**
JSMA	0.5847	0.7575	0.5248	**0.9869**
CW_2	0.716	0.9292	0.5239	**0.9842**

Crossing Attack Study: As an intuitive thought, we hoped the detector trained with one type of attack could be used to detect other types. Therefore, we studied the property of the detector's crossing attacks. We conducted the experiments on the CIFAR10 dataset and compared our method with the LID-based method [13], KD-based method [21], and NSS-based method [18] on different attacks. From Figure 4, we can observe that our method obtained better performance against crossing attacks than the other methods, which meant our method had the ability to detect unknown attacks. We speculated that it was because the IACS value was relatively stable on different attacks. To confirm our conjecture, we presented the IACS values at the penultimate layer on different attacks. As Figure 5 shows, the IACS values of normal examples distributed around about 0.85, and the adversarial examples around about 0.5. The results supported our conjecture.

Crossing Model Study: To further evaluate our method, we used a different model (ConvNet) with $3 \times 3 \times 32$ and $3 \times 3 \times 64$ convolutional layers on the CIFAR10 dataset to extract the features (it reported an accuracy of 84% on the test dataset). In other words, we detected adversarial examples generated by the different models. Table 4 reports the accuracy of the adversarial examples on different models and shows that the adversarial examples generated by DeepFool, JSMA, and CW_2 basically have no attack ability on Convnet. Then, we compared our method with the LID-based method [13] and the KD-based method [21], which rely on the target model to extract features. In Figure 6, we see that the performance did not change much and was even better on our method, while the performance of the other methods decreased significantly, especially the KD-based method [21]. We conjectured that the adjusted cosine similarity could better seize the intrinsic differences between adversarial examples and normal examples even though the adversarial examples had no attack effect.

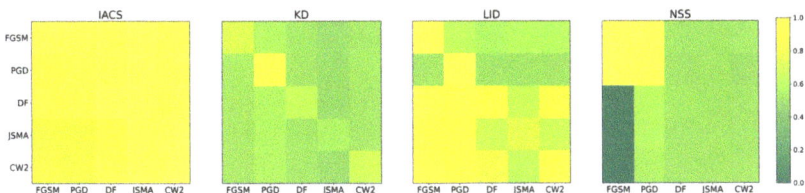

Figure 4. Crossing attacks performance: The horizontal axis represents the training set, and the vertical axis represents the test set. The closer the color is to yellow, the better the detector's performance is (DP refers to DeepFool).

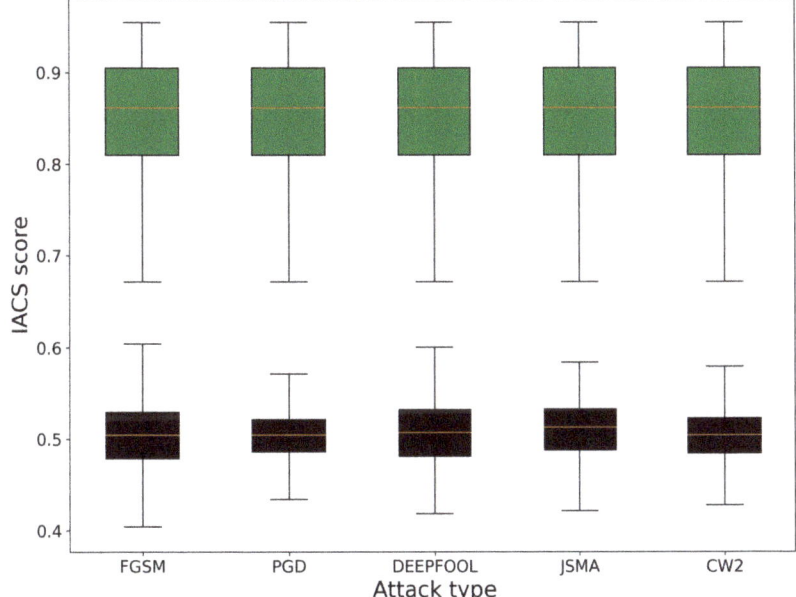

Figure 5. The IACS value with different attacks. Green denotes normal examples' box plots, and black denotes adversarial examples' box plots.

Table 4. The accuracy of different adversarial examples in Resnet (target model) and Convnet.

Model	Resnet	Convnet
FGSM	0.10	0.28
PGD	0.00	0.16
DeepFool	0.00	0.85
JSMA	0.09	0.81
CW_2	0.01	0.83

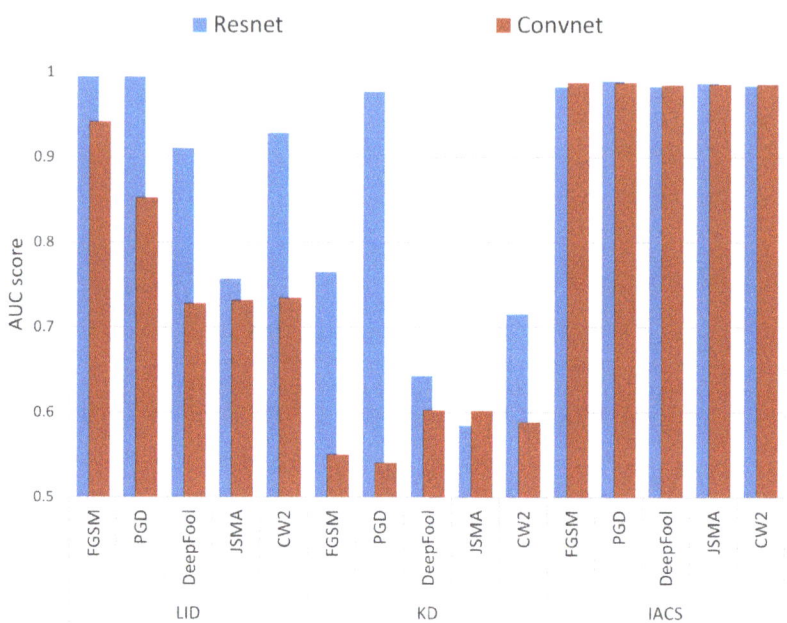

Figure 6. Crossing model performance: the green bars denote the AUC scores of the detector which extracts disentangled features from Resnet (target model), and the red bars denote the AUC scores of the detector based on Convnet.

5. Discussions

In order to figure out the reason why our method worked well, we further discuss the following problems.

Inner class: We performed an ablation study to analyze the contributions of the inner-class property. For comparison, we introduced the property of locality as a comparison with the property of inner class and leveraged K-nearest neighbors to capture the property of locality of the samples. We conducted comparative experiments on three datasets and five attacks with the local adjusted cosine similarity (LACS)-based method and the adjusted cosine similarity (ACS)-based method. For the LACS-based method, which is similar to our preliminary work [30], we averaged the adjusted cosine similarity in the k-nearest neighbors, but not within other normal samples with the same class as the sample. In the ACS-based method, we averaged the adjusted cosine similarity in a minibatch but the k-nearest neighbors or having the same label was not considered. As Table 5 shows, we found that without the inner-class property, the AUC score of the ACS-based method decreased significantly, and replacing the inner class with locality, the LACS-based method was not very efficient, especially for JSMA, DeepFool, and CW_2. These results meant that the label information played an important role. We speculated that this was because the label information predicted by the classifier limited the scope of comparison in the "same" class. As for why the ACS-based method had a low performance, we conjectured it was because the adjusted cosine similarity of adversarial examples was relatively low but the adjusted cosine similarity of normal examples with different classes' samples was also low. Therefore, the averages of the adjusted cosine similarity were close.

Table 5. A comparison of discrimination power (AUC score of a logistic regression classifier) among IACS, LACS, and ACS methods on the different datasets and with different attacks. The best results are highlighted in bold.

	MNIST			SVHN			CIFAR		
	IACS	LACS	ACS	IACS	LACS	ACS	IACS	LACS	ACS
FGSM	1.0000	0.9968	0.5938	**0.9987**	0.9920	0.6683	**0.9832**	0.9485	0.5838
PGD	1.0000	0.8075	0.5532	**0.9982**	0.9328	0.7188	0.9898	**0.9903**	0.6985
DeepFool	1.0000	0.9485	0.5864	**0.9996**	0.7690	0.5730	**0.9837**	0.8758	0.8652
JSMA	1.0000	0.9539	0.5165	1.0000	0.8854	0.6695	**0.9869**	0.6764	0.5385
CW$_2$	1.0000	0.9787	0.5910	1.0000	0.8689	0.5816	**0.9842**	0.9161	0.5680

Basic metric choice: In order to evaluate the contribution of the basic metric, the cosine similarity, we introduced the Euclidean distance as the basic metric, and we proposed the inner-class Euclidean distance (IED)-based method in which we averaged the Euclidean distance within the scope of the same predicted label. As Table 6 shows, the IACS had obvious advantages, especially for more complicated datasets. This further confirmed the advantages of the cosine similarity for high-dimensional data and that the direction of the adversarial perturbation mattered most.

Table 6. A comparison of discrimination power between IACS and IED method on the different datasets and with different attacks. The best results are highlighted in bold.

	MNIST		SVHN		CIFAR	
	IACS	IED	IACS	IED	IACS	IED
FGSM	1.0000	1.0000	0.9987	0.9920	0.9832	**0.9958**
PGD	1.0000	0.9541	0.9982	0.9328	**0.9898**	0.9546
DeepFool	1.0000	0.9878	0.9996	0.8690	**0.9837**	0.8759
JSMA	1.0000	0.9539	1.0000	0.7954	**0.9869**	0.7879
CW$_2$	1.0000	0.9614	1.0000	0.8689	**0.9842**	0.8125

6. Conclusions

In this paper, we proposed an adversarial examples detection method based on the inner-class adjusted cosine similarity. By introducing the predicted label information and leveraging the natural advantages of the cosine distance on high-dimensional data, it greatly improved the detection ability on adversarial examples. Extensive experiments were conducted and showed that our method could achieve a greater performance gain compared with other detection methods. Most importantly, our method could be extended to extract the disentangled features with different models other than the target model (the adversarial attack model) and could also detect adversarial examples from crossing attacks. Therefore, our method had a wider scope of application. Moreover, our method further confirmed that the direction of the adversarial perturbation mattered most. For future research, it would be meaningful to explore more datasets, especially more complicated datasets, such as ImageNet, and other fields such as video outlier detection.

Author Contributions: Conceptualization, D.G.; methodology, D.G.; software, D.G.; validation, D.G. and W.Z.; formal analysis, D.G.; investigation, D.G.; resources, D.G.; data curation, D.G.; writing—original draft preparation, D.G.; writing—review and editing, D.G.; supervision, W.Z.; project administration, W.Z. funding acquisition, W.Z. All authors have read and agreed to the published version of the manuscript.

Funding: This research was funded by the National Natural Science Foundation of China (No. U1811462).

Institutional Review Board Statement: Not applicable.

Informed Consent Statement: Not applicable.

Data Availability Statement: Data are available on request to the authors.

Conflicts of Interest: The authors declare no conflict of interest.

References

1. Szegedy, C.; Zaremba, W.; Sutskever, I.; Bruna, J.; Erhan, D.; Goodfellow, I.; Fergus, R. Intriguing properties of neural networks. *arXiv* **2014**, arXiv:1312.6199.
2. Yu, Z.; Zhou, Y.; Zhang, W. How Can We Deal With Adversarial Examples? In Proceedings of the 2020 12th International Conference on Advanced Computational Intelligence (ICACI), Yunnan, China, 14–16 March 2020; pp. 628–634.
3. Goodfellow, I.J.; Shlens, J.; Szegedy, C. Explaining and Harnessing Adversarial Examples. *arXiv* **2015**, arXiv:1412.6572.
4. Tanay, T.; Griffin, L. A Boundary Tilting Persepective on the Phenomenon of Adversarial Examples. *arXiv* **2016**, arXiv:1608.07690.
5. Miyato, T.; Maeda, S.i.; Koyama, M.; Nakae, K.; Ishii, S. Distributional Smoothing with Virtual Adversarial Training. *arXiv* **2016**, arXiv:1507.00677.
6. Tramèr, F.; Kurakin, A.; Papernot, N.; Goodfellow, I.; Boneh, D.; McDaniel, P. Ensemble Adversarial Training: Attacks and Defenses. *arXiv* **2020**, arXiv:1705.07204.
7. Madry, A.; Makelov, A.; Schmidt, L.; Tsipras, D.; Vladu, A. Towards Deep Learning Models Resistant to Adversarial Attacks. *arXiv* **2019**, arXiv:1706.06083.
8. Papernot, N.; McDaniel, P.; Wu, X.; Jha, S.; Swami, A. Distillation as a Defense to Adversarial Perturbations Against Deep Neural Networks. In Proceedings of the 2016 IEEE Symposium on Security and Privacy (SP), San Jose, CA, USA, 22–26 May 2016; IEEE: San Jose, CA, USA, 2016; pp. 582–597. [CrossRef]
9. Dong, Y.; Su, H.; Zhu, J.; Bao, F. Towards interpretable deep neural networks by leveraging adversarial examples. *arXiv* **2017**, arXiv:1708.05493.
10. Buckman, J.; Roy, A.; Raffel, C.; Goodfellow, I. Thermometer Encoding: One Hot Way to Resist Adversarial Examples. In Proceedings of the International Conference on Learning Representations (ICLR), Vancouver, BC, Canada, 30 April–3 May 2018; p. 22.
11. Dziugaite, G.K.; Ghahramani, Z.; Roy, D.M. A study of the effect of JPG compression on adversarial images. *arXiv* **2016**, arXiv:1608.00853
12. Das, N.; Shanbhogue, M.; Chen, S.T.; Hohman, F.; Chen, L.; Kounavis, M.E.; Chau, D.H. Keeping the Bad Guys Out: Protecting and Vaccinating Deep Learning with JPEG Compression. *arXiv* **2017**, arXiv:1705.02900.
13. Ma, X.; Li, B.; Wang, Y.; Erfani, S.M.; Wijewickrema, S.; Schoenebeck, G.; Song, D.; Houle, M.E.; Bailey, J. Characterizing Adversarial Subspaces Using Local Intrinsic Dimensionality. *arXiv* **2018**, arXiv:1801.02613.
14. Gondara, L. Detecting Adversarial Samples Using Density Ratio Estimates. *arXiv* **2017**, arXiv:1705.02224.
15. Wang, J.; Dong, G.; Sun, J.; Wang, X.; Zhang, P. Adversarial Sample Detection for Deep Neural Network through Model Mutation Testing. In Proceedings of the 2019 IEEE/ACM 41st International Conference on Software Engineering (ICSE), Montreal, QC, Canada, 25–31 May 2019; pp. 1245–1256. [CrossRef]
16. Katzir, Z.; Elovici, Y. Detecting Adversarial Perturbations Through Spatial Behavior in Activation Spaces. In Proceedings of the 2019 International Joint Conference on Neural Networks (IJCNN), Budapest, Hungary, 14–19 July 2019; IEEE: Budapest, Hungary, 2019; pp. 1–9. [CrossRef]
17. Yang, P.; Chen, J.; Hsieh, C.J.; Wang, J.L.; Jordan, M.I. ML-LOO: Detecting Adversarial Examples with Feature Attribution. *arXiv* **2019**, arXiv:1906.03499.
18. Kherchouche, A.; Fezza, S.A.; Hamidouche, W.; Déforges, O. Detection of adversarial examples in deep neural networks with natural scene statistics. In Proceedings of the 2020 International Joint Conference on Neural Networks (IJCNN), Glasgow, UK, 19–24 July 2020; pp. 1–7.
19. Grosse, K.; Manoharan, P.; Papernot, N.; Backes, M.; Cispa, P.M.; Campus, S.I.; University, P.S. On the (Statistical) Detection of Adversarial Examples. *arXiv* **2017**, arXiv:1702.06280
20. Gong, Z.; Wang, W.; Ku, W.S. Adversarial and Clean Data Are Not Twins. *arXiv* **2017**, arXiv:1704.04960.
21. Feinman, R.; Curtin, R.R.; Shintre, S.; Gardner, A.B. Detecting Adversarial Samples from Artifacts. *arXiv* **2017**, arXiv:1703.00410.
22. Papernot, N.; McDaniel, P.; Jha, S.; Fredrikson, M.; Celik, Z.B.; Swami, A. The Limitations of Deep Learning in Adversarial Settings. *arXiv* **2015**, arXiv:1511.07528.
23. Moosavi-Dezfooli, S.M.; Fawzi, A.; Frossard, P. DeepFool: A Simple and Accurate Method to Fool Deep Neural Networks. In Proceedings of the 2016 IEEE Conference on Computer Vision and Pattern Recognition (CVPR), Las Vegas, NV, USA, 27–30 June 2016; IEEE: Las Vegas, NV, USA, 2016; pp. 2574–2582. [CrossRef]
24. Carlini, N.; Wagner, D. Towards Evaluating the Robustness of Neural Networks. In Proceedings of the 2017 IEEE Symposium on Security and Privacy (SP), San Jose, CA, USA, 22–24 May 2017; IEEE: San Jose, CA, USA, 2017; pp. 39–57. [CrossRef]
25. Kurakin, A.; Goodfellow, I.J.; Bengio, S. Adversarial Machine Learning at Scale. *arXiv* **2016**, arXiv:arXiv:1611.01236.
26. Ross, A.S.; Doshi-Velez, F. Improving the Adversarial Robustness and Interpretability of Deep Neural Networks by Regularizing Their Input Gradients. In Proceedings of the AAAI Conference on Artificial Intelligence, New Orleans, LA, USA, 2–7 February 2018; p. 10.
27. Schwinn, L.; Nguyen, A.; Raab, R.; Bungert, L.; Tenbrinck, D.; Zanca, D.; Burger, M.; Eskofier, B. Identifying untrustworthy predictions in neural networks by geometric gradient analysis. In Proceedings of the Uncertainty in Artificial Intelligence, PMLR, Online, 27–29 July 2021; pp. 854–864.

28. Sarwar, B.; Karypis, G.; Konstan, J.; Reidl, J. Item-based collaborative filtering recommendation algorithms. In Proceedings of the Tenth International Conference on World Wide Web—WWW '01, Hong Kong, China, 1–5 May 2001; ACM Press: Hong Kong, China, 2001; pp. 285–295. [CrossRef]
29. Ling, X.; Ji, S.; Zou, J.; Wang, J.; Wu, C.; Li, B.; Wang, T. DEEPSEC: A Uniform Platform for Security Analysis of Deep Learning Model. In Proceedings of the 2019 IEEE Symposium on Security and Privacy (SP), San Francisco, CA, USA, 20–22 May 2019; IEEE: San Francisco, CA, USA, 2019; pp. 673–690. = [CrossRef]
30. Guan, D.; Liu, D.; Zhao, W. Adversarial Detection based on Local Cosine Similarity. In Proceedings of the 2022 IEEE International Conference on Artificial Intelligence and Computer Applications (ICAICA), Dalian, China, 24–26 June 2022; pp. 521–525.

Article

Multi-Granularity Semantic Collaborative Reasoning Network for Visual Dialog

Hongwei Zhang [1,*], Xiaojie Wang [1], Si Jiang [2] and Xuefeng Li [3]

[1] Center for Intelligence Science and Technology, Beijing University of Posts and Telecommunications, Beijing 100876, China
[2] School of Cyberspace Security, Beijing University of Posts and Telecommunications, Beijing 100876, China
[3] School of Artificial Intelligence and Big Data, Henan University of Technology, Zhengzhou 450001, China
* Correspondence: zsszhw@bupt.edu.cn

Abstract: A visual dialog task entails an agent engaging in a multiple round conversation about an image. Notably, one of the main issues is capturing the semantic associations of multiple inputs, such as the questions, dialog history, and image features. Many of the techniques use a token or a sentence granularity semantic representation of the question and dialog history to model semantic associations; however, they do not perform collaborative modeling, which limits their efficacy. To overcome this limitation, we propose a multi-granularity semantic collaborative reasoning network to properly support a visual dialog. It employs different granularity semantic representations of the question and dialog history to collaboratively identify the relevant information from multiple inputs based on attention mechanisms. Specifically, the proposed method collaboratively reasons the question-related information from the dialog history based on its granular semantic representations. Then, it collaboratively locates the question-related visual objects in the image by leveraging refined question representations. The experimental results conducted on the VisDial v.1.0 dataset verify the effectiveness of the proposed method, showing the improvements of the best normalized discounted cumulative gain score from 59.37 to 60.98 with a single model, from 60.92 to 62.25 with ensemble models, and from 63.15 to 64.13 with performing multitask learning.

Keywords: attention mechanisms; collaborative reasoning; multi-granularity; visual dialog

1. Introduction

With the many advances gleaned from the intersection of vision and language domains, several vision–language tasks (e.g., image captioning [1], visual question answering (VQA) [2], referring expression comprehension [3], and visual dialog [4]) have been introduced, attracting massive attention from the computer-vision community. Specifically, a visual dialog [4] is used to train an artificial intelligence (AI) agent to answer questions based on an image and its dialog history. This method offers practical benefits to society, such as in aiding visually impaired users to enjoy and appreciate electronic imagery.

To answer a question correctly and accurately, the AI agent first browses the dialog history to identify passages related to the question. Then, it locates the specific visual objects that match the explicit semantic intent of the question. Figure 1 shows an example of the visual dialog task, to predict the correct answer for the current question Q3, "Is it next to the apple and orange?", the agent traces the dialog history to obtain clues about the meaning of "it". In this case, "it" refers to a "bag", as mentioned in the Q2–A2 pairing. The agent must also determine if there are any mentions of the words "apple" and "orange". Then, it grounds these entities in the image. For this to work, the agent must comprehensively understand the current question and the past question–answer pairs. In this case, the agent learns that the spatial relationship should be correctly inferred between "it" and the apple and orange fruits pictured.

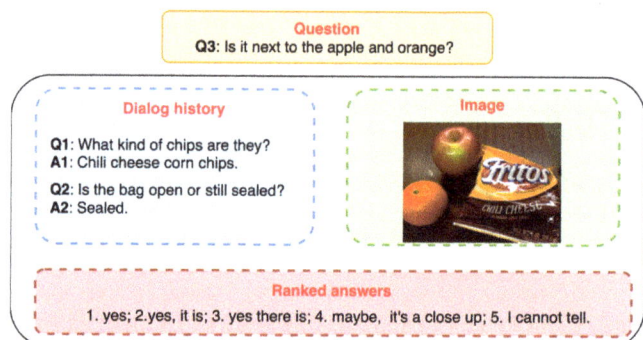

Figure 1. An example of visual dialog task. Based on an image and the dialog history, an agent needs to answer a series of questions by ranking a list of 100 candidate answers. Here, we only show top5 ranked answers.

To endow an AI agent with these capabilities, most methods use attention mechanisms to acquire clues from the dialog history and select the specific visual objects that are relevant to the question. Dual attention networks (DAN) [5] uses sentence-granularity semantic representations of the question and dialog history to compute the weight distributions of all of the question–answer rounds. Visual object identification then requires multi-head attention mechanisms. ReDAN [6] and dual-channel multi-hop reasoning model (DMRM) [7] use sentence-granularity semantic representations to iteratively capture information from the dialog history and the image via multiple reasoning steps. Using multi-head attention mechanisms, lightweight transformer for many inputs (LTMI) [8] and transformer-based modular co-attention (MCA) [9] apply token-granularity semantic representations to excavate token-to-token/visual object interactions and assign weight distributions. However, these methods only use single-granularity semantic representations of the question and dialog history to manage the multi-model intersections, which is intuitively insufficient to accurately answer the question. For example, when answering Q3 in Figure 1, the agent may infer incorrect answers if it only understands the semantic token information from the question while not understanding the question's semantic information. To compensate for this deficiency, the multi-view attention network (MVAN) [10] aggregates question-relevant historical information to facilitate visual object grounding by considering both token- and sentence-granularity semantic representations. MVAN overcomes the single-granularity modeling limitations, but it fails to explicitly and collaboratively explore the impact of different granularity semantic representations to locate the target visual objects.

Nevertheless, most of the approaches above are of single-granularity approaches, which exhibit limited reasoning skills and ignore latent information about relationships among the question, the dialog history, and the image. To this end, we propose a novel multi-granularity semantic collaborative reasoning network (MGSCRN) for visual dialog tasks. The model collaboratively extracts question-related information from the dialog history and original visual objects via multi-granularity encoding. Figure 2 shows a diagram for the proposed method. We first encode the question and dialog history into token- and sentence-granularity semantic representations. Then, we collaboratively assess the question-related information from the dialog history by considering the granular semantic representations of the question and dialog history. In this process, refined question representations with different granularities are obtained. Next, we perform collaborative reasoning to form an accurate alignment between visual objects and these two refined question representations. Finally, the target visual object representation-fused question features are delivered to the decoder for predicting answers.

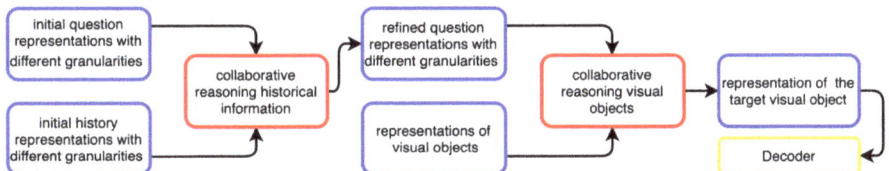

Figure 2. A diagram for proposed method MGSCRN.

The major contributions of this work are summarized as follows. First, the joint use of the multi-granularity semantic representations of the new question and the image's dialog history allows MGSCRN to successfully complete a visual dialog task. The collaborative reasoning of the coreference relations among multiple inputs capitalizes on the heterogeneous informative relationships (i.e., token- and sentence-granularity semantic information) in the history, including those of the various objects in the image. Second, our experiments with the VisDial v1.0 dataset demonstrate outstanding performance results that are superior to other state-of-the-art methods. Third, our qualitative analysis shows that MGSCRN infers both historical and visual question-related information more accurately than the other methods, which notably utilize single semantically granular representations.

This paper is structured in the following manner. The related works are shown in Section 2. Section 3 describes our proposed MGSCRN's model. In Section 4, we introduce our experimental setup and the visual dialog task measures. In Section 5, we empirically evaluate our proposed model using VisDial v1.0, VisPro, and VisDialConv datasets. Then, we report our experimental results and analyze the findings. This paper is concluded in Section 6.

2. Related Work

Visual Dialog

With the advancement of computing sciences and artificial intelligence, many researchers seek solutions to achieve truly intelligent artificial systems for complex real-life applications. Various machine learning techniques have been developed to tackle challenging real-world problems, such as unstructured data classification in social networks [11], sentiment analysis of Chinese short financial texts [12], and even energy conservation in communications [13].

Recently, many works have paid attention to handle multimodal cognitive problems by analyzing vision and language from real life. Compared to the majority of vision–language tasks, such as image captioning [1] and VQA [2], visual dialog [4], which involves multi-turn visual-grounded conversations, is more engaging and challenging. Two separate teams of researchers simultaneously published two types of visual dialog datasets. One is the Guess-what dataset that De Vries et al. [14] gathered. It is a goal-driven visual dialog. The questioner must correctly predict the visual target by asking a series of questions. The Oracle provides yes/no/n.a. answers. The other is a large-scale free-form visual dialog dataset called VisDial [4], where the questioner asks open-ended questions based on the image captions and dialog history to assist itself to better comprehend the visual contents; the answerer responds depending on the image and dialog history. In this paper, the second setting was used.

Various technical vantage points were used to study visual dialog tasks. Three baselines were initially proposed to go with the VisDial dataset: the hierarchical recurrent encoder (HRE) [4] included a dialog-RNN sitting on top of a recurrent block, which was capable of selecting relevant history from previous rounds; the late fusion (LF) [4] directly concatenated individual representations of the image, dialog history, and question followed by a linear transformation; and the memory network (MN) [4] treated each prior question–answer pair as a fact and used a SoftMax to obtain probabilities over the stored facts.

Most of the later techniques utilized various attention mechanisms. In each step of the reasoning process, ReDAN [6] calculated attention maps to the dialog history and visual objects. To manage the visual and textual information in a parallel and adaptive way, heterogeneous excitation-and-squeeze network (HESNet) [15] excavated multi-modal attention. Recursive visual attention (RvA) [16] updated visual attention while recursively browsing the dialog history to tackle the problem of visual reference resolution. In order to use the historical information in a balanced way, the reciprocal question representation learning network (RQRLN) [17] utilized transformer-based attention to reciprocally learn a new question representation through the intersections of two types of question representations with and without dialog history.

Other approaches concentrated on combining neural networks with graphical structural representation to capture the semantic dependence between cross-modal information. A visual graph was built by aligning vision–language for graph inference (AVLGI) [18]. It was used to enhance visual features guided by textual features, and a scene graph was then created to incorporate external knowledge. To discover the partially relevant contexts, a context-aware graph (CAG) [19] constructed a dynamic graph structure, which could be iteratively updated by applying an adaptive top-K message transmission mechanism.

Pretrained models were used in recent studies, and they performed admirably. Wang et al. [20] initialized the encoder with BERT to fuse of dialog history and visual contents in order to better leverage the pretrained language representations. Murahari et al. [21] utilized two-stream vision-language pretrained models (i.e., ViLBERT) for this task to improve the interactions between vision and language.

The aforementioned techniques address the problem of visual dialog from several angles. However, these extant methods employ a token or a sentence semantic granularity representation of multiple inputs, so the output still suffers from poor accuracy. In this study, we employ collaborative reasoning for the visual dialog task to gather more pertinent information from the semantic representation of each granularity.

3. Methodology

3.1. Problem Definition

The problem is stated as follows. Given image I, dialog history $H_t = \{C, h_1, \ldots, h_{t-1}\}$, and current question Q_t, where C is the caption describing the image, and $h_i = \left(Q_i, A_i^{gt}\right)$ is the ith question–answer pair after concatenating question Q_i and the ground-truth answer A_i^{gt}. The goal of the dialog agent is to infer the best answer to Q_t by discriminatively ranking a list of answer candidates: $A_t = \{A_1^t, \ldots, A_{100}^t\}$.

3.2. MGSCRN

In this section, we formally describe our proposed method. Figure 3 illustrates the overall structure of MGSCRN. It consists of three main modules. First, in the multi-granularity textual representation (MGTR) module, initial token- and sentence-granularity semantic representations are obtained by adopting self-attention mechanisms of transformers and convolutional neural networks (CNNs), respectively. Second, the question-aware attention-refer (QAAR) module leverages the transformer-based cross-attention. It aims to build intersections between the question and dialog history encoded using the same granular semantic representations, which are obtained by the MGTR module. In this process, two types of dialog history distributions are learned: one representing the attention weights of all of the tokens in the dialog history, and another representing the attention weights of all of the question–answer rounds. To more accurately identify the historical information associated with the question, one distribution is modified by the other during training. Consequently, even if one distribution deviates from the truth, the degree of deviation will be reduced via the corrections offered by the other distribution. Thus, the QAAR module collaboratively reasons the question-related information from the dialog history by leveraging different semantically granular representations of the current question and historical dialog. Third, the visual-aware attention alignment (VAAA) module learns the semantic alignments between the visual features and each representation of the question-fused history obtained from the previous module; hence, two types of distributions over visual objects are learned simultaneously. In the same way as the QAAR module, the VAAA module allows these two distributions to collaboratively assess the question-related visual objects. The following subsections discuss each module in detail.

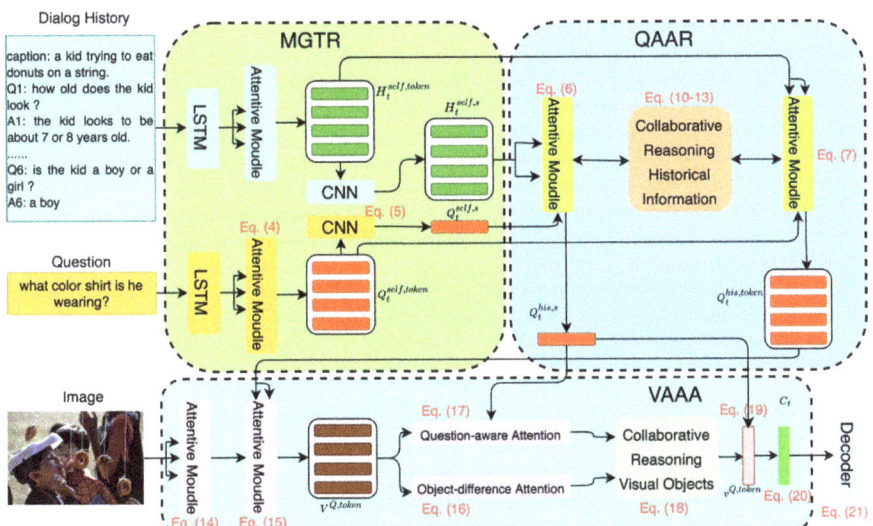

Figure 3. Model architecture of the multi-granularity semantic collaborative reasoning network (MGSCRN) for visual dialog task. CNN = Convolutional neural network; LSTM = Long short-term memory; MGTR = Multi-granularity textual representation; QAAR = Question-aware attention refer; VAAA = Visual-aware attention alignment.

3.2.1. MGTR

To collaboratively reason through the question and dialog history representations with different levels of granularity, the MGTR module adopts transformer-based attention mechanisms [22] and convolutional neural networks (CNN) to obtain the initial token- and sentence-granularity semantic representations. We first embed each token in the current question, Q_t, as $Q_t^G = \left[q_{t,1}^G, q_{t,2}^G, \ldots, q_{t,n_Q}^G\right]$, using a pretrained GloVe-embedding layer [23], where n_Q denotes the number of tokens in Q_t. We then use a long short-term memory (LSTM) to encode Q_t^G into a token sequence, $Q_t^{lstm} = \left[q_{t,1}^{lstm}, q_{t,2}^{lstm}, \ldots, q_{t,n_Q}^{lstm}\right]$. Next, we feed Q_t^{lstm} into a standard transformer attentive module (i.e., $AM(\cdot)$), which is carried out by a multi-head attention layer (i.e., $MutiHead(\cdot)$) and a position-wise fully connected feed-forward layer (i.e., $FFN(\cdot)$). Taking query matrix Q, key matrix K, and value matrix V as input, $AM(\cdot)$ is defined as follows:

$$AM(Q, K, V) = FFN(MultiHead(Q, K, V)), \qquad (1)$$

$$MultiHead(Q, K, V) = Concat(head_1, head_2, \ldots, head_{N_{head}})W^C, \qquad (2)$$

$$head_k = Softmax\left(\frac{QW_k^Q(KW_k^K)^T}{\sqrt{d}}\right)VW_k^V, \qquad (3)$$

where W^C and W_k^* are learned parameter matrices. By setting inputs Q, K, and V as Q_t^{lstm} in $AM(\cdot)$, we obtain the token-granularity semantic representation of the question, $Q_t^{self,token} = \left[q_{t,1}^{self}, q_{t,2}^{self}, \ldots, q_{t,n_Q}^{self}\right]$, after stacking L transformer blocks:

$$Q_t^{self,token} = AM\left(Q_t^{lstm}, Q_t^{lstm}, Q_t^{lstm}\right). \tag{4}$$

To more accurately obtain a sentence-granularity semantic representation of $Q_t^{self,s}$, we employ a CNN instead of average pooling the transformer or bidirectional encoder representations from transformers (BERT) token-embedding outputs, following the approach of [24]. We denote $Q_t^{self,s}$ as follows:

$$Q_t^{self,s} = CNN\left(Q_t^{self,token}\right). \tag{5}$$

Similarly, for the dialog history, H_t, we first concatenate the image caption and $t-1$ question–answer pairs, and then initialize them with the same embedding layer as in the question. Next, we utilize another LSTM and $AM(\cdot)$ to acquire a token-granularity semantic representation of the dialog history, $H_t^{self,token} = \left[h_{0,1}^{self}, \ldots, h_{0,n_H}^{self}, h_{1,1}^{self}, \ldots, h_{1,n_H}^{self}, \ldots, h_{t-1,1}^{self}, \ldots, h_{t-1,n_H}^{self}\right]$, where n_H denotes the number of tokens in each turn of the dialog history, H_t. Likewise, the sentence-granularity semantic representation of the dialog history, $H_t^{self,s} = \left[h_0^{self,s}, h_1^{self,s}, \ldots, h_{t-1}^{self,s}\right]$, is obtained by utilizing Equation (5).

3.2.2. QAAR

Differing from most of the extant works [4,6,17,19] that rely only on single-granularity representations to construct relationships between the question and dialog history, QAAR uses multi-granularity representations to collaboratively capture relevant historical information. To achieve this goal, the question guides the dialog history using sentence- and token-granularity semantic representations to fetch the history features relevant to the question. Two attention-weight distributions over the dialog history are computed simultaneously: all tokens and all rounds. Then, the distributions correct each other to accurately align the features of the question and dialog history items. To this end, we formally describe the QAAR module, which adopts $AM(\cdot)$ to learn the correlated history features while considering multi-granularity information. The equations are as follows:

$$Q_t^{his,s} = AM\left(Q_t^{self,s}, H_t^{self,s}, H_t^{self,s}\right), \tag{6}$$

$$Q_t^{his,token} = AM\left(Q_t^{self,token}, H_t^{self,token}, H_t^{self,token}\right), \tag{7}$$

where $Q_t^{his,s}$ and $Q_t^{his,token}$, respectively, represent the question with fused sentence- and token-granularity. When learning the representations of $Q_t^{his,s}$ and $Q_t^{his,token}$, the SoftMax function existing in $AM(\cdot)$ (i.e., Equations (6) and (7)) produce two types of distributions over the dialog history. Here, we take these out and denote them as $P_{his}^s = [p_0^s, p_1^s, \ldots, p_{t-1}^s] \in \mathcal{R}^t$ and $P_{his}^{token} = \left\{\left[p_{0,1}^{i,token}, \ldots, p_{0,n_H}^{i,token}, p_{1,1}^{i,token}, \ldots, p_{1,n_H}^{i,token}, \ldots, p_{t-1,1}^{i,token}, \ldots, p_{t-1,n_H}^{i,token}\right]\right\}_{i=1}^{n_Q} \in \mathcal{R}^{n_Q \times (t \times n_H)}$, respectively:

$$P_{his}^s = Softmax\left(\frac{Q_t^{self,s} W_k^{Q^s} \left(H_t^{self,s} W_k^{H^s}\right)^T}{\sqrt{d}}\right), \tag{8}$$

$$P_{his}^{token} = Softmax\left(\frac{Q_t^{his,token}W_k^{Q^{token}}\left(H_t^{self,token}W_k^{H^{token}}\right)^T}{\sqrt{d}}\right), \qquad (9)$$

where P_{his}^s represents the attention weights on t rounds of dialog history, and the ith row of P_{his}^{token} represents the attention weights on all of the tokens that possibly have dependencies to the ith token in the question.

To better explore the question-related tokens and sentences in the dialog history, we dynamically modify and update P_{his}^s and P_{his}^{token} under their mutual guidance during training. To make the deterministic distribution, P_{his}^s, more accurate, we first deal with the ith row of P_{his}^{token} by separately adding the weight values of tokens in the different rounds of the dialog history, which is defined as:

$$p_{his}^{i,token} = \left\{\sum_{k=1}^{n_H} p_{j,k}^{i,token}\right\}_{j=0}^{t-1} \in \mathcal{R}^t, \qquad (10)$$

where $p_{his}^{i,token}$ represents the different contributions of t rounds of dialog history to the ith token in the question. Thus, $P_{his}^{Q,token} = \left[p_{his}^{1,token}, p_{his}^{2,token}, \ldots, p_{his}^{n_Q,token}\right]^T \in \mathcal{R}^{n_Q \times t}$ indicates the different contributions of t rounds of dialog history to each token in the question. Then, we add the weight values of each column of $P_{his}^{Q,token}$ to obtain P_{his}^Q, followed by normalization:

$$P_{his}^Q = Norm\left(\sum_{k=1}^{n_Q} P_{his,k}^{Q,token}\right) \in \mathcal{R}^t, \qquad (11)$$

where P_{his}^Q represents the contribution of t rounds of dialog history to the whole question from the perspective of aggregating token-granularity semantic information. Next, as shown in Equation (12), we modify P_{his}^s with the help of P_{his}^Q:

$$P^s = Norm\left(P_{his}^s \odot P_{his}^Q\right) \in \mathcal{R}^t \qquad (12)$$

where $\odot$ denotes the elementwise product. To assist P_{his}^{token} in accurately assigning attention weights over all tokens from the dialog history, we also modify P_{his}^{token} under the constraint of P_{his}^s, which is calculated as follows:

$$P^{token} = \left\{Norm\left[P_{his,r}^s \cdot P_{rn_H:(r+1)n_H}^{i,token}\right]_{r=0}^{t-1}\right\}_{i=1}^{n_Q} \in \mathcal{R}^{n_Q \times (t \times n_H)}, \qquad (13)$$

where $[\cdot]$ denotes the scalar multiplication of vectors, and $p_{rn_H:(r+1)n_H}^{i,token}$ is the subset of the ith row of P_{his}^{token}, which represents the attention weights assigned by the ith token of the question to the all tokens in the rth round of the dialog history.

Finally, during collaborative reasoning, P_{his}^s and P_{his}^{token} are replaced with P^s and P^{token}, respectively, to obtain $Q_t^{his,s}$ and $Q_t^{his,token}$ via the interactive transmitting of information (i.e., P_{his}^s and P_{his}^{token}) in Equations (12) and (13). After stacking L transformer blocks as with the MGTR module, we obtain two types of question representations: $Q_t^{his,s}$ and $Q_t^{his,token}$, which are obtained by leveraging multi-granularity representations of the question and dialog history to collaboratively select question-aware historical information.

3.2.3. VAAA

Given the output representations of the QAAR module, $Q_t^{his,s}$ and $Q_t^{his,token}$, the VAAA module collaboratively aligns them with the visual features. Here, the initial features are extracted by employing the Faster-CNN [25], pretrained using Visual Genome [26], denoted as $V^0 \in \mathcal{R}^{m \times d_v}$, where m and d_v represent the overall number of visual objects in image I and the dimension size of each visual object, respectively. We use a linear projection to map the dimensions of the object embeddings to those of the token. Then, the visual features are denoted as $V \in \mathcal{R}^{m \times d}$. We first establish the latent connections among visual objects in image I via the $AM(\cdot)$:

$$V^{self} = AM(V, V, V) \in \mathcal{R}^{m \times d}. \tag{14}$$

Then, we apply the visual-aware attention mechanism to align visual objects with the visually related tokens from the question. The visual representation, V^{self}, queries the token-granularity semantic representation, $Q_t^{his,token}$, to obtain a new visual representation, $V^{Q,token}$, that fuses the attentive token-granularity question features. We utilize cross-attention to compute $V^{Q,token}$ as follows:

$$V^{Q,token} = AM\left(V^{self}, Q_t^{his,token}, Q_t^{his,token}\right) \in \mathcal{R}^{m \times d}. \tag{15}$$

However, the visual features, $V^{Q,token}$, are obtained by attending to the most relevant question-token features, which are not target visual features for adequately answering Q_t, then, an object-difference attention mechanism is used to locate the question-relevant visual objects in the corresponding image, formulated as:

$$P_V^{g,token} = Softmax\left(\left[V^{Q,token} \odot \left(V^{Q,token} - V^{self}\right)\right] W_V^{token}\right)^T \in \mathcal{R}^{g \times m}, \tag{16}$$

where $V^{Q,token} \odot \left(V^{Q,token} - V^{self}\right) \in \mathcal{R}^{m \times d}$ is the difference between $V^{Q,token}$ and V^{self} representations as guided by $V^{Q,token}$. $W_V^{token} \in \mathcal{R}^{d \times g}$ is the parameter matrix to be learned with g glimpses. The SoftMax function transforms g glimpse results into an attention weight over all visual objects. The final distribution on all of the visual objects is $P_V^{token} = 1/g \sum_{i=1}^{g} P_V^{g,token}$.

P_V^{token} is obtained by finding clues from question tokens and comparing visual object differences. Note that it may inaccurately identify relevant visual objects, depending on the token semantically granular representation of the question. Therefore, we also use the sentence-granularity semantic representation of $Q_t^{his,s}$ to align it with $V^{Q,token}$, which is implemented by the question-aware attention:

$$P_V^s = Softmax\left(\frac{Q_t^{his,s} W_k^{Q^{his,s}} \left(V^{Q,token} W_k^V\right)^T}{\sqrt{d}}\right). \tag{17}$$

The obtained P_V^{token} and P_V^s jointly decide an appropriate distribution, $P_V^{token+s}$, over the visual objects as follows:

$$P_V^{token+s} = Norm\left(P_V^{token} + P_V^s\right). \tag{18}$$

Based on distribution $P_V^{token+s}$, the weighted visual object representation is formulated as follows:

$$v^{Q,token} = \sum_{i=1}^{m} P_{V,i}^{token+s} V_i^{Q,token}. \tag{19}$$

Finally, we fuse the filtered visual feature, $v^{Q,token}$, with $Q_t^{his,s}$ (expressing the complete semantic intent of the question) as the final multimodal fused representation:

$$C_t = tanh\left(W_f \left[v^{Q,token}; Q_t^{his,s}\right]\right), \tag{20}$$

where [;] denotes the concatenation, yielding context representation C_t, which is then fed to the discriminative decoder to predict the answer to the current question, Q_t.

3.3. Discriminative Decoder

We encode the candidate answers in the same way as the questions and the dialog history. Here, we utilize the last hidden states of the LSTMs as candidate answer representations, $A_t = \{u_t^s\}_{s=1}^{100} \in \mathcal{R}^{100 \times d}$. The discriminative decoder ranks these candidates using the dot product operations on A_t and the context representation, C_t. A probability distribution over the candidates is then obtained using the SoftMax function, denoted as:

$$P_t^{Dis} = Softmax\left(C_t(A_t)^T\right). \tag{21}$$

Multi-class cross-entropy loss is employed as the discriminative objective function, which is formulated as follows:

$$\mathcal{L}_{Dis} = \sum_{i=1}^{100} y_i P_{t,i}^{Dis}, \tag{22}$$

where y_i is the one-hot encoded label vector of the ground-truth answer.

4. Experimental Setup and Metrics

4.1. Datasets

We conducted experiments on the VisDial v1.0 dataset [4] to verify our proposed model. The train, validation, and test splits contained 1.23 M, 20 K, and 44 K dialog rounds, respectively. For the train and validation splits, each image had a 10-round dialog, whereas in the test split, a random set of question–answer pairs and a current question were presented alongside each image. The train split included 123K images taken from the MS-COCO dataset [27], and 2 K and 8 K images from Flickr were used for the validation and test splits, respectively. Additionally, each question followed a list of 100 candidate answer options, one of which was the ground truth.

4.2. Evaluation Metrics

To evaluate our model, we used a number of retrieval metrics in the manner of Das et al. [4]. Specifically, *Recall@k* ($k \in \{1, 5, 10\}$) was used to determine where the ground-truth answer was located among the top k ranked responses. The mean rank (*Mean*) was the average rank of the ground-truth answer. It is defined as:

$$Mean = \frac{1}{|Q|} \sum_{i=1}^{|Q|} Rank_i, \tag{23}$$

where Q was the set of all the questions, and $Rank_i$ was the rank of the ground-truth answer.

The mean reciprocal rank (*MRR*) was the reciprocal rank of the ground-truth answer, which was computed as follows:

$$MRR = \frac{1}{|Q|} \sum_{i=1}^{|Q|} \frac{1}{Rank_i}. \tag{24}$$

The normalized discounted cumulative gain (*NDCG*) was introduced to penalize lower-ranked answers with high relevance. It should be note that the *NDCG* accepts many answers that are comparable as right, whereas other metrics only take the rank of a single answer. *NDCG* was defined as:

$$NDCG@k = \frac{DCG@k}{IDCG@k}, \tag{25}$$

where $DCG@k = \sum_{i=1}^{k} \frac{rel(i)}{log(2+i)}$ represents the submitted ranking, $IDCG@k = \sum_{i=1}^{k} \frac{2^{rel(i)}-1}{log(2+i)}$ represents the ideal ranking, the value k is the number of candidate answers with a relevance score greater than 0, and $rel(i)$ is the relevance score of ith candidate answer.

4.3. Training Details

The open-source code from Das et al. [4] was used to implement our model via PyTorch. The words that appeared at least five times in the training split were added to the vocabulary that we created. The question and dialog history were truncated to 20 and 40, respectively, or were padded. GloVe word vectors were used to initialize these words, and all LSTMs were set with a single layer and 512-dimensional hidden units. For the transformer blocks, we set the number of heads to eight, the hidden size to 512 dimensions, and the number of transformer layers to six by stacking the transformer blocks. On one single Titan RTX GPU, our model was trained using the Adam optimizer. The learning rate was increased from 1×10^{-4} to 5×10^{-4} during the first epoch, and then decreased by 0.2 at epochs 8 and 10. The batch size was set to 16.

5. Experimental Results and Analysis

5.1. Comparison with State-of-the-Art Methods

5.1.1. Results on Test-Standard v1.0 Split

Under the discriminative decoder setting using VisDial v1.0, we initially contrasted our proposed model with those of previously published results, such as LF [4], HRE [4], MN [4], co-reference neural module network (CorefNMN) [28], graph neural network (GNN) [29], factor graph attention (FGA) [30], dual visual attention network (DVAN) [31], RVA [16], dual encoding visual dialog (DualVD) [32], history-aware co-attention (HACAN) [33], synergistic network [34], DAN [5], textual-visual reference aware attention network (RAA-Net) [35], HESNet [15], CAG [19], LTMI-LG [36], VD-BERT [20], and MVAN [10].

Our MGSCRN model outperformed various techniques according to the NDCG metric, as shown in Table 1. Specifically, when compared with VD-BERT, our model improved the NDCG from 59.96 to 60.98 (+1.02) and achieved results that were competitive with most non-NDCG metrics. Note that VD-BERT employs pretrained BERT language models and unified discriminative and generative training settings to improve performance. We also noticed that the MGSCRN model performed better than MVAN across various metrics (e.g., NDCG from 59.37 to 60.98 (+1.61) and MRR from 64.84 to 65.40 (+0.56)).

To compare the MGSCRN to more sophisticated works, we provided our ensemble model's performance on the blind test-standard split of the VisDial v1.0 dataset in Table 2. For comparison, we chose the Synergistic [34], DAN [5], consensus dropout fusion (CDF) [37], and MVAN [10] models, because their ensemble performance results are available in the literature. For our ensemble model, we averaged the results (P_t^{Dis}) of the discriminative decoders after training six independent models with random initial seeds. Overall, our ensemble model outperformed all of the others (e.g., 62.25 NDCG compared to MVAN's 60.92, and 67.42 MRR compared to MVAN's 66.38). This further validates the effectiveness of our MGSCRN model as it focuses not only on predicting ground-truth answers (e.g., indicated by MRR) but also on selecting more plausible answers as correct ones (e.g., indicated by NDCG).

To improve task performance even more, we added multitask learning that combined the discriminative and generative decoders during training. For a fair comparison, we conducted multitask learning experiments (see Table 3). Following the visual dialog task [4], the generative decoder generates the next token, depending on the current one from the ground-truth answer, and utilizes log-likelihood scores to determine the rank of the candidates. We averaged the ranking outputs of the discriminative and generative decoders for evaluation. As shown in Table 3, our proposed model was effective in a multitask setting, achieving the best results on all metrics. Note that ReDAN [6] and MVAN [10] adopt the same method as ours, but the LTMI [8] uses only the discriminative decoder output for evaluation.

Table 1. Performance comparison on the test-standard dataset of VisDial v1.0 under the discriminative decoder setting. Better performance is indicated by "↑" when the value is higher and by "↓" when the value is lower. The best results are shown in bold.

Model	NDCG↑	MRR↑	R@1↑	R@5↑	R@10↑	Mean↓
LF	45.31	55.42	40.95	72.45	82.83	5.95
HRE	45.46	54.16	39.93	70.45	81.50	6.41
MN	47.50	55.49	40.98	72.30	83.30	5.92
CorefNMN	54.70	61.50	47.55	78.10	88.80	4.40
GNN	52.82	61.37	47.33	77.98	87.83	4.57
FGA	52.10	63.70	49.58	80.97	88.55	4.51
DVAN	54.70	62.58	48.90	79.35	89.03	4.36
RVA	55.59	63.03	49.03	80.40	89.83	4.18
DualVD	56.32	63.23	49.25	80.23	89.70	4.11
HACAN	57.17	64.22	50.88	80.63	89.45	4.20
Synergistic	57.32	62.20	47.90	80.43	89.95	4.17
DAN	57.59	63.20	49.63	79.75	89.35	4.30
RAA-Net	55.42	62.86	49.05	79.65	88.85	4.35
HESNet	57.13	63.21	49.70	80.63	89.65	4.19
CAG	56.64	63.49	49.85	80.63	90.15	4.11
LTMI-LG	58.55	64.00	50.63	80.58	90.20	4.12
VD-BERT	59.96	**65.44**	51.63	**82.23**	90.68	3.90
MVAN	59.37	64.84	51.45	81.82	90.65	3.97
MGSCRN	**60.98**	65.40	**51.92**	81.94	**90.89**	**3.88**

Table 2. Performance comparison of different methods of performing ensemble models on the test-standard dataset of VisDial v1.0. Better performance is indicated by "↑" when the value is higher and by "↓" when the value is lower. The best results are shown in bold.

Model	NDCG↑	MRR↑	R@1↑	R@5↑	R@10↑	Mean↓
Synergistic	57.88	63.42	49.30	80.77	90.68	3.97
DAN	59.36	64.40	59.90	81.18	90.40	3.99
CDF	59.49	64.40	50.90	81.18	90.40	3.99
MVAN	60.92	66.38	53.20	82.45	91.85	3.68
MGSCRN	**62.25**	**67.42**	**53.99**	**82.96**	**92.14**	**3.60**

Table 3. Performance comparison of different methods of performing multitask learning on the test-standard dataset of VisDial v1.0. Better performance is indicated by "↑" when the value is higher and by "↓" when the value is lower. The best results are shown in bold.

Model	NDCG↑	MRR↑	R@1↑	R@5↑	R@10↑	Mean↓
LTMI	60.92	60.65	47.00	77.03	87.75	4.90
ReDAN	61.86	53.13	41.38	66.07	74.50	8.91
MVAN	63.15	63.02	49.43	79.48	89.40	4.38
MGSCRN	**64.13**	**63.87**	**50.28**	**80.26**	**90.02**	**4.11**

5.1.2. Results on VisPro and VisDialConv Datasets

To examine our model's performance even further, we report the performance results of our model on those of VisPro and VisDialConv datasets [9], in which humans must browse the dialog history to predict the answers for the current question. We contrasted our model's outcomes with those of MCA-I-HGuidedQ, MCA-I-VGH, and MCA-I-H from the work of Agarwal et al. [9], who incorporated the dialog history in various ways using VisPro and VisDialConv datasets. Table 4 shows that our model continues to yield much better results over all of the metrics and that it has powerful generalization and prediction abilities owing to the collaborative intersection of semantic information with different granularity.

Table 4. Results comparison on VisPro and VisDialConv datasets. Better performance is indicated by "↑" when the value is higher and by "↓" when the value is lower. The best results are shown in bold.

Model	NDCG↑	MRR↑	R@1↑	R@5↑	R@10↑	Mean↓
VisPro dataset						
MCA-I-HGuidedQ	61.35	60.13	47.11	75.26	86.18	5.23
MCA-I-VGH	61.68	59.33	46.18	75.35	86.17	5.07
MCA-I-H	61.72	59.62	45.92	77.11	86.45	4.85
MGSCRN	**62.57**	**61.56**	**48.73**	**78.06**	**88.11**	**4.70**
VisDialConv dataset						
MCA-I-HGuidedQ	53.18	**62.29**	48.35	80.10	88.76	4.42
MCA-I-VGH	55.48	58.45	44.54	74.95	86.19	5.18
MCA-I-H	53.01	61.24	47.63	79.07	87.94	4.77
MGSCRN	**57.13**	61.87	**48.84**	**80.44**	**88.98**	**4.22**

However, the series of MCA methods mentioned above only modelled interactions at the token level of information granularity based on the question and dialog history, which may lead to yield lower scores in most metrics.

5.2. Ablation Study

To evaluate the influence of our model's key components, we performed an ablation study using the VisDial v1.0 validation split and the same discriminative decoder for all variations:

- Sentence-granularity semantic representations (SgSR): This model used only sentence-granularity semantic representations of the question and dialog history to acquire clues via the attentive module and CNN (Equations (4)–(6)). Then, it located question-related visual objects from V^{self} using the question-aware attention (Equation (17));
- Token-granularity semantic representations (TgSR): This model used only token-granularity semantic representations of the question and dialog history to acquire clues via the attentive module (Equations (4) and (7)). Then, it located question-related visual objects from $V^{Q,token}$: we first use the object-difference attention mechanism (Equation (16)) to obtain distribution on all of the visual objects, P_V^{token}; second, the final question-relevant visual representation is computed by $v^{Q,token} = \sum_{i=1}^{m} P_{V,i}^{token} V_i^{Q,token}$. The $v^{Q,token}$-fused question features is fed to the discriminative decoder to predict answers;
- TgSR without object-difference attention (TgSR w/o OdA): In this model, we performed self-attention instead of object-difference attention on $V^{Q,token}$ to locate question-related visual objects. The other settings matched those of TgSR;
- SgSR+TgSR: This model further combined SgSR and TgSR. Specifically, in the discriminative decoder, two probability distributions over answer candidates from SgSR and TgSR were added after normalization to produce the final distribution. It notably did not include collaborative reasoning historical information (CRHI) and collaborative reasoning visual object (CRVO) operations;
- MGSCRN: This is our full semantic model, which combined SgSR and TgSR with CRHI and CRVO operations.

Table 5 shows that both SgSR and TgSR considered only single-granularity semantic representations of the question and dialog history to reason the relational information among the question, dialog history, and visual objects. They achieved lower performance on all of the evaluation metrics compared with SgSR+TgSR. This illustrates the effectiveness of the joint decision-making that uses multi-granularity semantic information. Due to the introduction of CRHI and CRVO operations, MGSCRN bridged the multi-granularity semantic information to collaboratively reason the important information from the dialog history and visual objects, and it performed best on all of the metrics. This validates the method's provision of more accurate clues by leveraging a collaborative interaction strategy with multi-granularity semantic information. TgSR gained an extra boost over TgSR w/o OdA, which demonstrates the advantage of explicitly comparing different representations of the same visual object to identify those most related to the question.

Table 5. Ablation study of MGSCRN using the VisDial v1.0 validation split. Better performance is indicated by "↑" when the value is higher and by "↓" when the value is lower. The best results are shown in bold.

Model	NDCG↑	MRR↑	R@1↑	R@5↑	R@10↑	Mean↓
SgSR	59.85	63.65	50.88	80.76	89.20	4.37
TgSR	60.49	64.22	51.59	81.46	90.01	4.19
TgSR w/o OdA	59.58	63.45	50.93	80.80	89.38	4.32
SgSR+TgSR	61.25	64.98	52.68	81.92	90.46	4.01
MGSCRN	**62.28**	**66.44**	**53.42**	**82.80**	**91.21**	**3.83**

5.3. Qualitative Analysis

To further demonstrate the good interpretability of our model, we visualized the attention weights generated by QAAR and VAAA modules through the examples illustrated in Figure 4. For the QAAR module, two types of attention weights were shown. The first was attention weights P^s for t rounds of dialog history, obtained by modifying P^s_{his} with the assistance of P^Q_{his} (as in Equation (12)). Doing so revealed the relevance of different dialog history rounds. The second was $P^{token}_{i,j}$ over all tokens in the jth (selected by P^s) round of dialog history based on their relationship to the current question, where $P^{token}_{i,j} = Norm\left(P^{token}[i, jn_H : (j+1)n_H]\right)$. That is, $P^{token}_{i,j}$ reveals the relevance of different tokens in the jth round of dialog history to the ith token in the current question. Importantly, for both types of attention, we fetched the average attention weight over all heads computed by the topmost stacked transformer block. For the VAAA module, we first visualized the object-difference attention (P^{token}_V in Equation (16)) and question-aware attention (P^s_V in Equation (17)), and visualized the collaborative reasoning attention ($P^{token+s}_V$ in Equation (18)) computed by CRVO. Here, we display the top three ranked visual objects for each attention in the rightmost column of Figure 4.

Two QAAR visualization examples are presented, for which our model accurately captured question-aware historical information at both sentence- and token-granularities. In the top example, when answering the current question, "Is he wearing shorts?", to determine what "he" refers to, the model assigned a maximum weight to the "H1" round of dialog history (i.e., "Is the young boy a toddler? No."), in which the token, "boy", gained a relatively large weight by visualizing the weight of each token in "H1", given by the token "he" in the question. This observation validates that our model accurately obtains clues from historical information, thus clarifying the semantics of the question.

In the VAAA module's visualization, we observed that the question-related visual objects were located more accurately, benefiting from collaboratively reasoning the distribution over all of the visual objects using multi-granularity semantic representations of the question. For instance, when answering "Is there anything around the laptop?", the object-difference attention of the VAAA module grounded "cat," "screen", and "pens" as the top three target objects. However, "cat" and "screen" should not have been considered based on the complete semantics of the question. However, VAAA's question-aware attention located all of the visual objects around the laptop, including "table lamp", "keyboard", and "pens." Although the OdA did not produce an accurate distribution over the visual objects, the CRVO method that combined OdA and QaA determined a suitable distribution that still ranked "table lamp", "keyboard", and "pens" as the top three target objects. Similar phenomena were verified in the top example.

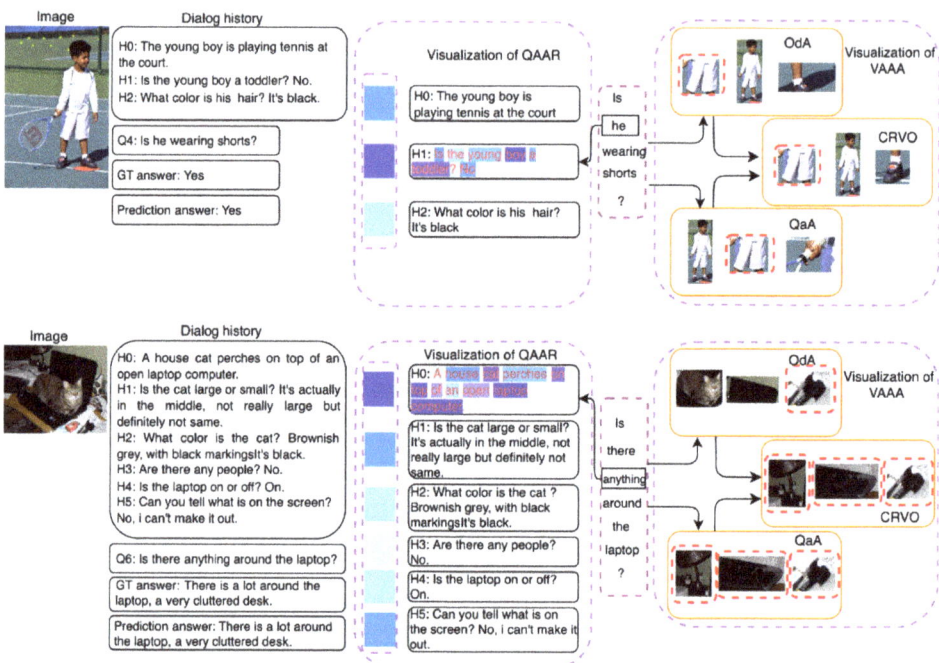

Figure 4. Example results on the VisDial v1.0 validation set. In the question-aware attention refer (QAAR) module, two types of attention weights are displayed. One is P^s on t rounds of dialog history, the other is $P^{token}_{i,j}$ over all tokens in the jth round of dialog history, which may have dependencies on the ith token in the question, Q_t. A darker blue square indicates a higher attention weight, and a lighter blue square indicates a lower attention weight. In the visual-aware attention alignment (VAAA) module, we visualized three types of attention: Object difference attention (OdA), question-aware attention (QaA), and collaborative reasoning visual object (CRVO), where top three visual objects for each attention are displayed. Visual objects in the red dotted box are the target objects for answering the question, Q_t.

To verify the effectiveness of the proposed model compared with the single granularity model, Figure 5 shows two example results on three models: SgSR model, TgSR model, and MGSCRN model. As observed in Figure 5, the MGSCRN model had a larger number of relevant answers on its list when compared to the top 10 ranking lists of the SgSR model. In the SgSR model, there were four and five relevant answers for each of the two cases. In contrast, our approach boosted the number to six and seven, respectively, which resulted in a higher NDCG score. Note that our model also outperformed the TgSR model in terms of NDCG. Additionally, in both cases, the ground-truth (GT) answer is ranked first by our proposed model. Through the above analysis, compared with a single granularity semantic representation model (SgSR/ TgSR), we validated our model performed better on the visual dialog task due to collaborative reasoning, using two different granularity semantic representations.

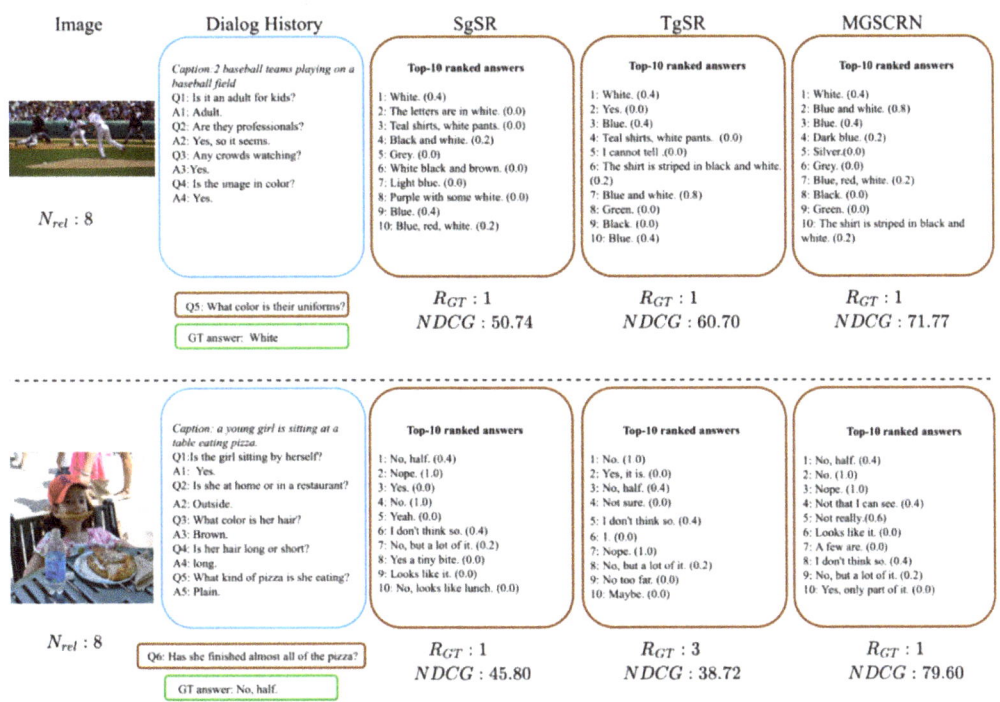

Figure 5. Example results from the validation set of VisDial v1.0 using the SgSR, TgSR, and MGSCRN models. The scores of the top 10 ranked answers are shown in brackets indicating the degree of relevance of the candidate answer options (out of 100) to the question. Here, R_{GT} denotes the rank of GT answer, NDCG is a score ranking of all relevant answers for the question, and N_{rel} is the number of candidate answer options with non-zero relevance.

5.4. Error Case Analysis

In this section, we illustrated some of the zero-scoring examples in terms of the R@10 metric. These can be categorized into three groups based on the reasons for the errors. The first is common-sense reasoning. That is, when faced with the question, "What material is the table?" or "How tall is the man?" (Figure 6a,b), common-sense reasoning is needed. However, our model was confused even after accurately locating the target visual objects. The second was related to the image not containing the target visual objects. Because "traffic lights" did not exist in the corresponding image (Figure 6c), our model picked the best candidate, which led to incorrect answers. The third type relates to keyword matching. When modeling the intersections between the question and dialog history, our model memorized the keywords from the dialog history (e.g., "eyes"), thus ranking answers with keywords at the top. For example, "No, his eyes are open" and "No, it's eyes are open" (Figure 6d). However, these answers are sometimes inaccurate.

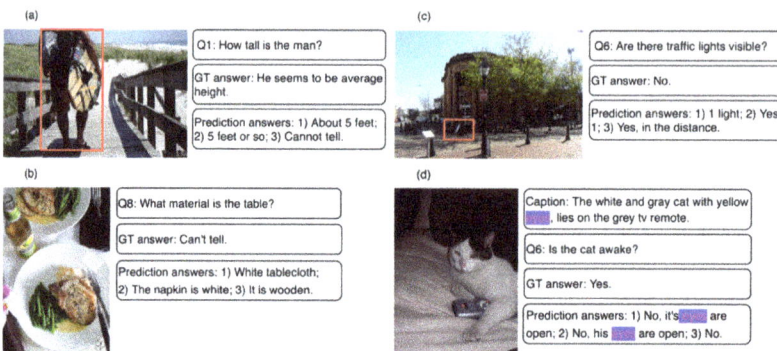

Figure 6. Examples of error cases from the VisDial v1.0 validation set. The visual objects in the red box are selected by the proposed model. The errors can be categorized into three groups based on the reasons: (1) common-sense reasoning, (2) the image not containing the target visual objects, and (3) keyword matching. (**a**,**b**) belong to the first group, (**c**) belong to second, and (**d**) belong to third, respectively.

6. Conclusions

In this paper, we proposed the MGSCRN visual dialog system. The QAAR module performs collaborative reasoning while leveraging a variety of granular semantic representations of a new question and an image's dialog history to capture and analyze historical information. The VAAA module performs collaborative reasoning to facilitate visual objects' grounding. We empirically validated our novel method on VisDial v1.0, VisPro, and VisDialConv datasets, achieving outstanding prediction accuracy and task efficacy.

In terms of applications, the proposed model is a magnificent implementation of AI-supported knowledge management and plays a crucial role in human–machine interaction, such as dealing with robotics applications and AI assistants. In terms of the limitations of the proposed method, our method fed the entire dialog history into the model, disregarding the dialog history's temporal relationship and discussed topic changes, when modeling the interactions between the current question and dialog history. By memorizing the topic changes, the model may capture the historical information more accurately as the conversation moves forward. Note that some of the questions can be answered by simply seeing the image without access to the history information. The proposed model may be unable to correctly answer these questions due to excessive memory of irrelevant history information. In addition, the accuracy of locating target objects may be decreased because our model does not consider the positional relationships between visual objects.

In the future, we will consider the temporal relationship between many rounds of the dialog history to determine how the discussed topic changes, so that the model can more precisely capture the background knowledge relevant to the current question. To answer questions that are unrelated to dialog history, how to use dialog history in a balanced manner is another challenge. Additionally, we expect to incorporate spatial relationships between visual objects by understanding the prior dialog history.

Author Contributions: Conceptualization, H.Z. and X.W.; methodology, H.Z.; software, H.Z. and S.J.; validation, H.Z. and S.J.; formal analysis, H.Z. and X.L; writing—original draft preparation, H.Z.; writing—review and editing, H.Z., X.W., X.L. and S.J. All authors have read and agreed to the published version of the manuscript.

Funding: This paper is supported by the National Natural Science Foundation of China (No. 62076032).

Institutional Review Board Statement: Not applicable.

Informed Consent Statement: Not applicable.

Data Availability Statement: Publicly available datasets were analyzed in this study. The data can be found here: https://visualdialog.org/data (accessed on 3 August 2022).

Conflicts of Interest: The authors declare no conflict of interest.

References

1. Guo, L.; Liu, J.; Tang, J.; Li, J.; Luo, W.; Lu, H. Aligning Linguistic Words and Visual Semantic Units for Image Captioning. In Proceedings of the 27th ACM International Conference on Multimedia, Nice, France, 21–25 October 2019; pp. 765–773.
2. Zhan, H.; Xiong, P.; Wang, X.; Wang, X.; Yang, L. Visual Question Answering by Pattern Matching and Reasoning. *Neurocomputing* **2022**, *467*, 323–336. [CrossRef]
3. Liu, J.; Wang, W.; Wang, L.; Yang, M.-H. Attribute-Guided Attention for Referring Expression Generation and Comprehension. *IEEE Trans. Image Process.* **2020**, *29*, 5244–5258. [CrossRef] [PubMed]
4. Das, A.; Kottur, S.; Gupta, K.; Singh, A.; Yadav, D.; Lee, S.; Moura, J.M.F.; Parikh, D.; Batra, D. Visual Dialog. *IEEE Trans. Pattern Anal. Mach. Intell.* **2019**, *41*, 1242–1256. [CrossRef] [PubMed]
5. Kang, G.-C.; Lim, J.; Zhang, B.-T. Dual Attention Networks for Visual Reference Resolution in Visual Dialog. In Proceedings of the 2019 Conference on Empirical Methods in Natural Language Processing and the 9th International Joint Conference on Natural Language Processing (EMNLP-IJCNLP), Hong Kong, China, 3–7 November 2019; Association for Computational Linguistics: Stroudsburg, PA, USA, 2019; pp. 2024–2033.
6. Gan, Z.; Cheng, Y.; Kholy, A.; Li, L.; Liu, J.; Gao, J. Multi-Step Reasoning via Recurrent Dual Attention for Visual Dialog. In Proceedings of the 57th Annual Meeting of the Association for Computational Linguistics, Florence, Italy, 28 July–2 August 2019; Association for Computational Linguistics: Stroudsburg, PA, USA, 2019; pp. 6463–6474.
7. Chen, F.; Meng, F.; Xu, J.; Li, P.; Xu, B.; Zhou, J. DMRM: A Dual-Channel Multi-Hop Reasoning Model for Visual Dialog. *AAAI* **2020**, *34*, 7504–7511. [CrossRef]
8. Nguyen, V.-Q.; Suganuma, M.; Okatani, T. Efficient Attention Mechanism for Visual Dialog That Can Handle All the Interactions between Multiple Inputs. In Proceedings of the European Conference on Computer Vision (ECCV), Glasgow, UK, 23–28 August 2020.
9. Agarwal, S.; Bui, T.; Lee, J.-Y.; Konstas, I.; Rieser, V. History for Visual Dialog: Do We Really Need It? In Proceedings of the 58th Annual Meeting of the Association for Computational Linguistics, Online, 5–10 July 2020; Association for Computational Linguistics: Stroudsburg, PA, USA, 2020; pp. 8182–8197.
10. Park, S.; Whang, T.; Yoon, Y.; Lim, H. Multi-View Attention Network for Visual Dialog. *Appl. Sci.* **2021**, *11*, 3009. [CrossRef]
11. Punia, S.K.; Kumar, M.; Stephan, T.; Deverajan, G.G.; Patan, R. Performance analysis of machine learning algorithms for big data classification: Ml and ai-based algorithms for big data analysis. *Int. J. E-Health Med. Commun. (IJEHMC)* **2021**, *12*, 60–75. [CrossRef]
12. Rao, D.; Huang, S.; Jiang, Z.; Deverajan, G.G.; Patan, R. A dual deep neural network with phrase structure and attention mechanism for sentiment analysis. *Neural Comput. Appl.* **2021**, *33*, 11297–11308. [CrossRef]
13. Niu, Y.; Zhang, H.; Gopal, G. Emerging Trends and the Importance of Network Evolution in Big Data. *Recent Adv. Comput. Sci. Commun. (Former. Recent Patents Comput. Sci.)* **2020**, *13*, 158.
14. De Vries, H.; Strub, F.; Chandar, S.; Pietquin, O.; Larochelle, H.; Courville, A. GuessWhat?! Visual Object Discovery through Multi-Modal Dialogue. In Proceedings of the 2017 IEEE Conference on Computer Vision and Pattern Recognition (CVPR), Honolulu, HI, USA, 21–26 July 2017; IEEE: Piscataway, NJ, USA, 2017; pp. 4466–4475.
15. Lin, B.; Zhu, Y.; Liang, X. Heterogeneous Excitation-and-Squeeze Network for Visual Dialog. *Neurocomputing* **2021**, *449*, 399–410. [CrossRef]
16. Niu, Y.; Zhang, H.; Zhang, M.; Zhang, J.; Lu, Z.; Wen, J.-R. Recursive Visual Attention in Visual Dialog. In Proceedings of the 2019 IEEE/CVF Conference on Computer Vision and Pattern Recognition (CVPR), Long Beach, CA, USA, 15–20 June 2019; IEEE: Piscataway, NJ, USA, 2019; pp. 6672–6681.
17. Zhang, H.; Wang, X.; Jiang, S. Reciprocal Question Representation Learning Network for Visual Dialog. *Appl. Intell.* **2022**. [CrossRef]
18. Jiang, T.; Shao, H.; Tian, X.; Ji, Y.; Liu, C. Aligning Vision-Language for Graph Inference in Visual Dialog. *Image Vis. Comput.* **2021**, *116*, 104316. [CrossRef]
19. Guo, D.; Wang, H.; Wang, M. Context-Aware Graph Inference with Knowledge Distillation for Visual Dialog. *IEEE Trans. Pattern Anal. Mach. Intell.* **2021**, 1. [CrossRef] [PubMed]
20. Wang, Y.; Joty, S.; Lyu, M.; King, I.; Xiong, C.; Hoi, S.C.H. VD-BERT: A Unified Vision and Dialog Transformer with BERT. In Proceedings of the 2020 Conference on Empirical Methods in Natural Language Processing (EMNLP), Online, 16–20 November 2020; Association for Computational Linguistics: Stroudsburg, PA, USA, 2020; pp. 3325–3338.
21. Murahari, V.; Batra, D.; Parikh, D.; Das, A. Large-Scale Pretraining for Visual Dialog: A Simple State-of-the-Art Baseline. In Proceedings of the European Conference on Computer Vision (ECCV), Glasgow, UK, 23–28 August 2020.
22. Vaswani, A.; Shazeer, N.; Parmar, N.; Uszkoreit, J.; Jones, L.; Gomez, A.N.; Kaiser, Ł.; Polosukhin, I. *Attention Is All You Need*; Guyon, I., Luxburg, U.V., Bengio, S., Wallach, H., Fergus, R., Vishwanathan, S., Garnett, R., Eds.; Curran Associates, Inc.: Long Beach, CA, USA, 2017; pp. 6000–6010.

23. Pennington, J.; Socher, R.; Manning, C. Glove: Global Vectors for Word Representation. In Proceedings of the 2014 Conference on Empirical Methods in Natural Language Processing (EMNLP), Doha, Qatar, 26–28 October 2014; Association for Computational Linguistics: Stroudsburg, PA, USA, 2014; pp. 1532–1543.
24. Choi, H.; Kim, J.; Joe, S.; Gwon, Y. Evaluation of BERT and ALBERT Sentence Embedding Performance on Downstream NLP Tasks. In Proceedings of the 2020 25th International Conference on Pattern Recognition (ICPR), Milan, Italy, 10–15 January 2021; IEEE: Piscataway, NJ, USA, 2021; pp. 5482–5487.
25. Ren, S.; He, K.; Girshick, R.; Sun, J. Faster R-CNN: Towards Real-Time Object Detection with Region Proposal Networks. *IEEE Trans. Pattern Anal. Mach. Intell.* **2017**, *39*, 1137–1149. [CrossRef] [PubMed]
26. Krishna, R.; Zhu, Y.; Groth, O.; Johnson, J.; Hata, K.; Kravitz, J.; Chen, S.; Kalantidis, Y.; Li, L.-J.; Shamma, D.A.; et al. Visual Genome: Connecting Language and Vision Using Crowdsourced Dense Image Annotations. *Int. J. Comput. Vis.* **2017**, *123*, 32–73. [CrossRef]
27. Lin, T.-Y.; Maire, M.; Belongie, S.; Hays, J.; Perona, P.; Ramanan, D.; Dollár, P.; Zitnick, C.L. Microsoft COCO: Common Objects in Context. In *Computer Vision—ECCV 2014, Proceedings of the 13th European Conference, Zurich, Switzerland, 6–12 September 2014*; Fleet, D., Pajdla, T., Schiele, B., Tuytelaars, T., Eds.; Lecture Notes in Computer Science; Springer International Publishing: Cham, Switzerland, 2014; Volume 8693, pp. 740–755. ISBN 978-3-319-10601-4.
28. Kottur, S.; Moura, J.M.F.; Parikh, D.; Batra, D.; Rohrbach, M. Visual Coreference Resolution in Visual Dialog Using Neural Module Networks. In *Computer Vision—ECCV 2018, Proceedings of the 15th European Conference, Munich, Germany, 8–14 September 2018*; Ferrari, V., Hebert, M., Sminchisescu, C., Weiss, Y., Eds.; Lecture Notes in Computer Science; Springer International Publishing: Cham, Switzerland, 2018; Volume 11219, pp. 160–178. ISBN 978-3-030-01266-3.
29. Zheng, Z.; Wang, W.; Qi, S.; Zhu, S.-C. Reasoning Visual Dialogs with Structural and Partial Observations. In Proceedings of the 2019 IEEE/CVF Conference on Computer Vision and Pattern Recognition (CVPR), Long Beach, CA, USA, 15–20 June 2019; IEEE: Piscataway, NJ, USA, 2019; pp. 6662–6671.
30. Schwartz, I.; Yu, S.; Hazan, T.; Schwing, A.G. Factor Graph Attention. In Proceedings of the 2019 IEEE/CVF Conference on Computer Vision and Pattern Recognition (CVPR), Long Beach, CA, USA, 15–20 June 2019; IEEE: Piscataway, NJ, USA, 2019; pp. 2039–2048.
31. Guo, D.; Wang, H.; Wang, M. Dual Visual Attention Network for Visual Dialog. In Proceedings of the Twenty-Eighth International Joint Conference on Artificial Intelligence, Macao, China, 10–16 August 2019; International Joint Conferences on Artificial Intelligence Organization: Brussels, Belgium, 2019; pp. 4989–4995.
32. Jiang, X.; Yu, J.; Qin, Z.; Zhuang, Y.; Zhang, X.; Hu, Y.; Wu, Q. DualVD: An Adaptive Dual Encoding Model for Deep Visual Understanding in Visual Dialogue. *AAAI* **2020**, *34*, 11125–11132. [CrossRef]
33. Yang, T.; Zha, Z.-J.; Zhang, H. Making History Matter: History-Advantage Sequence Training for Visual Dialog. In Proceedings of the 2019 IEEE/CVF International Conference on Computer Vision (ICCV), Seoul, Korea, 27 October–2 November 2019; IEEE: Piscataway, NJ, USA, 2019; pp. 2561–2569.
34. Guo, D.; Xu, C.; Tao, D. Image-Question-Answer Synergistic Network for Visual Dialog. In Proceedings of the 2019 IEEE/CVF Conference on Computer Vision and Pattern Recognition (CVPR), Long Beach, CA, USA, 15–20 June 2019; IEEE: Piscataway, NJ, USA, 2019; pp. 10426–10435.
35. Guo, D.; Wang, H.; Wang, S.; Wang, M. Textual-Visual Reference-Aware Attention Network for Visual Dialog. *IEEE Trans. Image Process.* **2020**, *29*, 6655–6666. [CrossRef]
36. Chen, F.; Chen, X.; Xu, C.; Jiang, D. Learning to Ground Visual Objects for Visual Dialog. *arXiv* **2021**, arXiv:2109.06013.
37. Kim, H.; Tan, H.; Bansal, M. Modality-Balanced Models for Visual Dialogue. In Proceedings of the AAAI Conference on Artificial Intelligence, New York, NY, USA, 7–12 February 2020.

Article

Smart Interactive Technologies in the Human-Centric Factory 5.0: A Survey

Davide Brunetti, Cristina Gena * and Fabiana Vernero

Computer Science Department, University of Turin, 10149 Turin, Italy
* Correspondence: cristina.gena@unito.it
† These authors contributed equally to this work.

Abstract: In this survey paper, we focus on smart interactive technologies and providing a picture of the current state of the art, exploring the way new discoveries and recent technologies changed workers' operations and activities on the factory floor. We focus in particular on the Industry 4.0 and 5.0 visions, wherein smart interactive technologies can bring benefits to the intelligent behavior machines can expose in a human-centric AI perspective. We consider smart technologies wherein the intelligence may be in and/or behind the user interfaces, and for both groups we try to highlight the importance of designing them with a human-centric approach, framed in the smart factory context. We review relevant work in the field with the aim of highlighting the pros and cons of each technology and its adoption in the industry. Furthermore, we try to collect guidelines for the human-centric integration of smart interactive technologies in the smart factory. In this wa y, we hope to provide the future designers and adopters of such technologies with concrete help in choosing among different options and implementing them in a user-centric manner. To this aim, surveyed works have been also classified based on the supported task(s) and production process phases/activities: access to knowledge, logistics, maintenance, planning, production, security, workers' wellbeing, and warehousing.

Keywords: user-centered AI; smart factory; smart interactive technologies

1. Introduction

In 2015, the introduction of emerging technologies such as wireless sensor networks, big data, cloud computing, embedded systems, and mobile Internet into the manufacturing environment determined the conditions for factories to enter the era of the fourth industrial revolution [1]. As a reaction to the massive installation of such technologies on modern factory floors, the strategic initiative called "Industry 4.0" was proposed and adopted as part of the "High-Tech Strategy 2020 Action Plan" of the German government [1]. Other main industrial countries followed, proposing strategies that were compliant with the idea of the just-born Industry 4.0. The Industry 4.0 concept describes production-oriented cyber-physical systems (CPS) which integrate production facilities, warehousing systems, logistics, and even social requirements to establish global value creation networks [2].

Various changes have been introduced by the smart factory vision in the ways production is conceived and workers operate on factory floors. Such changes involve different levels of the production process, introducing a number of disruptive technologies which enable the digitalization of the manufacturing sector, enclosed in different areas of expertise:

- Data, computational power, and connectivity;
- Analytic and artificial intelligence;
- Human–machine interaction;
- Digital-to-physical conversion.

In January 2021, the European Commission published a report called "Industry 5.0. Towards a sustainable, human-centric and resilient European industry" [3], presenting the need to speed up the transformation already underway, which uses digital and green technologies to heal the environment and the economy. Industry appears to be the pivot of this important transition, representing the only means to achieve wellbeing from a human- and a production-related point of view but also respecting the environment. Industry 5.0, which is intended to complement and extend the features of Industry 4.0, aims to make the industry sector more sustainable and resilient.

The report describes the main building blocks of the Industry 5.0 approach [3]:

- The industry must now become the accelerator and enabler of change and innovation;
- Digital technologies such as artificial intelligence (AI) and robotics can optimize human–machine interactions, underlining the added value human workers bring to the factory floor;
- By developing innovative technologies in a human-centric way, Industry 5.0 can support and empower, rather than replace, workers;
- Greening the economy will be successful with European industry taking a strong leadership role;
- Industry 5.0 will also have a transformative impact on society, especially for industry workers, who may see their role changed, requiring new skills.

In the Industry 5.0 vision, technologies can optimize both workplaces and worker's performances, favoring the interaction between human and machine, rather than replacing one with the other. The main difference between the fourth and fifth industry visions is that humans are envisioned to collaborate with robots and innovative technologies, which have to be designed in a human-centric way.

Industry 5.0 identifies the following six enabling technologies [4]:

1. Individualized human–machine interaction technologies that interconnect and combine the strengths of humans and machines.
2. Bio-inspired technologies and smart materials that allow materials to have embedded sensors and enhanced features while being recyclable.
3. Digital twins and simulations to model entire systems.
4. Data transmission, storage, and analysis technologies that are able to handle data and foster system interoperability.
5. Artificial intelligence to detect, for example, causalities in complex, dynamic systems, leading to actionable intelligence.
6. Technologies for energy efficiency, renewables, storage, and autonomy.

According to Xu et al. [5], Industry 5.0 is not a technology-driven revolution but a value-driven initiative that promotes technological transformation with a particular purpose, generating value by putting together economy (profitability, scalability, and business models), ecology (CO_2 reduction and circular economy), and society (societal challenges and human-centricity).

In this paper, keeping the context of the smart industry, we focus in particular on the dimension of smart interactive technologies, a term that can be traced back to that of intelligent user interfaces (IUIs), which aim at improving the symbiosis between humans and computers by merging artificial intelligence (AI) and human–computer interaction (HCI), including intelligent capabilities in the interface in order to improve performance, usability, and experience in critical ways (for more details, see [6]). This may also involve designing an interface that effectively leverages human skills and capabilities so that human performance with application improves. In addition, human-centered artificial intelligence (HCAI) has recently become a very popular term with a similar purpose and focuses on the need of bringing the human into the center of the AI design, thus creating systems that provide smart computations that are beneficial to humans, supporting them to achieve their objectives, *"focusing on enhancing human performance, making systems reliable, safe, and trustworthy"* [7]. Previously, researchers and developers directed their attention towards developing AI algorithms and systems with an emphasis on the machines' autonomy. Conversely, HCAI focuses on user experience design, putting human users at the center of design thinking. Researchers and developers of HCAI systems measure their success with human performance and satisfaction metrics [8]. Indeed, compared with traditional technologies, AI-based technologies elicit different expectations from a user's perspective. Besides the established principles of human-centered design, there are further important aspects that are peculiar to this type of system and that need to be considered during design, such as transparency, explainability, interpretability, user control vs. system autonomy, fairness, etc. [9].

In the following, we discuss some of the most relevant smart interactive technologies which will characterize the Industry 5.0 scenario. According to Sonntag [6], what defines an intelligent user interface, which is a term embracing smart interactive technologies as explained above, is the fact that the intelligence can be found:

- In the user interface(s) of the system, with the purpose of enabling an effective, natural, or otherwise appropriate interaction of users with the system. Examples are human-like communication methods such as speech or gesture [10], multimodal interfaces and smart environments (including IoT- and smart-object-based environments) [11,12], systems that personalize the modality of interaction to individual users taking into account her/his previous choices and preferences [13], etc.
- Behind the user interface, as, for instance, in personalized and not personalized recommender systems [14,15], i.e., systems that employ intelligent technology to support information retrieval; intelligent learning environments [16]; interface agents/robots that perform complex or repetitive tasks with some guidance from the user [17]; and situated assistance systems that monitor and support a user's daily activities [18], as, for instance, in IoT-based industrial environments [19], etc.

In Section 2, we propose examples that can be classified either in the first group, as large displays proposing touch and touchless interaction, virtual and augmented reality, and wearable, tangible user interfaces, or in the second group, as collaborative robots, or in both, as is the case for IoT and smart environments. In particular, we focus on the interactive part of such technologies, thus excluding from our discussion those cases where intelligence has a relatively limited impact on the type and modality of interaction.

2. Smart Interactive Technologies in the Human-Centric Smart Factory

The model of the smart industry, both in the 4.0 and 5.0 visions, is enabled by advanced digitalization and the spread of the Internet of Things (IoT), cyber-physical systems, and smart technologies on the factory floor, and it is destined to radically change the approach to work in today's industry. A significant increase in the demand for workforce is awaited at all levels, in order to manage complexity, abstraction, and problem solving processes. From the point of view of industrial workers, the introduction of opportunities and improvements in the quality of work are expected: a more stimulating work environment, greater autonomy, and opportunities for personal development. As a consequence, workers will be led to act on their own individual initiatives, acquiring excellent interaction and communication skills and organizing their workflows [20]. In this new vision, the term human-centred factory aims to define new social sustainable workplaces where the human dimension is a key cornerstone, highlighting the requirements for shifting from a traditional task-centric production to a worker-centric production [21]. Therefore, it becomes of fundamental importance to improve the interaction and collaboration between humans and intelligent machines, as with, for instance, cobots.

In Section 2.1, we present a set of relevant smart interactive technologies for the present and future of the Industry 5.0. For each of the presented technologies, we propose a brief definition and we summarize the main related work, and then we give examples in the smart factory. The reader should notice that the term smart factory includes in its definition what is understood as Industry 4.0 and 5.0, and we use it interchangeably We conclude the subsection with a brief summary of challenges and a set of guidelines.

2.1. Research Methodology

This article surveyed almost 100 papers and technical reports describing projects and visions related to a human-centered perspective in the smart factory, starting from 2018. At that time (2018–2021), we were involved in two national smart factory projects, HUMANS (human-centered manufacturing systems) (https://dmd.it/humans/en/humans/, accessed on 29 July 2022) and HOME (hierarchical open manufacturing Europe) (https://www.home-opensystem.org/index.php/en/home-3/, accessed on 29 July 2022) with the goal of leading research tasks focused on the interface and the interaction between man and machine through intelligent applications, as in wearable technologies, touchless interaction, interactive (production line) data visualization, etc. With regard to these aspects, we provided our industrial partners with requirements, guidelines, and specifications for the interaction between man and machine in the smart industry, as well as simulation demos on data visualization [22], gestural interaction [23,24], and wearable technologies.

Thus, we started collecting material on the basis of the requirements we had for the aforementioned projects. We reviewed the main survey papers ([25–33]) and company technical reports ([31,34,35]) in the field. Staring from the work referred to in the above-mentioned surveys, we extended our research using Researchgate as a platform and search keywords based on the main technologies we explored (collaborative robots, tangibles, wearables, large displays, IoT, gestural interaction, large displays, etc.), contextualized to the smart factory.

A fundamental inspiration for the technologies to consider and review, under a smart interaction perspective, came from the work by Romero, Stahre et al. [36], who developed the concept of Operator 4.0, which aims at expanding the capabilities of the industry worker with innovative technological means, rather than replacing the worker with robots. Operator 4.0 includes eight future projections of extended operators: the super-strength operator (operator + exoskeleton), the augmented operator (operator + augmented reality), the virtual operator (operator + virtual reality), the healthy operator (operator + wearable tracker), the smarter operator (operator + intelligent personal assistant), the collaborative operator (operator + collaborative robot), the social operator (operator + social networks), and the analytical operator (operator + big data analytics).

We decided to present the results of our analysis following a systematic frame, structured in such a way as to provide a set of guidelines for each chapter so that, in addition to reviewing the available technologies, we could offer a useful tool for understanding how to best implement them. The guidelines were derived from the works found for each chapter.

We believe that the analysis presented in this survey could provide hints for the design and development of new human-centered solutions in the factory of the future. According to our knowledge, no other comprehensive surveys have yet been published on smart technologies for Industry 5.0 that focus on their design in a human-centric way.

2.2. Large Displays

Large displays first appeared in the early 1960s, but they only began to be used in real installations at beginning of the new millennium. One of their advantages is their ability to bring users to collaborate and socialize, both in business and in entertainment contexts. While only touch interaction was supported at first, interaction through touchless gestures has become possible nowadays, due to the technological evolution. According to [37], large displays are characterized by five attributes or dimensions:

- Orientation (vertical, horizontal, diagonal, or at ground level);
- Display technology (monitor, front projection, or rear projection);
- Purpose (gaming, entertainment, productivity, social interactions, or advertising);
- Interaction methods (touch, touchless, tangible interfaces, or via an external device);
- Location (office/workplace, museums, universities, shops, or on the street).

In their study, the authors of [37] examined a number of scientific papers relating to large displays and observed that most displays are positioned vertically and are actually monitors, not simple screens with front- or rear-projected images. Such displays can be used at the workplace, where they can foster cooperation between staff members.

2.2.1. Touch and Touchless Interaction

Touch and touchless interaction modes can affect the type of applications to be used on large displays and have different pros and cons [37].

Touch interaction. Sambrooks and Wilkinson [38] carried out a study where they asked participants to perform a series of tasks where they had to select one or more elements through different interaction modes. Touch interaction was very precise and had a low margin of error. This positive result is probably partly due to the familiarity users have with touch interaction, albeit to a lesser extent than with mouse-based interaction, which proved to guarantee the highest level of precision. This primacy may depend on various factors, such as the size of the screen with which one interacts, the size of the icons or objects to be selected, or the problem of "fat fingers", whereby users with large fingers may encounter difficulty in using small screens.

It should be emphasized how touch interaction has evolved over time, passing from the simple touch of a single hand to multitouch, which allows the use of multiple fingers for actions such as zoom-in and zoom-out and moving elements on the screen and gives the possibility of using the human touch in combination with other interactive tools.

However, touch interaction can be uncomfortable in some occasions, especially in the business and industrial sectors and in cases where users need to wear gloves [23], or when there are other disturbing elements. Touchless interaction, where users do not have to touch any interface, can be used as an alternative to touch interaction on such occasions.

Touchless interaction. Different interaction styles can be classified under the "touchless" umbrella term, ranging from hand-based gestures recognized by a tracking device to the detection of the posture, position, or presence of the user's body to gaze tracking and facial expression recognition. The best known examples of touchless interaction are often associated with the gaming and entertainment domain, where both the tracking of body movements (e.g., using Kinect (https://developer.microsoft.com/en-us/windows/kinect/, accessed on 29 July 2022)) and the detection of gestures executed via an external, dedicated device (e.g., PlayStation Move) are used. Other dedicated devices, such as the Myo bracelet, can be worn by users. Touchless interaction exploits the natural language of the body, which is why it could significantly reduce the distance between the user and the interface, especially if used in the business context. However, it is essential to design the interface and user experience (UX) so that they are as fluid and engaging as possible, in order to accompany the user in the interactive process and not cause frustration. In this regard, the UX design will also have to take into account the effort that repeating actions with gestural interfaces can determine, much higher than when approaching a touch- or mouse-based interface. In addition, designers will have to make sure that the interface layout allows a precise execution of actions, also considering the fact that many potential users are not very familiar with this type of interaction. For this reason, it is also necessary to ensure that the interface returns clear feedback and makes any error reversible, including those caused by involuntary body movements and incorrectly detected gestures (*immersion syndrome*).

2.2.2. The Smart Factory Context

Smart factories are proving increasingly capable of absorbing and proposing solutions from a wide range of disciplines. The introduction of gestural interfaces and the widespread use of large screens is a case in point. Gestural interfaces, similarly to other technologies and approaches in the HCI field, are part of the *physical layer* in [39]'s classification of technologies for the smart factory.

Applications: 3D object manipulation and task browsing. Ref. [40] investigated the interaction between people and 3D objects shown on public displays in an urban planning scenario. Participants were asked to perform tasks through spontaneously produced handgestures and phone-gestures. The process led to the definition of two sets of user-defined gestures. Although the proposed study limited its investigations to the area of urban planning, similar studies which identify sets of user-defined gestures can be conducted in various fields, including the industry sector.

For example, Ref. [24] proposed the use of large displays in combination with touchless gestural interaction on the shop floor (Figure 1). In the context of a smart industry project, the project consortium developed a smart armband which allows us to detect gestures from movement and muscle biosignals, while a machine learning library allows us to calibrate and recognize task-specific gestures (Figure 2). The definition of an appropriate set of gestures underwent several steps, including a guessability study [23]. The proposed application was tested in small industry specialized in sheet metal fabrication, where welders frequently switch between their workbench and a nearby desktop computer to browse the tasks they have been assigned and visualize 3D models of the final product. Results were generally positive and the participants were favorable to our solution and willing to use it in their everyday work activities.

Figure 1. User interacting with a large display via a touchless gestural interface [23,24].

Figure 2. User testing a smart armband for gestural interaction on the shop floor [23,24].

Tackling usability issues. An interesting solution, specifically focusing on the context of smart factories and combining gesture recognition with augmented reality (see Section 2.4) to address usability issues, was given by [41]. To overcome the hard and time-taking learning curves in switching from an industrial device to another, Ref. [41] investigated the development of a universal interaction device, capable of communicating with various field devices and plant modules of an industrial facility via common wireless communication standards. Their aim is the creation of a platform that has one user interface for all purposes, is nonproprietary, and can be designed individually dependent on its owner's requirements. Merging together gestures recognition with augmented reality, they offer intuitive interactions, freeing the operator from the constraints of manipulating hand-held objects.

2.2.3. Challenges and Guidelines for Touch and Touchless Interaction with Large Displays in the Smart Factory

Based on our analysis of the relevant literature, it is apparent that free-form gestural interaction is useful in contexts where touch-based interaction is not possible (e.g., because users wear gloves) or not advisable (e.g., because of hygiene policies). On the other hand, performing gestures might be physically demanding and, especially in the case of touchless interaction, some specific issues may emerge, such as:

- Unintentional gestures might be misinterpreted by the system as intentional gestures.
- Carrying out tasks that require precision through free-form gestures might be problematic.

More specifically, Garzotto and Valoriani [42] proposed the following guidelines for the design of gestural interaction, based on previous work by [10,43,44].

- **Guideline 1: Semantic intuitiveness.** Gestures should have a clear cognitive association with the semantic functions they perform and the effects they achieve.
- **Guideline 2: Minimize fatigue.** Gestural communication involves more muscles than keyboard interaction or speech. Gestural commands must therefore be concise and quick and minimize the user's effort and physical stress.
- **Guideline 3: Learnability.** It must be easy for the user to learn how to perform and remember gestures, minimizing the mental load of recalling movement trajectories and associated actions. The gestures that are most natural and easy to learn and are immediately assimilated by the user are those that belong to everyday life or involve the least physical effort. These gestures should be associated with the most frequent interactions.
- **Guideline 4: Intentionality (immersion syndrome).** Users can perform unintended gestures, i.e., movements that are not meant to communicate with the system they are interacting with. These are usually evoked when the user is communicating simultaneously with other devices or people, or just resting his or her body. Immersion syndrome occurs if every movement is interpreted by the system, whether or not it was intended, which may determine interaction effects against the user's will. The designer must identify well-defined means to detect the intention of the gestures, as distinguishing useful movements from unintentional ones is not easy.
- **Guideline 5: Precision.** Tasks that require precise interaction, e.g., the fine selection of a specific value in a large set of alternatives presented on the screen, may be problematic: when operating at a distance, we cannot obtain a good resolution because of the intrinsic instability of movements in free space. Touchless gestural input or control should be carefully designed with a special attention to precision.
- **Guideline 6: Feedback.** Appropriate feedback indicating the effects and correctness of the gesture performed is necessary for successful interaction, to improve users' confidence in the system, to allow users to learn the appropriate manner of performance, and to help users understand what was wrong with their actions.
- **Guideline 7: Provide reversible actions.** Commands must be easy to undo to easily cancel any unintended action, and "backward navigation" must be supported, to allow user return to previously seen objects or revise previous choices.

2.3. Virtual Reality

Virtual reality (VR) implies a complete immersion in a digitally built world. VR first appeared in the late 1980s, but it took another thirty years before it became actually available [45]. Only in the last few years affordable devices such as virtual reality cases for smartphones appeared on the market (Figure 3), providing potential enhancements in the field of manufacturing also for smaller businesses. Similarly to augmented reality (see Section 2.4), VR is often combined with gestural interfaces. Since VR creates immersive experiences, meaningful interactions with the virtual environment enabling users to touch, move, and interact with virtual objects via standardized gestures must be supported. Thus, studies and enhancements in the field of gestural interfaces do have an impact also in this area, determining significant advancements in the perceived naturalness of the virtual environment.

Figure 3. Virtual reality visor with smartphone case.

On the other hand, one cannot ignore the effects that technologies such as VR have on users in terms of the modification of their consciousness and perception. Ref. [46] pointed out how web or mobile interfaces can potentially, in specific cases, disconnect users from the physical world, increasing the risk of user alienation and lowering the user experience. It is thus trivial to think how such risks grow exponentially with the adoption of systems that provide a full immersion in a digitally built world. Ref. [47] explored the effects of virtual reality on the modification of the consciousness of users and the pathological implications that arise in such systems. It highlighted the risks and pointed out the need for serious scientific study in the field in order to gain the best from the adoption of such technologies in environments such as the smart factories' shop floors, limiting potential side effects.

2.3.1. The Smart Factory Context

VR can be adopted at many stages of the production process.

Applications: (remote) factory layout planning. Factory layout planning, for example, is a long standing area in production engineering that could potentially benefit from VR integration to allow workers and equipment to be more productive. Facility layout techniques and, particularly, factory layout planning, apply to the case where several physical means have to be located in a certain area, aiming at developing an efficient and effective plant layout for all the available resources [48]. Ref. [49] proposed a modeling approach for VR-supported layout planning (VLP) tasks. The authors identified three methods for modeling the virtual environment:

1. Using cameras or scanners along with algorithms to automatically convert image and video data into spatial data.
2. Modeling facilities entirely using computer-aided-design software (CAD) or virtual reality modeling languages (VRML).
3. Combining the previous two as a hybrid approach.

Collaboration between users situated in different locations could also be supported by immersive virtual reality user interfaces (VRUIs). Ref. [50] described a VR-based approach to factory planning, aimed to allow the simultaneous visualization, investigation, and analysis of data by multiple connected users. The authors classified interactions into human–human interactions and human–machine interactions, to analyze and assign them, taking into account the needs of factory planning in a virtual environment. They structured the whole factory planning process into three fields: target planning, conceptual planning, and realization planning. Moreover, they explained how to speed up actions within the planning process by implementing collaborative factory planning tools realized by interconnected but spatially distributed VR systems.

Applications: virtual commissioning and digital twins. Related to VR factory planning is virtual commissioning (VC), namely, evaluating a production line in a virtual environment before the physical production line is constructed [51]. According to Lee and Park [52], a virtual manufacturing system (namely, virtual commissioning) is a computer-based environment that simulates individual manufacturing processes in an efficient way. Indeed, virtual commissioning enables the full verification of a manufacturing system by performing a simulation involving a virtual plant and a real controller. This requires the virtual plant model to be fully described at the level of sensors and actuators. Although virtual commissioning can significantly reduce the time and effort required at the real commissioning stage, there are obstacles to the implementation of virtual commissioning. Since a virtual plant needs to communicate with a real controller, the virtual devices should be modeled at the level of sensors and actuators, which is not easy for control engineers who do not have in-depth knowledge on modeling and simulation [52]. Closely related to VC is the emerging digital twin (DT) technology, commonly referred to as one of the key enabling technologies of Industry 5.0. A DT can be defined as an evolving digital profile of the historical and current behavior of a physical object or process that helps optimize business performance (https://www2.deloitte.com/content/dam/Deloitte/kr/Documents/insights/deloitte-newsletter/2017/26_201706/kr_insights_deloitte-newsletter-26_report_02_en.pdf, accessed on 29 July 2022). As VC can be defined as the validation of automated industrial production systems before any physical commissioning is made, a DT could be intended as a virtual model of a physical industrial production system being constantly updated and updating the physical object through a real-time and bidirectional data exchange (a DT usually includes data streams in both directions between the physical and virtual objects). According to Lidell et al. [52], the increased use of VC and DTs is important for many reasons, such as "*increasing safety for operators through minimising harmful situations and correctly validating safety systems, as well as allowing improved working conditions, as described in the second interview. Furthermore, increased use of VC and DTs should also make the process of designing and developing production systems more cost-effective. More use of VC and DTs could also minimize wastes, such as ordering wrong components and machines, and to optimize the resource and energy usage through simulations and using DTs*".

Tools for the creation of VR environments. Finally, Ref. [45] investigated the problem of teaching how to create industry-themed virtual reality environments to mechanical engineers and faced the absence of tools that would fulfill the purpose without requiring complicated coding. The authors created their own framework, using a game engine called a source engine and enriching it with a library of textures, models, and scripts called DigiTov, later adapted also for Unity3D.

2.3.2. Challenges and Guidelines for Virtual Reality Interfaces in the Smart Factory

Our review shows that VR technologies can prove useful in a smart factory context, especially when it comes to planning and enhancing collaboration. On the other hand, there may be negative physical side effects such as nausea, seizures, or eye soreness; in addition, users' consciousness may be affected, bringing a loss of spatial awareness, dizziness, and disorientation (https://www.classvr.com/health-and-safety/, accessed on 29 July 2022).

Ref. [53] reviewed the relevant literature and identified the following guidelines to support the development of VR applications:

- **Guideline 1:** The degree of freedom should be minimal;
- **Guideline 2:** Avoid sickness related to brightness, acceleration, and the unnecessary use of images;
- **Guideline 3:** Create the sense of a 3D environment by using depth cues;
- **Guideline 4:** The correct use of user interface (UI) elements;
- **Guideline 5:** A user guide that helps to start the 3D environment;
- **Guideline 6:** Use a minimum number of controls, which helps the user to remember the controls;
- **Guideline 7:** Virtual objects should be made from real-world objects;
- **Guideline 8:** Try to use Gestalt principles such as similarity, proximity, and hierarchy;
- **Guideline 9:** Try to give feedback to the user when they interact with any virtual object;

- **Guideline 10:** Use audio to help the user experience the real world in a virtual environment.

2.4. Augmented and Mixed Reality

Augmented reality (AR) allows the user to interact with a real-world environment where objects are enhanced by computer-generated virtual projections of data and information, sometimes making use of multiple sensory modalities such as visual, auditory, haptic, somatosensory, and olfactory.

Widely adopted in combination with gestural interfaces to obtain immersive experiences, augmented reality requires that some device is used to display the aforementioned projections and therefore allow users to interact with them. Apart from large displays (see Section 2.2), many augmented reality applications take advantage of mobile phones and tablets. A more specific solution is represented by smart glasses and head-mounted displays. The latter in particular are already used in the military and engineering fields and consist of devices that allow the display to be positioned in front of the user's eyes using a helmet or headbands, allowing total freedom of movement. Such a display can be monocular (i.e., for one eye only), biocular (i.e., two displays are used, showing the same image), or binocular (i.e., for stereoscopic images). Designing an augmented reality experience requires that several factors are taken into account, such as the device chosen for the projections, which can influence their effectiveness, and the surrounding environment, which may have unsuitable surfaces (too bright, reflective, or transparent) for displaying augmented data and may not be suitable for performing certain gestures for safety reasons.

Largely discussed in the literature, a standard definition of mixed reality is yet to be established. Rather than presenting a radically different paradigm, the concept of mixed reality refers to the different levels of the distortion of the real environment, from pure augmented reality to fully virtual reality. Ref. [54] defined a model with two extrema: a fully real environment, the real world, and a fully virtual environment. Each level in between represents the different levels of what the author calls mixed reality.

Ref. [55] reported on a literature survey of 68 papers as well as interviews with AR/VR experts, aimed at understanding the state of the art of mixed reality technologies and related theories. The feature which distinguishes mixed reality from augmented reality is the creation of fully explorable, 3D images in a real-world environment. As described in Section 2.4, in fact, augmented reality only enriches the real environment with 2D elements such as markers, information panels, etc.

2.4.1. The Smart Factory Context

Applications: assembly and maintenance. Augmented reality is applied to guide and help workers in processes such as the assembly and maintenance of complex objects and quality checking, thus decreasing users' cognitive load and improving efficiency [56,57]. More specifically, workers can interact with three-dimensional projections of the objects they are going to assemble and have the possibility of making any necessary checks before starting to build objects in the physical world.

Applications: access to technical data. Another possibility regards the provisioning of technical data, such as manuals, component availability, and maintenance history [57]: augmented reality can improve efficiency in providing relevant information in time as well as geo-located at the appropriate place. In this respect, Ref. [58] identified a series of principles to improve user experience:

- Interoperability, i.e., the use of the same standards for texts and visual elements, so as to facilitate human–machine interaction through documentation;
- Virtualization, i.e., paper documents are virtually copied on the machine, which can thus monitor user actions;
- Decentralization, i.e., documents are divided into sections, so that the machine can show users only the part they need at a certain time;

- Real-time functionality, i.e., the system must be able to analyze the collected data in real time in order to detect any errors. Similarly, technical documentation must be updated in real time, if necessary;
- Service orientation, i.e., carrying out each procedure as if it were a service (e.g., remote maintenance operations);
- Modularity, i.e., the adoption of a modular structure which allows greater flexibility, for example, when new technical procedures must be included within the existing documentation.

An example of real-time technical data delivery is given by [59], who developed an app aimed to provide the operators on the shop floor with technical manuals, operating diagrams, maintenance history, and components availability in the warehouse, connected to the smart manufacturing software.

Available devices and technologies. As far as specific devices and technologies are concerned, augmented reality smart glasses (ARSG) are widely used in the context of smart factories [60]. However, they are prone to give rise to privacy-related issues, in that the use of cameras and other sensors could affect users' behavior and decision making. According to [61], in fact, privacy concerns are one of the factors influencing consumers' decisions to adopt ARSGs.

Regarding mixed reality, the first commercial solution was Microsoft HoloLens, developed due to a collaboration between NASA and Microsoft. Devised at first for gaming purposes, Microsoft HoloLens were then widely applied on the shop floors of smart factories. A real-world application scenario was provided by Fifthingenium (https://fifthingenium.com/, accessed on 29 July 2022), a company specialized in hybrid reality solutions for the industry sector: the Holo Prototype Viewer allows workers to interact with 3D models in their physical environments, creating a mixed reality experience.

2.4.2. Challenges and Guidelines for Augmented and Mixed Reality Interfaces in the Industry 4.0 Context

Our review shows that augmented and mixed reality applications can be helpful when it comes to dynamically providing information to operators on production lines, as well as interactive manuals to be used in the assembly and maintenance areas. However, in addition to privacy-related issues, implementation can be difficult:

- The smooth motions of augmented contents can be hard to obtain with ordinary mobile devices or tablet gyroscopes.
- Depending on the projection surface texture and location, limitations may arise which may hinder an accurate understanding of the surface itself.

To sum up these observations and inspire augmented and mixed reality implementation in the smart factory, we report the following guidelines, based on the work of [60]:

- **Guideline 1: Selection.** An accurate choice of the device that will support the implementation of AR on the shop floor can affect its effectiveness on the production process and must be perpetrated through a step-by-step evaluation of the market.
- **Guideline 2: Compliance.** Privacy policies must be examined and choices on the technologies to be adopted must follow such requirements in order to avoid inapplicable decisions.

Further attention points, which mainly take an implementation-oriented perspective, are included in Google Augmented Reality Design Guidelines (https://designguidelines.withgoogle.com/ar-design/augmented-reality-design-guidelines/introduction.html, accessed on 29 July 2022):

- **Guideline 3: Environment.** Surfaces where augmented reality contents will appear must have correct light exposure and adequate textures. Dim lighting, extremely bright environments, and transparent or reflective surfaces can compromise an accurate understanding of surfaces.

- **Guideline 4: Movements.** When designing the AR experience, exploit the interaction possibilities given by the 360-degree virtual world and encourage users to use movements to dynamically explore the environment
- **Guideline 5: Safety.** The immersive experience provided by the AR must not divert the operator's consciousness from the real world around. Movements must be designed accordingly, to prevent them from unconsciously performing dangerous actions.

2.5. Internet of Things

While the concept of a network of smart devices was discussed as early as 1982, it is only in the last two decades that the increasing possibility of embedding sensors and Internet connectivity into physical devices has led to the definition of an entirely new interaction paradigm, the Internet of Things, which has rapidly brought radical enhancements to fields as diverse as home automation and industry.

A full Internet of Things definition dates back to what Ashton wrote in 2009 [62]: if we had computers that knew everything there was to know about things, using the data they collected without our help, we could track and count everything and significantly reduce waste, losses, and costs. We would know when products needed to be replaced, repaired, or recalled from store warehouses and what the percentage of their wear and tear was. We must enable computers to use their own means of collecting information so that they can see, hear, and feel the world's trends in all their beauty. RFID and sensor technologies enable computers to observe, detect, and understand the world without the constraints of human input.

A further complete definition was given by Rand Europe (https://www.rand.org/randeurope/research/projects/internet-of-things.html, accessed on 29 July 2022), a non-profit research institute: *The "Internet of Things comes from today's Internet, by creating a pervasive and self-organizing network of interconnected, identifiable, and addressable physical entities to enable application development across key vertical industries through embedded chips, sensors, actuators, and inexpensive miniaturization".* Finally, Ref. [63] defined the Internet of Things as follows: *"The Internet of Things (IoT) is an emerging concept quickly gaining ground in the modern wireless telecommunications landscape. The underlying idea of this paradigm lies in the ever-present around us of a multitude of things or objects, for example, radio frequency identification tags (RFID), sensors, actuators, smartphones, etc., capable of mutually interact and cooperate with one another to pursue shared objectives through common addressing patterns."*

According to Skobelev and Borovik [64], the implementation of IoT technologies assumes a transfer of computation to the virtual world (cloud) where each virtual twin of objects in the real world acts according to the selected algorithm and rules. For communication in the real and virtual worlds, intelligent agents may be used. They can perceive information from the real world, make decisions, and coordinate them with other objects or users in real time. At the same time, real objects can work independently or be parts of more complex objects (household things, flexible production lines, groups of drones, etc.).

Moving to a real-world example, modern automobiles, where sensor-gathered data are used to enhance the driving experience, are a case in point: for example, sensors can monitor tire pressure to prevent wheels from locking up or collect information on specific parts of the engine. While in this example data are kept within the system itself, when technologies such as the GPS come into play, a whole new set of possibilities is presented, where information can travel from the vehicle to other external systems. The vehicle therefore becomes smart, exploiting internal data to communicate with other entities and enhance its potential.

Similarly to what has happened in the world of augmented reality with the introduction of ARSGs (Section 2.4), the Internet of Things has also caused the occurrence of privacy issues. In fact, as the Internet of Things is evolving into a decentralized system of cooperating smart objects, such decentralization has a great impact on the way personal information generated and consumed by smart objects should be protected [65]. To address this issue, Ref. [65] proposed a framework which allows users to specify privacy preferences based on a three-level taxonomy of object "smartness", i.e., the object capability of sensing and processing individual data. Starting from the idea that, due to the complexity of data flows among different devices, it is easy for users to lose control of the way their data are distributed and processed, the model implements privacy preferences which allow users not only to pose conditions on which portion of their data can be collected, for what purpose, for how long, and by whom but also to limit the way data can be elaborated to derive new information.

2.5.1. The Smart Factory Context

The Internet of Things is expected to have a huge impact on smart and connected factories.

Applications: access to sensor-gathered data. Almost any existing object or device can be linked to back-end services and become capable to gather and analyze data, elaborate them, and display additional information obtained through physical analytics, possibly leveraging on augmented reality techniques [66]. For example, in a production plant, IoT devices can be used to monitor parameters such as temperature and pressure and to consequently switch different production processes on and off. They can also be employed to monitor hazards such as harmful gas leaks and activate countermeasures such as the ringing alarms meant to alert human operators [67]. Beyond its application on the shop floor, the Internet of Things has already brought changes to the whole product lifecycle: in fact, sensor-gathered data can not only be used to show additional information to the user but also to foster research and therefore enhance the services provided. Likewise, IoT technologies can be exploited, in combination with machine learning models which run on sensor-collected data, to test product quality, thus reducing the time and cost of testing [68]. Furthermore, the Internet of Things can help enhance the supply chain infrastructure so as to improve internal and external connectivity with suppliers and customers. Among other things, IoT devices can be used to track storage conditions throughout the supply chain and to facilitate product traceability [69].

Applications: energy efficiency. In the context of a smart factory project, Ref. [22] proposed the pervasive installation of sensors on production lines to solve consumption management issues. To perform efficient data monitoring with the aim to manage consumption within the context of the smart factory and thus promote a more sustainable approach, all of the ever-changing fields that bring innovation to the fourth industrial revolution were involved in the projects. Technological advancements in information visualization techniques allowed fluent interactions between end-users and big amounts of data; enhancements in machine learning (see [70] for more details) and artificial intelligence engineering made the extraction of valuable information easier to perform on retrieved data; the distance between the digital and physical worlds has been shortened by the pervasive installation of sensors on the production lines and by a participatory approach to the design of the overall cyber-physical system. In a similar vein, Ref. [71] suggested to exploit an IoT layer to make industrial systems more energy-efficient. Loads with such variations that can compromise power quality and increase energy usage are monitored in real time using a sensor-area network, and a central processing server is in charge of deciding which actions to take in order to optimize power consumption.

Enabling technologies. Indoor positioning systems are systems which use wireless communication networks (short- to long-range) [72] and can be easily adapted to address challenges in asset management, people tracking, security, or warehouses [73], thus having direct implications in the developments of the smart factory. Many are the technologies adopted to implement localization systems, from optical sensors to sound waves sensors to electromagnetic field sensors. Among them, radio-frequency-based systems represent a key enabler technology. To this purpose, Ref. [74] provided a state-of-the-art review on one particular type of radio frequency system, the radio frequency identification system commonly known as RFID that represents one of the most suitable choices due to its cost-effectiveness and energy efficiency.

2.5.2. Challenges and Guidelines for Internet of Things in the Industry 4.0 Context

The Internet of Things solutions can have several benefits in the smart factory, such as:

- Large amounts of data can be gathered through connected objects.
- The automation of the network can be enhanced.
- Machine-to-machine communication, as well as human–machine interaction, can be improved.

Such enhancements, however, come at a price. As the amount of data gathered grows, so does the risk of cyber-attacks, making security issues of foremost importance [68]. In addition, the wide use of Internet of Things technologies may negatively impact energy consumption costs.

Cicibaş and Demir [75] proposed a series of guidelines which address both technical and social issues for manufacturing companies. We report here an extract of the guidelines tackling social issues which specifically focus on IoT acceptance and stakeholder involvement and were formulated based on previous work [76,77]:

- **Guideline 1: User acceptance**. Seek ways to achieve user acceptance. Pay special attention to conferences, trainings, and other types of information-sharing activities.
- **Guideline 2: Privacy and ethics**. Inform users and let them adjust privacy settings for private data collections using IoT devices.
- **Guideline 3: Education and training**. Develop and conduct an effective training program for the users.
- **Guideline 4: Stakeholder management**. Identify all stakeholders and pay attention to stakeholder management.

Other guidelines which also clearly embrace a human perspective are more commercial in their nature, such as those which can be drawn from https://www.uxmatters.com/mt/archives/2022/05/designing-for-the-internet-of-things-iot.php, accessed on 29 July 2022:

- **Guideline 5: The importance of UX research**. During the initial phases of design, it is always a good idea to think about what value an IoT device would offer to the users and must deliver to the business.
- **Guideline 6: Taking a holistic view**. Ideally, IoT solutions consist of multiple devices that have various capabilities—both digitally and physically. One must take a holistic approach to designing an IoT device, looking across the whole system, which needs to *work seamlessly* together to create a meaningful experience for users.
- **Guideline 7: Safety and security**. IoT solutions are not purely digital. Once the IoT is placed into a real-world context, the consequences can be severe when something goes wrong.

Other guidelines are proposed by https://www.iotforall.com/designing-user-experience-iot-products, accessed on 29 July 2022:

- **Guideline 8: Simplified onboarding**. The first step of introducing a new system to users can also be the hardest. In the case of multidevice interaction, it often implies repeated authentications, gateway processes that differ from device to device, and switching to additional services, such as email, for verification. Simplified onboarding—secure but effortless authentication with code verification instead of passwords—is a promising beginning.
- **Guideline 9: Smooth cross-device design and interaction**. The key to a consistent user experience across multiple IoT products is in the cloud. Cloud-based apps and connected devices allow us to keep all the parts of the system constantly up-to-date. As a result, it provides users with *seamless transitions* between system elements with minimum effort, adaptation, and wasted time.

- **Guideline 10: One-space experience**. One of the most problematic tasks in UX design for IoT is minimizing the gaps between the physical world of connected devices and creating a smooth experience across all system elements. [...] The challenge of a seamless experience is to integrate diverse independent components into a one-stop solution while saving its functionality and reliability.
- **Guideline 11: New interfaces**. [...] Today, the designers of consumer-oriented IoT products already focus on voice and audio, with more and more digital assistants seen in the home. However, voice is not the only new interface. The future of smooth user experience becomes more contextual and natural [...].

Finally, we report some further guidelines from andrei-klubnikin.medium.com/, https://andrei-klubnikin.medium.com/5-steps-to-great-iot-user-experience-5913955587f1, accessed on 29 July 2022:

- **Guideline 12: Provide the ultimate user experience**. As general as it sounds, the Internet of Things' user experience design principles still revolve around usability, accessibility, utility, and desirability. [...] There are several factors that affect the Internet of Things' user experience, including high power consumption, the lack of a display, the accuracy of sensor data, and device interoperability, and these issues should be addressed during the proof of concept stage.
- **Guideline 13: Do not take Internet connectivity for granted**. Although the key idea behind every IoT project is to connect either consumer electronics or initially dumb objects to the Internet and enable "things" to exchange data over a network, a smart device should perform basic functions even in an offline mode.
- **Guideline 14: Keep interoperability in mind**. Without open-source APIs, reliable device management platforms and unified communication protocols (ZigBee, for instance), IoT is just a bunch of objects connected to the cloud and mobile apps. What we need is a global interconnected environment where products created by different vendors interact with each other.
- **Guideline 15: Embrace accessibility.** The Internet of Things can potentially remove the barriers people with special needs face on a daily basis. [...] That is why forward-thinking vendors enhance their connected solutions with voice recognition and even eye-tracking technologies, thus raising the quality of life for special consumers.

2.6. Wearable

Among smart objects (see Section 2.5) are all those devices which can be woven or otherwise incorporated into clothing or worn as accessories. Many examples of wearable devices have been developed at an experimental level, while some of them have actually made it to the market and eventually become accepted as everyday objects. Popular examples of wearable devices range from fashion items (also known as fashion electronics or fashion technology) to activity trackers or healthcare solutions, able to keep track of body values via specific sensors.

Smart glasses (see Section 2.4), which can be potentially coupled with graduated lenses, also fall into this category. Notice that, while augmented reality mainly involves superimposing interactive computer graphics onto physical objects in the real world, smart glasses have mainly been designed for microinteractions [78]. Being designed for mobility, hands-free interactions such as gestures, voice recognition, and eye-tracking are all good candidates for possible interactions with these devices.

Definitely more common than smart glasses are smartwatches (Figure 4). New guidelines, such as the WatchOs Human Interface Guidelines by Apple (https://developer.apple.com/design/human-interface-guidelines/watchos/overview/themes/, accessed on 29 July 2022), have been defined to support the design of appropriate interfaces for such small devices. On the other hand, Ref. [79] investigated around-device interaction modalities using electric field sensing: more specifically, the authors explored gestural interactions going beyond the boundaries of screens, introducing a new concept of tangible user interfaces and enabling a spontaneous binding between physical objects and digital functions.

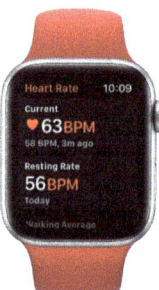

Figure 4. The smartwatch, one of the most widely used wearable devices.

2.6.1. The Smart Factory Context

Due to the interaction modalities they allow, enabling hand-held, touch, and touchless inputs [78], smart glasses can play an important role in smart factories.

Applications: access to knowledge. Ref. [80] studied an application used to document knowledge about assembly and maintenance processes using video recording with smart glasses (namely, Google Glasses). In particular, the application profiles and identifies not only users but also the working context, taking advantage of QR codes or barcodes placed on the machines, thus allowing users to retrieve or post videos from and to a repository. This application was evaluated by administering a survey to a few experienced workers, who were firstly instructed about the usage and interaction modalities of the technology and then proceeded with an on-site test, thus allowing the authors to assess the application on real shop floors. The same evaluation was performed in two different companies: in the first one, the system was used to document standardized tasks within assembly processes in the automotive sector, while in the second case the challenge was to document maintenance tasks. Results showed significant improvements in efficiency and reliability, in comparison with the usual documentation modalities already in use.

Applications: workers' wellbeing. Focusing on the wellbeing of workers, the HuManS (human-centered manufacturing system) project (https://dmd.it/humans/en/humans/, accessed on 29 July 2022) experimented with the creation of a hardware/software Internet of Things architecture for monitoring, analyzing, and controlling the posture of users. Sensorized shirts were designed to be worn by workers on the shop floors, along with an app capable of receiving and elaborating data from the wearable devices. Workers were supposed to log into the app and provide personal data such as their weight and height. They could then monitor their movements in real time during the progress of their daily work and use the app to examine various figures showing the history of their movements during the work shifts.

2.6.2. Challenges and Guidelines for Wearable Devices in the Industry 4.0 Context

Being closely connected to Internet of Things smart objects, most of the pros, cons, and guidelines discussed in Section 2.5.2 can be applied to wearable devices. In addition, as we have seen, wearable devices can improve workers' mobility and support the implementation of other technologies into everyday objects (think of smart glasses with augmented reality). As discussed in Section 2.4.1, however, privacy issues may arise when wearable devices are adopted on the shop floor.

In addition to the guidelines proposed for Internet of Things solutions (see Section 2.5.2), a few more points are worth mentioning (https://developer.apple.com/design/human-interface-guidelines/watchos/overview/apps/, accessed on 29 July 2022):

- **Guideline 1: Glanceability.** Make sure the user interface is organized so that people can quickly and easily find the information they need and perform actions.

- **Guideline 2: Privacy.** Obscure personal information that users would not want casual observers to see, such as health data. In connection with Guideline 1, make sure other types of information remain glanceable, to ease task completion.

2.7. Tangible User Interfaces

In the Internet of Things era, most devices still provide only web or mobile interfaces. Ref. [46] argued that constant interaction with such interfaces could decrease user experience and possibly lead to user alienation from the physical world, these being interactions disconnected from tangible reality. On the contrary, allowing the binding between physical objects and digital functions, augmented reality and other interactive mediums open up to the world of tangible user interfaces (TUI). Tangibles are a particular type of user interface where a person interacts with digital information through the physical environment by touching, displacing, rotating, sliding, or generally interacting in different ways with physical objects that provide inputs to a system and feedback to the user [81].

The analysis carried out by [46] summarizes the current trends in tangible interaction and extrapolates eight properties that could be exploited for designing tangible user interfaces for IoT objects. Such properties range from the ability to leverage natural human skills such as haptic and peripheral interactions to the possibility of integrating tangible interactions with IoT objects in users' daily routines.

In a similar vein, Ref. [82] studied how taxonomies and design principles for tangible interaction should be mapped into the new landscape of IoT systems, investigating parallels between the properties of IoT systems and tangible interactions and therefore envisaging a shift from the world of IoT (Internet of Things) to that of IoTT (Internet of Tangible Things).

2.7.1. The Smart Factory Context

Applications: assembly. Focusing on the smart factory context, Ref. [83] explored the concept of user-defined tangibles: users can turn any physical object at their workplace into a tangible control, thus spontaneously binding it to digital functions. As far as supporting technologies are concerned, the authors found that, in manual assembly workplaces, projection is more suitable than surface-computers, since it cannot be affected by the accidental drop of materials, which is a common event in such a scenario. Consequently, Ref. [83] designed a system which combines a top-mounted Kinect and a top-mounted projector to enable touch interaction, the highlighting of objects, and the display of controls, along with a bottom-mounted leap motion aimed at capturing the user's gestures.

Enabling technologies. Many techniques can be adopted to track objects and enable interactions, among which are RFID tags, capacitive systems, cameras, and magnets. Envisaging the assembly line in factories as a possible application scenario, Ref. [84] explored radar sensing as a way to support tangible interaction with six sensing mechanisms: counting, ordering, and identifying the objects and tracking the orientations, movements and distances of these objects. The authors showed that miniature radar sensing is accurate even with minimal training and that it can support new forms of tangible interaction.

2.7.2. Challenges and Guidelines for Tangible User Interfaces in the Industry 4.0 Context

As discussed in our review, tangible interfaces can help to make interaction more natural and engaging. In particular, tangibles can:

- Stimulate users to interact with the concrete world around them, thus contrasting the sense of alienation which may arise from continuous exposure to screen-based devices.
- Provide immediate feedback in the real world, instead of exploiting a graphical interface which provides a representation of reality.

On the other hand, acceptability issues may arise when digital objects incorporating tangibles replace the everyday objects operators are used to [85]. In addition, tangibles can be hardly standardized, which implies that operators might be required to make a substantial effort to learn how to use each of them [85]. Similarly, the use of some tangibles might be restricted to specific environments [85].

In addition to the guidelines proposed for Internet of Things solutions (see Section 2.5.2) based on the work of [83], we report a brief list of guidelines for the implementation of tangible user interfaces in the smart factory which specifically focus on the above-mentioned issues:

- **Guideline 1: Codesigning.** Whenever possible, try to involve users in the design process of tangibles, in order to avoid unexpected acceptability issues.
- **Guideline 2: Learning.** Consider the possibility of undertaking training sessions to allow operators to build the mental models required to operate TUIs.

2.8. Collaborative Robots

Another paradigm that changed with the development of the Smart Industry 4.0 and 5.0 is surely that of human–robot interaction (HRI), leading to the modification and enhancement of the acceptance level of collaborative robots on the shop floors.

The first collaborative robot was devised in 1996 by James E. Colgate and Michael A. Peshkin, who defined it as *"an apparatus and a method for direct physical interaction between a person and a generic manipulator controlled by a computer"* [86]. The term "cobot" was later listed among the new terms by the Wall Street Journal, meaning a collaborative robot designed to help workers in their businesses rather than replace them [87]. Today, more than twenty years after its invention, the concept of collaborative robotics has commonly taken on the meaning of work sharing. Collaboration is manifested through human access to the robotic system and the workspace to perform functionally related actions [88]. The collaboration between humans and robots aims to combine human skills and flexibility with the benefits associated with robotic systems. This allows an increase in productivity and product quality while reducing ergonomics-related risks for operators [89].

Different levels of collaboration between humans and robots are possible [90]. Conventionally, in the factory, the robot is located inside protected areas that are not accessible to humans; access to the workspace is only allowed when the robot is stationary to carry out maintenance or programming operations. This is the first level of collaboration and is characterized by a *strict separation* between workspaces. The second level can be called *coexistence*: in this case there is no sharing of the workspace, but a physical barrier is missing. At this level, humans can access the robot's work area, but human presence is detected by a safety system that causes the robot to regulate the power and speed of movements. The third level is that of *synchronized operations*, in which worker and robot share the same workspace, but at different times; therefore, there is a condition of temporal separation. A fourth level is that of *cooperation*, in which spatial and temporal separation are reduced and man and robot are allowed to occupy the same work area at the same time, remaining separate, however, because of the lack of joint activities. Finally, at the highest level of collaboration, man and robot *work on a common task* without any temporal or spatial separation of the work area, but rather a voluntary contact between man and machine can be envisaged.

The collaboration between humans and robots in charge to assist their work deserves to be investigated and improved. HRI being a sub-branch of HCI that is rapidly emerging and creating its own standards, a stand-alone research sector deserves to be considered for this category of interactions and the relevant studies that were carried out.

2.8.1. The Smart Factory Context

Given a clear definition of what a collaborative robot is and what its components are, we move one step forward in our examination of the current state of the art, providing real-world examples of their implementation on shop floors.

Applications: e-waste management. Ref. [91] put forward the adoption of collaborative robots to solve e-waste management problems, optimizing the recycling process of electronic equipment. Companies are always more subject to public and government pressure to reduce their environmental impact. When dealing with e-waste, manual operations can be financially prohibitive and full automation is not easy to implement due to the lack of uniformity of devices. It is trivial to notice how this is clearly a scenario where a collaboration between humans and robots may bring enhancements to the process. Alvarez-de-los Mozos and Renteria [91] examined the e-waste management techniques and the limitations of fully automated techniques for waste electrical and electronic equipment (WEEE) and then proposed a solution for WEEE recycling that involves the use of collaborative robots. The authors brought a real-world example discussing the use of Liam, a collaborative robot developed by Apple to effectively disassemble the iPhone (Figure 5). The authors also pointed out that one of the main problems that can possibly arise when dealing with bigger electronic appliances is that of identifying cables, flexible parts, or components which are usually difficult to recognize. This represents a point in the process where collaboration can happen and a skilled operator could carry out the job, leaving the tedious and potentially dangerous tasks of operating the materials to the robots.

Figure 5. Liam, a collaborative robot by Apple which disassembles iPhones.

Applications: assembly. Ref. [92] investigated the combination of sensors, embedded in wearable devices with gestures recognition, to propose an HRI framework applicable in assembly operations, where collaborative robots can assist workers, delivering tools and parts and holding objects. The aim of this and many other investigations in the field is that of exploiting the best abilities of robots, such as accuracy or repetitive work, and the best abilities of humans, for example, cognition and management, in order to reach a collaborative scenario where the most is made out of every available resource. Moreover, we should consider that mobile robots and exoskeletons have the potential to make certain tasks less physically demanding, see Spada at al. for more details [93]. This may allow women to take on tasks that were previously reserved for men due to the required physical strength. A vast range of further opportunities will arise by the further digitalization of the workforce [36].

Focusing on the general context of performing dull tasks on production lines, a study aimed at enhancing the effectiveness of already existing robots was carried out by [94]. The authors started their work from the assumption that collaborative robots are more useful when they can be displaced at a level of easiness that makes them "mobile" [94]. They investigated a system to enable robots such as Baxter and Sawyer by Rethink Robotics to smartly perform movements within the shop floor, sensing persons or obstacles and moving safely throughout the space. A downward-facing QR code camera was used for the precise placement of the robot at a work station and, when not assigned to a specific cobot, the platform can be used as a general-purpose automated guided vehicle.

Worker–robot interaction and collaboration. A key feature of collaborative robots is their ability to partner with human operators in mixed teams. They need to coordinate their actions to engage in joint activities and to coordinate their behavior to human behaviors at different levels: semantic, contextual, temporal, and more, see [95], which investigated the cognitive systems that build the awareness needed to obtain such interactions. The authors provided a tool for addressing this problem by using the notion of deep hybrid representations and the facilities that this common state representation offers for the tight coupling of planners on different layers of abstraction. According to Villani et al. [17], the main challenges related to cobots are: safety issues, intuitive user interfaces, so that human operators can easily interact with the robot, and design methods, which mean control laws, sensors and task allocation, and planning approaches, which allow the human operator to safely stand close to the robot, actively sharing the working area and tasks and providing the interaction system with the required flexibility [17]. In particular, regarding the worker–robot interaction, the use of collaborative robots in industrial processes proves beneficial also given the fact that they can be managed and taught through intuitive systems, based on augmented reality [96], walk-through programming [97], or programming by demonstration [98].

As far as intuitive user interfaces are concerned, differently from instructing a (skilled) human worker on how to carry out a task, programming a robot requires providing the robot with explicit motion-oriented instructions, detailing the points and trajectories that the robot has to follow. Nonetheless, the goal is that of explicitly instructing the robot in a human-friendly manner and without negatively affecting the productivity of the system. To this purpose, Villani at al. [17] proposed to use novel approaches as walk-through programming, programming by demonstration, and the use of multimodal interfaces and augmented/virtual reality, which are characterized by high intuitiveness since they constitute instances of natural and tangible user interfaces (NUIs and TUIs, respectively). For instance, NUIs allow users to directly manipulate and interact with robots rather than instruct them to do so by typing commands. Techniques used include, for instance, speech, gestures, eye tracking, facial expression, and haptics, in addition to the traditional ones, namely, keyboard, mouse, monitor, touchpad, and touchscreen.

It is worthwhile mentioning control techniques and approaches aimed at improving the safety and ergonomics of operators interacting with robots. Typical control problems related to safety include collision avoidance, collision detection, motion planning, and safety-oriented control system design. Similarly, for ergonomics, they include scheduling and ergonomics-oriented control system design, as well as the common area of motion planning. Several approaches are available to tackle each problem: for example, considering the ergonomics area, these include: biological and nonbiological trajectory optimization, minimum jerk trajectory planning (motion planning), mixed-integer linear programs, stochastic Petri-nets, cognitive load optimization, feedforward/feedback optimization, decision making models (for scheduling), haptic assistance, optimal control, whole-body control, game theory, gesture-based control, admittance control, learning-based control, and reinforcement learning (for ergonomics-oriented control system design). See for more details the comprehensive survey by Proia et al. [99].

Cobots as autonomous systems. Ref. [100] carried out a useful study on the paradigm of the smart factory, focusing on the role of cobots in this context. The authors explored how cobots are defined and highlighted how learning processes can be carried out by such robots, through the adoption of artificial intelligence techniques, in order to enhance productivity and the quality of manufactured goods and thus create a smart factory. Examining the nine pillars of Industry 4.0, Ref. [100] discussed the role of collaborating robots in the scope of the first pillar: autonomous systems. They defined cobots as automated systems, including sensors, actuators, and controllers, capable of performing tasks continuously and designed to be applied in the industrial field [100]. Two types of autonomous systems were shown in their study, i.e., multiagent systems and intelligent industrial robots.

This second category of autonomous systems is particularly interesting for this section. Ref. [100] provided a standard definition of what such robots are, in terms of their characteristic components: an *"intelligent industrial robot is a useful combination of a manipulator arm, sensors, and intelligent controllers, which replaces a human worker and can complete tasks and resolve the problems. Eventually, it will be able to learn from humans at first. The use of these machines in industrial automation can improve productivity and product quality, creating smart industry"*.

2.8.2. Challenges and Guidelines for Collaborative Robots in the Industry 4.0 Context

As highlighted by [101], industrial collaborative robotics is one of the most promising technologies of the smart industry. In particular, human–robot collaboration in assembly will be particularly interesting for manufacturing companies. In this context, the interaction between humans and robots opens new possibilities:

- The elimination of repetitive or dangerous tasks from human operators' concerns, to allow the human resources to focus on those tasks that better suit human minds.
- A reduction in risks on the shop floor.

However, there are also challenges:

- A long learning curve, to allow a smooth interaction between operators and robots.
- Possible difficulties in the realization of a smooth inclusion into mixed teams, due to the difficulties in creating coordinated behaviors in such robots.
- Eventual slowdowns of the production process may result from faults, especially if many tasks are assigned to collaborative robots.

Ref. [95] proposed a set of guidelines for the adoption of collaborative robots on factory floors:

- **Guideline 1: Selection.** A correct distinction between the tasks that should be carried out by human operators and those that better belong to robots must be conducted before planning the work and tasks must be assigned accordingly.
- **Guideline 2: Behavior.** When designing collaborative robots or when making decisions on which solution to adopt, their ability to coordinate their behavior with that of humans must be taken into consideration as a priority.
- **Guideline 3: Safety.** ISO/TS 15066:2016 specifies safety requirements for collaborative industrial robot systems and the work environment and must be taken into consideration when adopting such solutions within a smart factory (https://www.iso.org/standard/62996.html, accessed on 29 July 2022).

Ref. [102] proposed and then validated [101] new design guidelines for systems integrator designers to develop safe and ergonomic collaborative assembly workstations. We report the most general ones:

- **Guideline 4**. Minimize specific mechanical hazards related to the entrapment of human body parts.
- **Guideline 5**. Minimize specific mechanical hazards related to collisions with human body parts.
- **Guideline 6**. Minimize specific mechanical hazards related to robot system parts falling.
- **Guideline 7**. Minimize the biomechanical overload of upper limbs related to repetitive tasks.
- **Guideline 8**. Minimize the biomechanical overload of the whole body related to the manual lifting/lowering of objects.
- **Guideline 9**. Minimize the biomechanical overload of head/neck/trunk/upper or lower limbs related to static or awkward working postures.
- **Guideline 10**. Maximize operator psychological wellbeing and satisfaction.
- **Guideline 11**. Maximize the efficiency of manual and robot assembly activities.

3. Discussion and Conclusions

Smart interactive technologies are revolutionizing workers' activities on the factory floor. While throughout our survey we have adopted a technology-driven perspective, illustrating the changes and possibilities enabled by the emerging technologies in the Industry 4.0 and 5.0 visions, Table 1 summarizes the contributions of surveyed work which specifically falls into the smart factory context by highlighting the problems and phases they address in the production process. As we can see, most transformations regard the production phase and access to knowledge.

Communication, learning, and knowledge-sharing. When examining the modifications brought or suggested by the fourth and later fifth industrial revolutions (respectively: smart manufacturing, smart mass production, smart products, smart working, smart supply chain, and system(s) optimization; sustainability, environmental stewardship, human-centricity, and social benefit, see for more details [103]) to the modern factory floors, however, one cannot ignore how such changes are influencing the way communication happens between operators and coworkers and between operators and machines. For example, Ref. [30] investigated the mutual human–machine learning in smart factories, with the ultimate goal to identify new learning patterns in such environments. They defined mutual learning as *a bidirectional process involving reciprocal exchange, dependence, action, or influence within human and machine collaboration, which results in creating new meaning or concepts, enriching the existing ones, or improving skills and abilities in association with each group of learners*, and distinguished three groups of tasks that can be carried out within the smart factory: those assigned specifically to humans, those dispensed for machines, and the shared ones, where exchange and thus mutual learning occurs. Ref. [30] then illustrated a conceptual model for mutual learning, based on the model of hybrid learning proposed by Zitter and Hoeve [104]. All their results have been applied and tested in a real-world context, at the TU Wien Pilot Factory.

All in all, we can state that learning processes within the smart factory are and will be increasingly more affected by the process of digitalization. In this vein, Ref. [105] reviewed virtual training systems with a focus on their teaching styles and identified new research directions in the field of adaptive training systems.

Benefits derived from the changes introduced with the Industry 4.0 and 5.0 extend to activities carried out outside the factory walls. A case in point is the work of [20], which examined knowledge sharing solutions based on Industry 4.0 to improve mobile service technicians' daily work performance and work satisfaction. The authors started a human-centered design process that led to the creation of the Mobile Service Technician 4.0 concept: it utilizes industrial internet, virtual, and augmented reality as well as wearable technologies to improve both the user experience of workers within the examined field and the quality of their work.

Human-centricity. The Industry 5.0 paradigm reinserts proactively humans back into the automation chain [106], and this means that technology used in manufacturing should be "*adapted to the needs, and diversity of industry workers, instead of having the worker continuously adapt to ever-evolving technology. The worker is more empowered and the working environment is more inclusive. To achieve this, workers are to be closely involved in the design and deployment of new industrial technologies, including robotics and AI*" [3]. Hence, approaches such as codesign and prototyping should be adopted in this new vision, also helped by new technologies, such as virtual and augmented reality, that allow prototype simulations before the actual realization.

Table 1. Surveyed works classified based on the supported task(s) and production process phases/activities: access to knowledge (K), logistics (L), maintenance (M), planning (Pl), production (Pr), security (S), workers' wellbeing (We), and warehousing (Wa).

Technology	Reference	Supported Tasks	K	L	M	Pl	Pr	S	We	Wa
Large displays	[24]	Task browsing; interaction with 3D objects		X			X			
	[39]	Interaction with information systems								
	[40]	Interaction with 3D objects				X				
	[41]	Access to different specialized devices (monitoring, diagnosis, and maintenance)			X					
Virtual reality	[45]	Create virtual factory environments				X				
	[49]	Factory layout planning				X				
	[50]	Factory layout planning				X				
	[51]	Virtual commissioning				X				
	[52]	Virtual commissioning and performance optimization				X				
Augmented reality	[56]	Product assembly and maintenance; quality checking	X		X		X			
	[57]	Product assembly and maintenance; quality checking; access to technical data	X		X		X			
	[58]	Access to technical data	X							
	[59]	Access to technical data			X		X			
	[60]	Assembly, maintenance, quality control, and material handling					X			
	[61]	Product engineering, employee coaching, warehousing, and logistics		X						X
Internet of Things	[22]	Energy consumption data monitoring	X							
	[66]	Access and analyze technical data	X							
	[67]	Process monitoring, hazard reduction	X							
	[68]	Product testing		X			X	X		
	[69]	Product traceability; monitoring of storage conditions	X				X			X
	[71]	Energy consumption optimization					X			
	[73]	Asset management, people tracking, security, or warehouse		X				X		X
	[74]	Asset management, people tracking, security, or warehouse		X				X		X
Wearable	[80]	Document knowledge about assembly and maintenance processes	X							
	HuManS (https://dmd.it/humans/en/humans/) accessed on 29 July 2022	Monitoring, analyzing, and controlling the posture of user							X	
Tangible user interfaces	[83]	Assembly					X			
	[84]	Assembly					X			
Collaborative robots	[17]	Collaboration in manufacturing tasks					X			
	[36]	Strength-demanding tasks					X			
	[91]	e-waste management			X		X			
	[92]	Assembly					X			
	[94]	Performing dull tasks on production lines					X			
	[95]	Supporting human operators in mixed teams					X			
	[99]	Collaboration in manufacturing tasks; posture improvement					X		X	
	[100]	Assembly					X			

In addition to that, Industry 5.0 also emphasizes human-centricity through the use of AI-based technologies to empower the worker's performance and capacity. In this regard, wearable devices that boost cognitive and operational capacities are increasingly being utilized and improved in manufacturing industries [107]. Exoskeletons, i.e., augmenter equipment that give extra strength and physical capabilities to protect the operator from the adverse effects of heavy workloads [93], are a case in point. According to Jafari et al. [108], virtual technologies such as smart AR glasses, spatial AR projectors, etc., are viable and novel gadgets that facilitate flexible operations and technical guidance through information transmission and virtualization. Moreover, wearables could open new channels for alerting workers and their general practitioners about critical health conditions, both physical and mental, as well as supporting workers in adopting healthy behaviors in the workplace [3].

However, these improvements in working conditions cannot be conducted at the expense of workers' fundamental rights of privacy, security, autonomy, and human dignity. According to our vision, it is essential that future HCI and HCAI specialists become aware of the potential ethical and practical issues of smart interactive technologies, also considering the smart factory context and the new, central role of workers, see for more details Longo et al. [107].

Sustainability. Another relevant Industry 5.0 concept, also emphasized in the 2021 European Commission's report [3] and highlighted in Section 2.5.2, is the one of environmental sustainability. According to Akundi et al. [103] Industry 5.0 *"recognizes the capacity of industry to fulfill social objectives beyond employment and development, to become a sustainable source of development, by making production regard the limitations of our planet and prioritizing employee health first"*. Sustainability is closely related also to the promotion of a circular economy, i.e., the idea of developing circular processes that reuse, repurpose, and recycle natural resources, reducing waste and environmental impact [3]. One of the enabling technologies for reaching sustainability goals is certainly IoT. Drawing from the IoT Guidelines for Sustainability produced by the World Economic Forum (https://www.weforum.org/, accessed on 29 July 2022) [109], we recall a set of points which specifically refer to the sustainability and impact measurement area. Firstly, along with all the valuable data they may collect with IoT systems, smart factories should make sure to measure and process energy usage data, so as to minimize costs, increase savings, and reduce waste (*consumption*). Then, smart factories should embrace a sustainability awareness culture to respond to new generational demand, enhancing brand reputation and attracting top talent (*culture*). Furthermore, potential impact should be evaluated and results measured based on some ad hoc framework, such as the United Nations Sustainable Development Goals (https://sdgs.un.org/goals, accessed on 29 July 2022) (*impact*). When planning an Internet of Things project, potentially addressable sustainable development goals and targets should be identified and incorporated into the commercial design (*goals*). Finally, RFID or GPS sensors monitoring should be implemented both to track products in the delivery process and to track inventory items within the warehouse and the production lines (*monitoring*).

Further challenges. Along with all the enhancements and improvements brought by the Industry 4.0 and 5.0 to production processes and to the workers' performances, unavoidably there come new risks for both individuals and organizations that can directly affect productivity and translate into financial risks. Herrmann [110] gave an overview of the technical components of a smart factory, raising the awareness of this manufacturing trend in terms of risks evaluation. The author focused on the topics of standardization, information security, the availability of the IT structures, the availability of fast Internet, complex systems, as well as organizational and financial risks in the scope of the fourth industrial revolution. He pointed out how investigation in the field must be pushed parallel to the development progress and highlighted the need for further research in order to provide a complete overview of the smart factory and its status. Last but not least, in the definition of Industry 5.0 we found the concept of resilience, referring to the need to develop a higher degree of robustness in industrial production, arming it better against disruptions and ensuring it can provide and support critical infrastructure in times of crisis.

The future industry needs to be resilient enough to swiftly navigate the (geo-)political shifts and natural emergencies [3], as sadly witnessed by the recent events of COVID-19 and the war between Russia and Ukraine.

In this paper, we have provided a picture of the current state of the art of smart interactive technologies on the factory floor, and we have also explored the way new technologies are changing the relations between workers and operations. On the one hand, we wanted to emphasize the fact that smart factories provide a challenging and stimulating environment, where workers are required to be resourceful and possess excellent communication, organization, and collaboration skills, in order to manage complexity and abstraction in problem solving processes, as also highlighted by [111]. On the other hand, we wanted to provide some practical examples of the use of intelligent technologies in the smart factory, also proposing guidelines to design interactions that should be human-centered.

Intelligent system components may have unexpected and biased behavior, due to the success and large use of probabilistic approaches such as machine learning, neural networks, deep learning, etc., based on the data collected in large data sets which may have some latent bias (see, for instance, [112]) and thus confuse users, erode their confidence, and lead to the abandonment of AI technology. High-profile reports of failures (see for example: https://spectrum.ieee.org/ai-failures, accessed on 29 July 2022, and https://www.ftc.gov/news-events/news/press-releases/2022/06/ftc-report-warns-about-using-artificial-intelligence-combat-online-problems, accessed on 29 July 2022) range from humorous and embarrassing mistakes (e.g., autocompletion errors, misunderstandings in conversational agents, etc.) to more serious circumstances in which users cannot effectively handle an AI system (e.g., driving a semiautonomous car). These factors show that designers and developers need knowledge on proper methodologies to create effective human-centered intelligent systems. *User in control* is one of the pillars of human-centered design: this can be achieved by granting transparency in system behavior, i.e., in the form of the explainability of the AI decision making process empowering the end-users to configure and adapt such behavior (for more details, see [9]). This example shows how important it is to consider human factors and human perspectives in intelligent systems, which need to be designed and implemented in a user-centered/human-centric way. We hope that with the discussions, examples, and guidelines reported in this survey paper, we have made a small but relevant advance with respect to this goal.

Author Contributions: Conceptualization, D.B., C.G. and F.V.; Investigation, D.B., C.G. and F.V.; Methodology, D.B., C.G. and F.V.; Writing—original draft, D.B., C.G. and F.V. All authors have read and agreed to the published version of the manuscript.

Funding: This research was partially funded by Regione Piemonte, grant number 319-50 (Programma Operativo Regionale POR-FESR 2014/2020, HOME project).

Institutional Review Board Statement: Not applicable.

Informed Consent Statement: Not applicable.

Data Availability Statement: Not applicable.

Conflicts of Interest: The authors declare no conflict of interest.

References

1. Wang, S.; Wan, J.; Li, D.; Zhang, C. Implementing Smart Factory of Industrie 4.0: An Outlook. *Int. J. Distrib. Sens. Netw.* **2016**, *2016*, 1–10. [CrossRef]
2. Frazzon, E.; Ehm, J.; Makuschewitz, T.; Scholz-Reiter, B. Towards Socio-Cyber-Physical Systems in Production Networks. *Procedia CIRP* **2013**, *7*, 49–54. [CrossRef]
3. Directorate-General for Research and Innovation European Commission; Breque, M.; De Nul, L.; Petridis, A. *Industry 5.0: Towards a Sustainable, Human-Centric and Resilient European Industry*; Publications Office: Luxembourg, 2021. [CrossRef]
4. Directorate-General for Research and Innovation European Commission; Müller, J. *Enabling Technologies for Industry 5.0—Results of a Workshop with Europe's Technology Leaders*; Publications Office: Luxembourg, 2020.

5. Xu, X.; Lu, Y.; Vogel-Heuser, B.; Wang, L. Industry 4.0 and Industry 5.0—Inception, conception and perception. *J. Manuf. Syst.* **2021**, *61*, 530–535. [CrossRef]
6. Sonntag, D. Intelligent User Interfaces—A Tutorial. *arXiv* **2017**, arXiv:1702.05250.
7. Shneiderman, B. Human-Centered Artificial Intelligence: Reliable, Safe & Trustworthy. *Int. J. Hum. Comput. Interact.* **2020**, *36*, 495–504. [CrossRef]
8. Shneiderman, B. Bridging the Gap Between Ethics and Practice: Guidelines for Reliable, Safe, and Trustworthy Human-Centered AI Systems. *ACM Trans. Interact. Intell. Syst.* **2020**, *10*, 26. [CrossRef]
9. Costabile, M.F.; Gena, C.; Matera, M.; Paternò, F.; Tortora, G.; Zancanaro, M. Teaching HCI for AI: Co-Design of a Syllabus. Final Report by the Workshop Organizers. 1999. Available online: http://sigchitaly.eu/en/hci4ai-syllabus-it/workshop-results/ (accessed on 29 July 2022).
10. Baudel, T.; Beaudouin-Lafon, M. Charade: Remote Control of Objects Using Free-hand Gestures. *Commun. ACM* **1993**, *36*, 28–35. [CrossRef]
11. Blumendorf, M.; Feuerstack, S.; Albayrak, S. Multimodal User Interfaces for Smart Environments: The Multi-Access Service Platform. In Proceedings of the Working Conference on Advanced Visual Interfaces, AVI'08, Napoli, Italy, 28–30 May 2008; Association for Computing Machinery: New York, NY, USA, 2008; pp. 478–479. [CrossRef]
12. Gianotti, M.; Riccardi, F.; Cosentino, G.; Garzotto, F.; Matera, M. Modeling Interactive Smart Spaces. In Proceedings of the Conceptual Modeling, Vienna, Austria, 3 November 2020; Dobbie, G.; Frank, U.; Kappel, G., Liddle, S.W., Mayr, H.C., Eds.; Springer International Publishing: Cham, Switzerland, 2020; pp. 403–417.
13. Jameson, A.; Gabrielli, S.; Kristensson, P.O.; Reinecke, K.; Cena, F.; Gena, C.; Vernero, F. How can we support users' preferential choice? In Proceedings of the International Conference on Human Factors in Computing Systems, CHI 2011, Extended Abstracts Volume, Vancouver, BC, Canada, 7–12 May 2011; Tan, D.S., Amershi, S., Begole, B., Kellogg, W.A., Tungare, M., Eds.; ACM: New York, NY, USA, 2011; pp. 409–418. [CrossRef]
14. Jannach, D.; Zanker, M.; Felfernig, A.; Friedrich, G. *Recommender Systems—An Introduction*; Cambridge University Press: Cambridge, UK, 2010.
15. Gena, C.; Grillo, P.; Lieto, A.; Mattutino, C.; Vernero, F. When Personalization Is Not an Option: An In-The-Wild Study on Persuasive News Recommendation. *Information* **2019**, *10*, 300. [CrossRef]
16. Desmarais, M.C.; Baker, R.S. A review of recent advances in learner and skill modeling in intelligent learning environments. *User Model.-User-Adapt. Interact.* **2012**, *22*, 9–38. [CrossRef]
17. Villani, V.; Pini, F.; Leali, F.; Secchi, C. Survey on human–robot collaboration in industrial settings: Safety, intuitive interfaces and applications. *Mechatronics* **2018**, *55*, 248–266. [CrossRef]
18. Corcella, L.; Manca, M.; Nordvik, J.E.; Paternò, F.; Sanders, A.; Santoro, C. Enabling personalisation of remote elderly assistance. *Multim. Tools Appl.* **2019**, *78*, 21557–21583. [CrossRef]
19. Hajjaji, Y.; Boulila, W.; Farah, I.R.; Romdhani, I.; Hussain, A. Big data and IoT-based applications in smart environments: A systematic review. *Comput. Sci. Rev.* **2021**, *39*, 100318. [CrossRef]
20. Kaasinen, E.; Aromaa, S.; Väätänen, A.; Mäkelä, V.; Hakulinen, J.; Keskinen, T.; Elo, J.; Siltanen, S.; Rauhala, V.; Aaltonen, I.; et al. Mobile Service Technician 4.0: Knowledge-Sharing Solutions for Industrial Field Maintenance. *IxD&A* **2018**, *38*, 6–27.
21. MAY, G.; Taisch, M.; Bettoni, A.; Maghazei, O.; Matarazzo, A.; Stahl, B. A New Human-centric Factory Model. *Procedia CIRP* **2015**, *26*, 103–108. [CrossRef]
22. Benedetto, F.; Brunetti, D.; Gena, C.; Lai, M.; Meo, R.; Vernero, F. Intelligent monitoring applications for Industry 4.0. In Proceedings of the IUI '20: 25th International Conference on Intelligent User Interfaces, Cagliari, Italy, 17–20 March 2020; ACM: New York, NY, USA, 2020; pp. 67–68. [CrossRef]
23. Andolina, S.; Ariano, P.; Brunetti, D.; Celadon, N.; Coppo, G.; Favetto, A.; Gena, C.; Giordano, S.; Vernero, F. Experimenting with Large Displays and Gestural Interaction in the Smart Factory. In Proceedings of the 2019 IEEE International Conference on Systems, Man and Cybernetics, SMC 2019, Bari, Italy, 6–9 October 2019; pp. 2864–2869. [CrossRef]
24. Andolina, S.; Ariano, P.; Brunetti, D.; Celadon, N.; Coppo, G.; Favetto, A.; Gena, C.; Giordano, S.; Vernero, F. Introducing Gestural Interaction on the Shop Floor: Empirical Evaluations. In Proceedings of the Human-Computer Interaction-INTERACT 2021-18th IFIP TC 13 International Conference, Bari, Italy, 30 August–3 September 2021; Ardito, C., Lanzilotti, R., Malizia, A., Petrie, H., Piccinno, A., Desolda, G., Inkpen, K., Eds.; Springer: Berlin/Heidelberg, Germany, 2021; Volume 12936, pp. 451–455. [CrossRef]
25. Krupitzer, C.; Müller, S.; Lesch, V.; Zufle, M.; Edinger, J.; Lemken, A.; Schäfer, D.; Kounev, S.; Becker, C. A Survey on Human Machine Interaction in Industry 4.0. *arXiv* **2020**. [CrossRef]
26. Golightly, D.; Sharples, S.; Patel, H.; Ratchev, S. Manufacturing in the cloud: A human factors perspective. *Int. J. Ind. Ergon.* **2016**, *55*, 12–22. [CrossRef]
27. Spasojević-Brkić, V.; Putnik, G.; Shah, V.; Castro, H.; Veljkovic, Z. Human-Computer Interactions and User Interfaces for Remote Control of Manufacturing Systems. *FME Trans.* **2013**, *41*, 250–255.
28. Gorecky, D.; Schmitt, M.; Loskyll, M.; Zühlke, D. Human-machine-interaction in the industry 4.0 era. In Proceedings of the 2014 12th IEEE International Conference on Industrial Informatics (INDIN), Porto Alegre, Brazil, 27–30 July 2014; pp. 289–294.
29. Aehnelt, M.; Klamma, R.; Pammer, V. Human Computer Interaction Perspectives on Industry 4.0. *Interact. Des. Archit.* **2019**, *38*. Available online: https://www.researchgate.net/publication/332466625_Human_Computer_Interaction_Perspectives_on_Industry_40_Interaction_Design_Architectures_Vol_38 (accessed on 29 July 2022). [CrossRef]

30. Ansari, F.; Erol, S.; Sihn, W. Rethinking Human-Machine Learning in Industry 4.0: How Does the Paradigm Shift Treat the Role of Human Learning? *Procedia Manuf.* **2018**, *23*, 117–122. [CrossRef]
31. Deneen, K. Human-Machine Interface Technologies: What Impact on Industry 4.0? Available online: https://medium.com/astercapital/human-machine-interface-technologies-what-impact-on-industry-4-0-6a105f97529d/ (accessed on 29 July 2022).
32. Klumpp, M.; Hesenius, M.; Meyer, O.; Ruiner, C.; Gruhn, V. Production logistics and human-computer interaction—State-of-the-art, challenges and requirements for the future. *Int. J. Adv. Manuf. Technol.* **2019**, *105*, 3691–3709. [CrossRef]
33. Ras, E.; Wild, F.; Stahl, C.; Baudet, A. Bridging the Skills Gap of Workers in Industry 4.0 by Human Performance Augmentation Tools: Challenges and Roadmap. In Proceedings of the 10th International Conference on PErvasive Technologies Related to Assistive Environments, PETRA'17, Island of Rhodes, Greece, 21–23 June 2017; Association for Computing Machinery: New York, NY, USA, 2017; pp. 428–432. [CrossRef]
34. Merz, E. Industrie 4.0: Usability Design for the Industry of Tomorrow. Available online: https://www.amnytt.no/getfile.php/2875229.2265.rtbaurxvwt/Usability_Industrie_4.0_EN.pdf (accessed on 29 July 2022).
35. Deloitte. The Industry 4.0 Paradox. Available online: https://www2.deloitte.com/global/en/pages/energy-and-resources/articles/the-industry-4-0-paradox.html (accessed on 29 July 2022).
36. Romero, D.; Stahre, J.; Wuest, T.; Noran, O.; Bernus, P.; Fast-Berglund, Å.; Gorecky, D. Towards an Operator 4.0 Typology: A Human-Centric Perspective on the Fourth Industrial Revolution Technologies. In Proceedings of the International Conference on Computers & Industrial Engineering (CIE46), Tianjin, China, 29–31 October 2016; pp. 1–11.
37. Ardito, C.; Buono, P.; Costabile, M.F.; Desolda, G. Interaction with Large Displays: A Survey. *ACM Comput. Surv.* **2015**, *47*, 46:1–46:38. [CrossRef]
38. Sambrooks, L.; Wilkinson, B. Comparison of Gestural, Touch, and Mouse Interaction with Fitts' Law. In Proceedings of the 25th Australian Computer-Human Interaction Conference: Augmentation, Application, Innovation, Collaboration, OzCHI'13, Adelaide, Australia, 25–29 November 2013; Association for Computing Machinery: New York, NY, USA, 2013; pp. 119–122. [CrossRef]
39. Lacueva-Pérez, F.J.; Khakurel, J.; Brandl, P.; Hannola, L.; Gracia-Bandrés, M.A.; Schafler, M. Assessing TRL of HCI Technologies Supporting Shop Floor Workers. In Proceedings of the 11th PErvasive Technologies Related to Assistive Environments Conference, PETRA '18, Corfu, Greece, 26–29 June 2018; ACM: New York, NY, USA, 2018; pp. 311–318. [CrossRef]
40. Du, G.; Degbelo, A.; Kray, C.; Painho, M. Gestural interaction with 3D objects shown on public displays: An elicitation study. *Interact. Des. Archit.* **2018**, *2018*, 184–202. [CrossRef]
41. Meixner, G.; Petersen, N.; Koessling, H. User Interaction Evolution in the SmartFactoryKL. In Proceedings of the 24th BCS Interaction Specialist Group Conference, BCS '10, Dundee, UK, 6–10 September 2010; British Computer Society: Swinton, UK, 2010; pp. 211–220.
42. Garzotto, F.; Valoriani, M. Touchless Gestural Interaction with Small Displays: A Case Study. In Proceedings of the Biannual Conference of the Italian Chapter of SIGCHI, Trento, Italy, 16–20 September 2013; ACM: New York, NY, USA, 2013; pp. 26:1–26:10. [CrossRef]
43. Kuikkaniemi, K.; Jacucci, G.; Turpeinen, M.; Hoggan, E.; Müller, J. From Space to Stage: How Interactive Screens Will Change Urban Life. *Computer* **2011**, *44*, 40–47. [CrossRef]
44. Norman, D.A. Natural User Interfaces Are Not Natural. *Interactions* **2010**, *17*, 6–10. [CrossRef]
45. Horejsi, P.; Polcar, J.; Rohlíková, L. *Digital Factory and Virtual Reality: Teaching Virtual Reality Principles with Game Engines*; IntechOpen Limited: London, UK, 2016. [CrossRef]
46. Angelini, L.; Mugellini, E.; Abou Khaled, O.; Couture, N. Internet of Tangible Things (IoTT): Challenges and Opportunities for Tangible Interaction with IoT. *Informatics* **2018**, *5*, 7. [CrossRef]
47. Lutsenko, Y.V. Shine and Poverty of Virtual Reality. *Polythematic Online Sci. J. Kuban State Agrar. Univ.* **2016**, *124*. [CrossRef]
48. Khurana Rohit Monga, V. Facility Layout Planning: A Review. *Int. J. Innov. Res. Sci. Eng. Technol.* **2015**, *04*, 976–980. [CrossRef]
49. Gong, L.; Berglund, J.; Fast-Berglund, Å.; Johansson, B.; Wang, Z.; Börjesson, T. Development of virtual reality support to factory layout planning. *Int. J. Interact. Des. Manuf.* **2019**, *13*, 935–945. [CrossRef]
50. Menck, N.; Yang, X.; Weidig, C.; Winkes, P.; Lauer, C.; Hagen, H.; Hamann, B.; Aurich, J. Collaborative Factory Planning in Virtual Reality. *Procedia CIRP* **2012**, *3*, 317–322. [CrossRef]
51. Lidell, A.; Ericson, S.; Ng, A. The Current and Future Challenges for Virtual Commissioning and Digital Twins of Production Lines. *Adv. Transdiscipl. Eng.* **2022**, *21*, 508–519. [CrossRef]
52. Lee, C.G.; Park, S.C. Survey on the virtual commissioning of manufacturing systems. *J. Comput. Des. Eng.* **2014**, *1*, 213–222. [CrossRef]
53. Rasheed, G.; Khan, M.; Malik, N.; Akhunzada, A. Measuring Learnability through Virtual Reality Laboratory Application: A User Study. *Sustainability* **2021**, *13*, 10812. [CrossRef]
54. Milgram, P.; Takemura, H.; Utsumi, A.; Kishino, F. Augmented reality: A class of displays on the reality-virtuality continuum. *Telemanipulator Telepresence Technol.* **1994**, *2351*, 282–292. [CrossRef]
55. Speicher, M.; Hall, B.; Nebeling, M. What is Mixed Reality? In Proceedings of the 2019 CHI Conference on Human Factors in Computing Systems (CHI '19), Glasgow, Scotland, UK, 4–9 May 2019; Association for Computing Machinery: New York, NY, USA, 2019; pp. 1–15. [CrossRef]

56. Perez, F.J.L.; Brandl, P.; Bandres, M.A.G. *Technolgy Monitoring: Report on Information Needed for Workers in the Smart Factory*; Technical Report; FACTS4WORKERS (Worker Centric Workplace in Smart Factories): Graz, Austria, 2006.
57. Perey, C.; Wild, F.; Helin, K.; Janak, M.; Davies, P.; Ryan, P. Advanced manufacturing with augmented reality. In Proceedings of the 2014 IEEE International Symposium on Mixed and Augmented Reality (ISMAR), Munich, Germany, 10–12 September 2014; p. 1. [CrossRef]
58. Gattullo, M.; Scurati, G.W.; Fiorentino, M.; Uva, A.E.; Ferrise, F.; Bordegoni, M. Towards augmented reality manuals for industry 4.0: A methodology. *Robot. Comput. Integr. Manuf.* **2019**, *56*, 276–286. [CrossRef]
59. Deac, C.; Popa, C.L.; Ghinea, M.; Cotet, C. Using Augmented Reality in Smart Manufacturing. In Proceedings of the 28th DAAAM International Symposium, Zadar, Croatia, 8–11 November 2017; DAAAM International: Vienna, Austria, 2017; pp. 727–732. [CrossRef]
60. Syberfeldt, A.; Danielsson, O.; Gustavsson, P. Augmented Reality Smart Glasses in the Smart Factory: Product Evaluation Guidelines and Review of Available Products. *IEEE Access* **2017**, *5*, 9118–9130. [CrossRef]
61. Rauschnabel, P.; He, J.; Ro, Y.K. Antecedents to the adoption of augmented reality smart glasses: A closer look at privacy risks. *J. Bus. Res.* **2018**, *92*, 374–384. [CrossRef]
62. Ashton, K. That Internet of Things Thing: In the Real World Things Matter More than Ideas. Available online: https://www.rfidjournal.com/that-internet-of-things-thing (accessed on 29 July 2022).
63. Atzori, L.; Iera, A.; Morabito, G. The Internet of Things: A survey. *Comput. Netw.* **2010**, *54*, 2787–2805. [CrossRef]
64. Skobelev, P.; Borovik, S. On the way from Industry 4.0 to Industry 5.0: From digital manufacturing to digital society. *Sci. Tech. Union Mech. Eng.* **2021**, *2*, 307–311.
65. Sagirlar, G.; Carminati, B.; Ferrari, E. Decentralizing Privacy Enforcement for Internet of Things Smart Objects. *arXiv* **2018**, arXiv:1804.02161.
66. Wehle, H.D. Augmented Reality and the Internet of Things (IoT)/Industry 4.0. Available online: https://www.researchgate.net/profile/Hans-Dieter-Wehle/publication/288642701_Augmented_Reality_and_the_Internet_of_Things_IoT_Industry_40_en/links/5682703308aebccc4e0df03f/Augmented-Reality-and-the-Internet-of-Things-IoT-Industry-40-en.pdf (accessed on 29 July 2022).
67. Zhang, C.; Chen, Y. A Review of Research Relevant to the Emerging Industry Trends: Industry 4.0, IoT, Block Chain, and Business Analytics. *J. Ind. Integr. Manag.* **2019**, *5*, 165–180. [CrossRef]
68. Fatima, Z.; Tanveer, M.H.; Waseemullah; Zardari, S.; Naz, L.F.; Khadim, H.; Ahmed, N.; Tahir, M. Production Plant and Warehouse Automation with IoT and Industry 5.0. *Appl. Sci.* **2022**, *12*, 2053. [CrossRef]
69. Ben-Daya, M.; Hassini, E.; Bahroun, Z. Internet of things and supply chain management: A literature review. *Int. J. Prod. Res.* **2019**, *57*, 4719–4742. [CrossRef]
70. Vaccaro, L.; Sansonetti, G.; Micarelli, A. An Empirical Review of Automated Machine Learning. *computers* **2021**, *10*, 11. [CrossRef]
71. Mahmud, B. Internet of Things (IOT) for Manufacturing Logistics on SAP ERP Applications. *J. Telecommun. Electron. Comput. Eng. (JTEC)* **2017**, *9*, 43–47.
72. Xiao, J.; Zhou, Z.; Yi, Y.; Ni, L.M. A Survey on Wireless Indoor Localization from the Device Perspective. *ACM Comput. Surv.* **2016**, *49*, 25:1–25:31. [CrossRef]
73. Fang, Y.; Cho, Y.; Zhang, S.; Perez, E. Case Study of BIM and Cloud–Enabled Real-Time RFID Indoor Localization for Construction Management Applications. *J. Constr. Eng. Manag.* **2016**, *142*, 05016003. [CrossRef]
74. Alsinglawi, B.; Elkhodr, M.; Nguyen, Q.V.; Gunawardana, U.; Maeder, A.J.; Simoff, S.J. RFID Localisation For Internet Of Things Smart Homes: A Survey. *Int. J. Comput. Netw. Commun.* **2017**, *9*, 81–99. [CrossRef]
75. Cicibaş, H.; Demir, K. Integrating Internet of Things (IoT) into Enterprise Organizations: Socio-Technical Issues and Guidelines. *J. Manag. Inf. Syst.* **2016**, *1*, 106–117.
76. Demir, K. A Survey on Challenges of Software Project Management. In Proceedings of the 2009 International Conference on Software Engineering Research & Practice, Las Vegas, NV, USA, 13–16 July 2009; CSREA Press: Sterling, VA, USA, 2009; Volume 2, pp. 579–585.
77. European Commission; EPoSS. Internet of Things in 2020 Roadmap for the Future. Available online: https://docbox.etsi.org/erm/Open/CERP%2020080609-10/Internet-of-Things_in_2020_EC-EPoSS_Workshop_Report_2008_v1-1.pdf (accessed on 29 July 2022).
78. Lee, L.; Hui, P. Interaction Methods for Smart Glasses: A Survey. *IEEE Access* **2018**, *6*, 28712–28732. [CrossRef]
79. Zhou, J.; Zhang, Y.; Laput, G.; Harrison, C. AuraSense: Enabling Expressive Around-Smartwatch Interactions with Electric Field Sensing. In Proceedings of the 29th Annual Symposium on User Interface Software and Technology, Tokyo, Japan, 16–19 October 2016; Association for Computing Machinery: New York, NY, USA, 2016; pp. 81–86. [CrossRef]
80. Quint, F.; Loch, F.; Weber, H.; Venitz, J.; Gröber, M.; Liedel, J. Evaluation of Smart Glasses for Documentation in Manufacturing. In Proceedings of the Mensch und Computer 2016—Workshopband, Aachen, Germany, 4–7 September 2016.
81. Ishii, H. Tangible Bits: Designing the Seamless Interface between People, Bits, and Atoms. In Proceedings of the 8th International Conference on Intelligent User Interfaces, IUI'03, Miami, FL, USA, 12–15 January 2003; Association for Computing Machinery: New York, NY, USA, 2003; p. 3. [CrossRef]

82. Angelini, L.; Mugellini, E.; Lechelt, Z.; Hornecker, E.; Marshall, P.; Liu, C.; Brereton, M.; Soro, A.; Couture, N.; Abou Khaled, O. Internet of Tangible Things: Workshop on Tangible Interaction with the Internet of Things. In Proceedings of the Extended Abstracts of the 2018 CHI Conference on Human Factors in Computing Systems, Montreal, QC, Canada, 21–26 April 2018; Association for Computing Machinery: New York, NY, USA, 2018; pp. 1–8. [CrossRef]
83. Funk, M.; Korn, O.; Schmidt, A. An Augmented Workplace for Enabling User-Defined Tangibles. In Proceedings of the CHI'14: CHI Conference on Human Factors in Computing Systems, Toronto, ON, Canada, 26 April–1 May 2014. [CrossRef]
84. Yeo, H.S.; Minami, R.; Rodriguez, K.; Shaker, G.; Quigley, A. Exploring Tangible Interactions with Radar Sensing. *Proc. ACM Interact. Mob. Wearable Ubiquitous Technol.* **2018**, *2*, 1–25. [CrossRef]
85. Wallbaum, T.; Matviienko, A.; Heuten, W.; Boll, S. Challenges For Designing Tangible Systems. In Proceedings of the 3rd European Tangible Interaction Studio (ETIS 2017), Luxembourg, 19–23 June 2017; pp. 21–23.
86. Colgate, J.; Wannasuphoprasit, W.; Peshkin, M. Cobots: Robots for collaboration with human operators. In Proceedings of the 1996 ASME International Mechanical Engineering Congress and Exposition, Atlanta, GA, USA, 17–22 November 1996; pp. 433–439.
87. Silverman, R.E. The Words of Tomorrow. Available online: https://www.wsj.com/articles/SB944517141695981261/ (accessed on 29 July 2022).
88. Vicentini, F. *La Robotica Collaborativa. Sicurezza e Flessibilità Delle Nuove Forme di Collaborazione Uomo-Robot*; Tecniche Nuove: Milano, Italy, 2017.
89. Linsinger, M.; Sudhoff, M.; Lemmerz, K.; Glogowski, P.; Kuhlenkötter, B. Task-based Potential Analysis for Human-Robot Collaboration within Assembly Systems. In *Tagungsband des 3. Kongresses Montage Handhabung Industrieroboter*; Springer: Berlin/Heidelberg, Germany, 2018. [CrossRef]
90. Oberc, H.; Prinz, C.; Glogowski, P.; Lemmerz, K.; Kuhlenkötter, B. Human Robot Interaction—Learning how to integrate collaborative robots into manual assembly lines. *Procedia Manuf.* **2019**, *31*, 26–31. [CrossRef]
91. Alvarez-de-los Mozos, E.; Renteria, A. Collaborative Robots in e-waste Management. *Procedia Manuf.* **2017**, *11*, 55–62. [CrossRef]
92. Neto, P.; Simão, M.; Mendes, N.; Safeea, M. Gesture-based human-robot interaction for human assistance in manufacturing. *Int. J. Adv. Manuf. Technol.* **2018**, *101*, 119–135. [CrossRef]
93. Spada, S.; Ghibaudo, L.; Gilotta, S.; Gastaldi, L.; Cavatorta, M.P. Analysis of exoskeleton introduction in industrial reality: Main issues and EAWS risk assessment. In *International Conference on Applied Human Factors and Ergonomics*; Springer: Berlin/Heidelberg, Germany, 2017; pp. 236–244.
94. Fleck, P. Collaborative Robot. Available online: www.researchgate.net/publication/313665204_Collaborative_Robots?channel=doi&linkId=58a21ad545851598babae8ff&showFulltext=true (accessed on 29 July 2022). [CrossRef]
95. Manso, L.; Bustos, P.; Bandera, J.; Romero-Garcés, A.; Calderita, L.; Marfil, R.; Bandera, A. *Deep Representations for Collaborative Robotics*; Springer: Berlin/Heidelberg, Germany, 2016; Volume 10087, pp. 179–193. [CrossRef]
96. Lambrecht, J.; Kruger, J. Spatial programming for industrial robots based on gestures and Augmented Reality. In Proceedings of the 2012 IEEE/RSJ International Conference on Intelligent Robots and Systems, Vilamoura-Algarve, Portugal, 7–12 October 2012; pp. 466–472. [CrossRef]
97. Landi, C.T.; Ferraguti, F.; Secchi, C.; Fantuzzi, C. Tool compensation in walk-through programming for admittance-controlled robots. In Proceedings of the IECON 2016-42nd Annual Conference of the IEEE Industrial Electronics Society, Florence, Italy, 23–26 October 2016; pp. 5335–5340.
98. Billard, A.G.; Calinon, S.; Dillmann, R. Learning from Humans. In *Springer Handbook of Robotics*; Siciliano, B., Khatib, O., Eds.; Springer International Publishing: Cham, Switzerland, 2016; pp. 1995–2014. [CrossRef]
99. Proia, S.; Carli, R.; Cavone, G.; Dotoli, M. Control Techniques for Safe, Ergonomic, and Efficient Human-Robot Collaboration in the Digital Industry: A Survey. *IEEE Trans. Autom. Sci. Eng.* **2022**, *19*, 1798–1819. [CrossRef]
100. Benotsmane, R.; Dudás, L.; Kovács, G. Collaborating robots in Industry 4.0 conception. *IOP Conf. Ser. Mater. Sci. Eng.* **2018**, *448*, 012023. [CrossRef]
101. Gualtieri, L.; Rauch, E.; Vidoni, R. Development and validation of guidelines for safety in human-robot collaborative assembly systems. *Comput. Ind. Eng.* **2022**, *163*, 107801. [CrossRef]
102. Gualtieri, L.; Rauch, E.; Vidoni, R.; Matt, D.T. Safety, Ergonomics and Efficiency in Human-Robot Collaborative Assembly: Design Guidelines and Requirements. *Procedia CIRP* **2020**, *91*, 367–372. [CrossRef]
103. Akundi, A.; Euresti, D.; Luna, S.; Ankobiah, W.; Lopes, A.; Edinbarough, I. State of Industry 5.0—Analysis and Identification of Current Research Trends. *Appl. Syst. Innov.* **2022**, *5*, 27. [CrossRef]
104. Zitter, I.; Hoeve, A. *Hybrid Learning Environments: Merging Learning and Work Processes to Facilitate Knowledge Integration and Transitions*; OECD Education Working Papers; OECD Publishing: Paris, France, 2012. [CrossRef]
105. Loch, F.; Böck, S.; Vogel-Heuser, B. Teaching Styles of Virtual Training Systems for Industrial Applications? A Review of the Literature. *IxD&A* **2018**, *38*, 46–63.
106. Demir, K.A.; Döven, G.; Sezen, B. Industry 5.0 and Human-Robot Co-working. *Procedia Comput. Sci.* **2019**, *158*, 688–695. [CrossRef]
107. Longo, F.; Padovano, A.; Umbrello, S. Value-oriented and ethical technology engineering in industry 5.0: A human-centric perspective for the design of the factory of the future. *Appl. Sci.* **2020**, *10*, 4182. [CrossRef]

108. Jafari, N.; Azarian, M.; Yu, H. Moving from Industry 4.0 to Industry 5.0: What Are the Implications for Smart Logistics? *Logistics* **2022**, *6*, 26. [CrossRef]
109. World Economic Forum. Internet of Things Guidelines for Sustainabilitydeas. Available online: https://www3.weforum.org/docs/IoTGuidelinesforSustainability.pdf (accessed on 29 July 2022).
110. Herrmann, F. The Smart Factory and Its Risks. *Systems* **2018**, *6*, 38. [CrossRef]
111. Kaasinen, E.; Schmalfuß, F.; Özturk, C.; Aromaa, S.; Boubekeur, M.; Heilala, J.; Heikkilä, P.; Kuula, T.; Liinasuo, M.; Mach, S.; et al. Empowering and engaging industrial workers with Operator 4.0 solutions. *Comput. Ind. Eng.* **2020**, *139*, 105678. [CrossRef]
112. Peng, A.; Nushi, B.; Kiciman, E.; Inkpen, K.; Kamar, E. Investigations of Performance and Bias in Human-AI Teamwork in Hiring. In Proceedings of the AAAI Conference on Artificial Intelligence, Palo Alto, CA, USA, 22 February–1 March 2022.

Article

Important Features Selection and Classification of Adult and Child from Handwriting Using Machine Learning Methods

Jungpil Shin [1,*], Md. Maniruzzaman [1], Yuta Uchida [1], Md. Al Mehedi Hasan [1], Akiko Megumi [2], Akiko Suzuki [2] and Akira Yasumura [3]

1. School of Computer Science and Engineering, The University of Aizu, Aizuwakamatsu 965-8580, Fukushima, Japan; d8232112@u-aizu.ac.jp (M.M.); m5261176@u-aizu.ac.jp (Y.U.); mehedi@u-aizu.ac.jp (M.A.M.H.)
2. Graduate School of Social and Cultural Sciences, Kumamoto University, 2-40-1 Kurokami Chuo-ku, Kumamoto City 860-8555, Kumamoto, Japan; megmakko@gmail.com (A.M.); spyx8fu9@flute.ocn.ne.jp (A.S.)
3. Faculty of Humanities and Social Sciences, Kumamoto University, 2-40-1 Kurokami Chuo-ku, Kumamoto City 860-8555, Kumamoto, Japan; yasumura@kumamoto-u.ac.jp
* Correspondence: jpshin@u-aizu.ac.jp

Abstract: The classification of different age groups, such as adult and child, based on handwriting is very important due to its various applications in many different fields. In forensics, handwriting classification helps investigators focus on a certain category of writers. This paper aimed to propose a machine-learning (ML)-based approach for automatically classifying people as adults or children based on their handwritten data. This study utilized two types of handwritten databases: handwritten text and handwritten pattern, which were collected using a pen tablet. The handwritten text database had 57 subjects (adult: 26 vs. child: 31). Each subject (adult or child) wrote the same 30 words using Japanese hiragana characters. The handwritten pattern database had 81 subjects (adult: 42 and child: 39). Each subject (adult or child) drew four different lines as zigzag lines (trace condition and predict condition), and periodic lines (trace condition and predict condition) and repeated these line tasks three times. Handwriting classification of adult and child is performed in three steps: (i) feature extraction; (ii) feature selection; and (iii) classification. We extracted 30 features from both handwritten text and handwritten pattern datasets. The most efficient features were selected using sequential forward floating selection (SFFS) method and the optimal parameters were selected. Then two ML-based approaches, namely, support vector machine (SVM) and random forest (RF) were applied to classify adult and child. Our findings showed that RF produced up to 93.5% accuracy for handwritten text and 89.8% accuracy for handwritten pattern databases. We hope that this study will provide the evidence of the possibility of classifying adult and child based on handwriting text and handwriting pattern data.

Keywords: handwritten text; handwritten pattern; adult and child classification; machine learning; sequential forward floating selection approach

1. Introduction

In recent years, the field of handwriting has attracted interest from various aspects, such as biometrics [1] and the medical field [2]. In addition, handwritten characters can be obtained from a variety of sources such as paper documents, images, touch screens, and other devices. This makes the data easy to collect and suitable for classification. Furthermore, since handwriting is something that everyone uses every day in school, it is a method that is less stressful for people. There are few studies on handwriting classification for adults and children, and most of the studies are on the classification of face recognition [3], age groups [4], age, gender, and nationality [5], gender [6,7], gender and handedness [8], detection of alcohol [9], and Parkinson's disease (PD) [10] based on handwriting images.

There are two types of handwriting data: offline and online. The input data collected using a scanning machine are called "offline", whereas input data obtained using a pen tip are called "online" [11]. In our research work, we used the online-based handwritten database. Moreover, a single writer's handwriting may be unique or differ slightly, but the handwriting of a child and adult must always be different. Most forensic handwriting analysis is based on the inspection of specific character shapes, character ligatures, size, pen lift, pen pressure, speed, letter spacing, etc., to identify a suspected person. Age group detection will be a great solution before detecting the actual suspected person in forensic analysis. It will give additional evidence about the suspected person's age. Currently, there are many applications of handwritten recognition, for example, signature authentication used in industrial applications [2], authenticating of criminal investigations in a court of justice [12,13], document examinations [14], and so on. The most difficult aspects of handwriting identification are distortions and pattern variations; feature extraction is of supreme importance. Handwritten forensic analysis or handwriting recognition using machine learning (ML) algorithms can be a great solution to classify adults and children based on their handwritten text and handwritten pattern.

Ahmad et al. (2004) proposed support vector machine (SVM) with some kernels for online handwritten recognition [15]. They showed that at the character level, the SVM recognition rate was dramatically better due to the use of maximizing boundaries in the decision function. The only problem with this algorithm was storing large support vector for a huge training character that requires a larger memory size. Babu et al. (2014) proposed k-nearest neighbors (k-NN) for recognizing handwritten digits based on structural features, which does not require thinning operation and size normalization approach [16]. Ramzan et al. (2018) implemented neural networks (NN) and their variants to recognize handwritten digits. The survey details some existing techniques implemented for handwritten digit recognition (HWDR) being carried out [17]. Baldominos et al. (2019) [18] also proposed convolutional neural networks (CNNs) to distinguish previous work to recognize handwritten characters using some data augmentation from works using the original dataset out-of-the-box [19,20]. They provided the most extensive and updated survey of the MNIIST and EMNIST datasets and achieved the lowest error rate.

Poon et al. (2019) [2] applied logistic regression to predict PD based on handwritten recognition. They utilized the publicly available PD database and extracted secondary kinematic handwriting features from the dataset. It is being studied not only for personal identification but also in the medical field. The limitation of their proposed model was that they used small sample size of the dataset and lacked control in the study design. As for all the limitations of handwriting recognition, Japanese handwritten character recognition is complex due to the various types of writing styles, characters, and confusion among similar characters. One of the major causes of the inefficient classification of Japanese characters is a large number of letters. However, many methods have been developed to recognize Japanese handwriting as text images for several applications, but there are few studies on the classification of adults and children based on Japanese handwritten recognition.

Nisimura et al. (2004) [21] suggested a discriminating strategy based on statistical learning and extracted linguistic features from speech or voice data to classify adults and children. They applied SVM and found that it performed with better classification accuracy than the Gaussian mixture model [22]. The disadvantage of this strategy is that it has a trait in common with both labels. Makihara et al. (2010) [23] proposed a method to classify gender and age using video-based gait feature analysis with a large-scale multi-view gait database. They adopted the k-NN classifier to classify gender and age. They used three databases (HumanID, Soton, and CASIA) that contained over 100 subjects. These datasets have their particular limitations, such as the small view images in the HumanID dataset; also, single view images in the Soton dataset, and maximum subjects in the CASIA dataset included in the 20's or 30's. Faghel-Soubeyrand classified adult and child based on faces [24]. In this study, we propose a new approach for the classification of adults and children based on their handwritten text and pattern recognition. Our proposed method can achieve more than 89% classification accuracy, implying that classification accuracy with handwritten characters can be expected.

The organization of this paper is as follows: Section 2 presents materials and methods, including proposed ML-based framework; description of datasets, feature extraction, feature selection, classifiers along with their performance evaluation metrics are discussed in this section. The experimental results and discussion are discussed in Section 3. Finally, the conclusion is discussed in Section 4.

2. Materials and Methods

In this section, we summarize the proposed ML-based framework. Next, two databases used in this research work are described. We also describe feature extraction, feature selection, and two classification methods along their performance evaluation metrics in this section.

2.1. Proposed ML-Based Framework

The goal of this work is to propose an ML-based model for predicting adult and child based on their handwritten texts and handwritten patterns. The proposed ML-based framework is presented in Figure 1. First, we divide the handwriting (text and pattern) dataset into two phases: the training phase and the testing phase. We take 80% of the dataset in the training phase and the remaining 20% of the dataset for the test phase. The second step is to preprocess the handwriting data. After preprocessing handwriting data, we extract 30 features and then select an optimal subset of the features using sequential forward floating selection (SFFS). We applied two ML-based algorithms, SVM and RF, for the classification of adult and child. We tuned the hyperparameters of the classifiers (SVM and RF) using a grid search method and trained SVM and RF-based classifiers with five-fold cross-validation protocol. After training, classifiers (SVM and RF) are used in the testing phase for the classification of adult and child. Accuracy, recall, precision, f1-score, and area under the curve (AUC) are used to evaluate the performance of the classifiers.

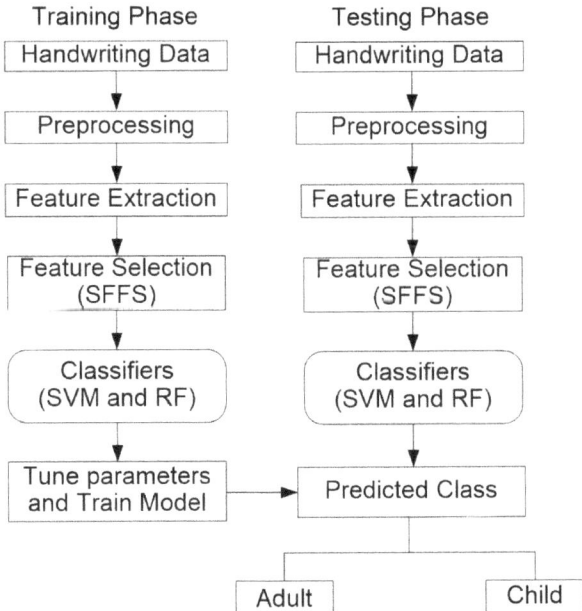

Figure 1. Proposed ML-based classification model.

2.2. Dataset

2.2.1. Device for Data Collection

Handwriting data were recorded using a pen tablet system (Cintiq Pro 16, Wacom Co., Ltd., Saitama, Japan). The tablet was connected to a laptop PC running Windows 10. Figure 2 illustrates the coordinates of the parameters generated by the pen tablet. The screen size of the pen tablet was 15.6 inches, and the resolution size was 2560 × 1440 pixels.

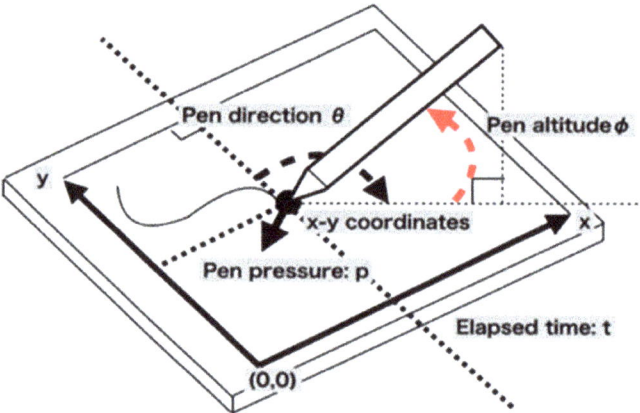

Figure 2. Pen tablet device.

2.2.2. Handwritten Text

We developed a new dataset to evaluate our proposed method where adult and child handwriting-based text data were collected using a pen tablet. A total of 57 participants were taken for this work, consisting of 26 adults (aged 19–59 years) with handwriting and 31 children (aged 12–13 years). Each subject (child or adult) was asked to write the same 30 words (tasks) using hiragana characters only on the pen tablet using a dedicated stylus pen. Each word contains a minimum of 2 characters and a maximum of 7 characters. A summary of the handwritten text dataset is described in Table 1.

Table 1. Summary of the handwritten text dataset.

Group Name	Age (Years)	No. of Subjects	No. of Words	Total Samples
Adult	19–59	26	30	780 (26 × 30)
Child	12–13	31	30	930 (31 × 30)

2.2.3. Handwritten Pattern

Handwriting-based pattern data were also collected from 81 subjects using a pen tablet system. The dataset had 39 children and 42 adults. In this study, we adopted two patterns. One was drawing a continuous zigzag line, essentially a continuous set of triangles without a base. Another was drawing a continuous periodic line pattern (PL) that was repeated squares and triangles sequentially without a base. The trace and predict conditions were used for each pattern. Each subject was asked to draw these four patterns on the pen tablet using a dedicated stylus pen and each drawing pattern was repeated 3 times. The traced over the sample zigzag lines are presented in Figure 3a, and the data are written on a blank sheet of paper after memorizing the sample. The traced over the sample PL lines are also presented in Figure 3b, and the data were derived from memorizing the sample and writing it on a blank sheet of paper. The data were collected by separating the zigzag line and the PL line, taking data for 30 s, resting for 20 s, taking data again for 30 s, and resting for 20 s, and so on, until six data were collected. The reason for the intervals was to let the brain rest. A summary of the handwritten pattern dataset is described in Table 2.

Table 2. Summary of the handwritten pattern dataset.

Group Name	Age (Years)	No. of Subjects	No. of Task	No. of Repeat	Total Samples
Adult	19–43	42	4	3	504 (42 × 4 ×3)
Child	8–13	39	4	3	468 (39 × 4 ×3)

Figure 3. Handwritten pattern dataset for (**a**) zigzag line sample and painted; (**b**) PL line sample and painted.

2.3. Feature Extraction

The handwriting data contained six pieces of information, including the time of writing, pen pressure, x-coordinate and y-coordinate of the writing position, angle of the horizontal component of the pen, and angle of the vertical component of the pen. To classify adults and children based on their handwriting, 30 feature parameters are evaluated for each task. These feature parameters only required the localization of primary features of handwritten text images, namely, the width, height, speed, peak, different types of grip angle, and various types of pressure, which are given in detail in Table 3.

Table 3. Extracted feature names and their description.

SN	Feature	Description
1	Width	Max (X)–Min (X)
2	Height	Max (Y)–Min (Y)
3	Length	The total length of the drawing
4	Velocity	(Length)/(total) drawing time
5	PIVH	The maximum speed recorded at any time point
6	PIVL	Minimum speed recorded at any time point (PMS > 0)
7	PIAH	Maximum acceleration recorded at any time point
8	PIAL	Minimum acceleration recorded at any time point (PMA > 0)
9	GripAngleMeanW	Mean of grip angle values for the entire drawing task (Horizontal)
10	GripAngleMeanL	Mean of grip angle values for the entire drawing task (Vertical)
11	GripAngleSDW	SD of grip angle values for the entire drawing task (Horizontal)
12	GripAngleSDL	SD of grip angle values for the entire drawing task (Vertical)
13	PressureMean	Mean of recorded pressure values for the entire task
14	PressureSD	SD of recorded pressure values for the entire task
15	PCAvgPos	Mean increase in pressure between two-time points
16	PCSDPos	SD of increase in pressure between two-time points
17	PCMax	The maximum increase in handwriting pressure between two-time points
18	PCAvgNeg	Mean decrease in pressure between two-time points
19	PCSDNeg	SD of decrease in pressure between two-time points
20	PCMin	Maximum reduction in handwriting pressure between two-time points
21	Error	Number of outliers and triangle square errors based on angles
22	PeakPresMean	Mean pressure at minima
23	ErrorStopTime	Mean stuck time at the starting minima point just before the error
24	AngleMean	Mean of angles at maxima and minima
25	AngleVar	The variance of angles at maxima and minima
26	ReglineSlope	The slope of the regression line
27	ReglineIntercept	The intercept of the regression line
28	LoopCount	Time spent writing divided by the number of peaks
29	AngleSpeed	Mean of velocities at the edge of the peaks and valleys
30	ErrorRate	(Error)/(All Peaks) Error rate

2.4. Feature Normalization

Data normalization is a technique that minimizes redundancy and improves the efficiency of the data. Mathematically, it is defined as follows:

$$z = \frac{X - \mu}{\sigma} \quad (1)$$

where X is the original feature vector; μ is the mean of that feature vector, and σ is its standard deviation. The value of z lies between 0 to 1.

2.5. Feature Selection

Feature selection is the process of removing irrelevant features to improve the efficiency of the model. We have used SFFS for feature selection, which is an extension of sequential forward selection (SFS), to reduce the initial d-dimensional feature space into a k-dimensional feature subspace ($k < d$) [25]. Let $Y = \{y_1, y_2, \ldots, y_d\}$ be a set of all features and $X_k = \{x_j | j = 1, 2, \ldots, k; x_j \in Y\}$, where $k \in (0, 1, 2, \ldots, d)$ and X_k is a subset of Y. We start the algorithm with $X_o = \varnothing, k = 0$. The steps of SFFS are described as follows:

Step 1: $x^+ = \operatorname{argmax} J(X_k + x)$, where $x \in Y - X_k$, J is an evaluation index and x^+ is the feature with the highest evaluation when it chooses.

Step 2: $X_{k+1} = X_k + x^+$. The feature with the highest evaluation by selecting is used.

Step 3: $k = k + 1$.

Step 4: **Step 1** to **Step 3** is repeatedly iterating. Then, x^+ when k reaches the specified number which is the set of the most appropriate features obtained.

SFFS is performed up to **Step 3** of SFS, and a process for searching for features to be deleted is added. At first, **Step 1** to **Step 4** are performed starting from $X_0 = \varnothing, k = 0$, as in the SFS.

Step 5: $x^- = \operatorname{argmax} J(X_k - x)$, where $x \in X_k$ and x^- is the feature with the best performance when the feature is deleted.

Step 6: If $J(X_k - x) > J(X_k)$:
$X_{k-1} = X_k - x^-$
$k = k - 1$
Go to **Step 1**.

In **Step 1**, we capture the features that best improve the performance of the feature subset from the feature space. Then, we proceed to **Step 2**. In **Step 2**, remove features only if they improve the performance of the resulting subset. In this study, the Sequential Feature Selector in mlxtend library was used and implemented [26].

2.6. Classifiers

2.6.1. Support Vector Machine

Support vector machine (SVM) [27,28] is supervised learning that is used for both classification and regression problems. In this study, we implemented SVM in Scikit-learn support vector classification (SVC) [29]. SVM is classified on the largest hyperplane up to the nearest training data point of the class. A highly accurate model can be obtained with a small amount of data, and the accuracy of identification can be kept even when the number of features increases. The main objective of SVM is to find the hyperplane in the feature space that can easily separate the classes, which needs to solve the following constraint problem:

$$\max \alpha \sum_{i=1}^{n} \alpha_i - \frac{1}{2} \sum_{i=1}^{n} \sum_{j=1}^{n} \alpha_i \alpha_j y_i y_j K(x_i, x_j) \quad (2)$$

Subject to

$$\sum_{i=1}^{n} y_i^T \alpha_i = 1, \ 0 \leq \alpha_i \leq C, i = 1, \ldots, n \ \forall \ i = 1, 2, 3, \ldots, n \quad (3)$$

The final discriminate function takes the following form:

$$f(x) = \sum_{i=1}^{n} \alpha_i K(x_i, x_j) + b \quad (4)$$

where, b is the bias term.

2.6.2. Random Forest

Random forest (RF) [30] is one type of ensemble learning used for classification, regression, etc. It is a model in which decision trees are created in parallel and predictions are made by calculating the majority vote of the output results of each learning machine. Random learning enables fast learning and identification even for high-dimensional features, and the random selection of training data makes it strong against noise. Therefore, it is possible to build an overall good model. In this study, we also implemented RF with random forest classifier in Scikit-learn [29].

2.7. Performance Evaluation Metrics

To evaluate the performance of the classification model, we adopted five evaluation metrics: classification accuracy (ACC), recall (Rec), precision (Pre), f1-score, and AUC. The evaluation metrics of accuracy, recall, precision, and f1-score are computed based on true positive (t_p), false positive (f_p), true negative (t_n), and false negative (f_n), which are briefly explained as follows:

$$\text{ACC (\%)} = \frac{t_p + t_n}{t_p + f_p + t_n + f_n} \times 100 \quad (5)$$

$$\text{Rec (\%)} = \frac{t_p}{t_p + f_n} \times 100 \quad (6)$$

$$\text{Pre (\%)} = \frac{t_p}{t_p + f_p} \times 100 \quad (7)$$

$$\text{f1-score (\%)} = 2 \times \frac{(\text{Pre} \times \text{Rec})}{\text{Pre} + \text{Rec}} \times 100 \quad (8)$$

3. Experimental Results and Discussion

3.1. Experimental Setup

To perform the classification of adult and child, 80% of the dataset was utilized for training sets and 20% of the dataset for testing sets. For all statistical analysis, Python version 3.9 and Scikit-learn version 1.0.2 were used. We used Windows 10 21H1 (build 19043.1151) 64-bit with an Intel (R) Core (TM) i5-10400 processor and 16 GB of RAM.

3.2. Baseline Characteristics of Adult and Child

The baseline characteristics of adults and children for the handwritten text and pattern datasets are presented in Table 4. For the handwritten text dataset, the prevalence of adult and child was 45.6% and 54.4%. Among them, 42.7% and 59.3% of adult and child were female. The average ages of adult and child for the handwritten text dataset were 27.3 ± 10.5 and 12.5 ± 0.3 years.

For the handwritten pattern dataset, the average ages of adult and children were 23.9 ± 4.9 and 11.8 ± 1.6 years. The overall prevalence of females was 59.3%. Approximately 64.6% and 35.4% of adult and child were female. It was observed that age and gender (except gender for handwritten text data) were significantly associated with adult and child for both handwritten text and pattern dataset (p-value < 0.05).

Table 4. Baseline characteristics of adult and child.

Variables	Overall	Adult	Child	p-Value [1]
Handwritten text dataset				
Total, n (%)	57	26 (45.6)	31 (54.4)	–
Gender, Female, n (%)	27 (47.4)	11 (42.7)	16 (59.3)	0.483
Age (year), Mean ± SD	19.2 ± 10.2	27.3 ± 10.5	12.5 ± 0.3	0.001
Handwritten pattern dataset				
Total, n (%)	81	42 (51.9)	39 (48.1)	–
Gender, Female, n (%)	48 (59.3)	31 (64.6)	17 (35.4)	0.0025
Age (year), Mean ± SD	17.8 ± 7.9	23.9 ± 4.9	11.8 ± 1.6	0.001

n is the total number of subjects; [1] p-value is obtained from t-test for age and chi-square test for gender.

3.3. Hyperparameter Tuning

For the classification tasks, we set the following hyperparameters for SVM as cost (C) = [0.0001, 0.001, 0.01, 0.1, 1, 10, 100, 1000]; kernel = ["rbf", "linear", "poly", "sigmoid"]; and gamma: [0.00001, 0.0001, 0.001, 0.01, 0.1, 1]. We also set the hyperparameters for RF as max_depth = [2, 3, 5, 10], n_estimators = [50, 100, 200, 300, 400], min_samples_split = [2, 3, 10], min_samples_leaf = [1, 3, 10], bootstrap = [True, False], and criterion = ["gini", "entropy"]. We implemented grid search algorithms to tune these hyperparameters. We choose the hyperparameters that will provide the highest classification accuracy.

3.4. Experiment-1: Evaluation for Handwritten Text Dataset

In this experiment, we used different types of handwritten texts and then extracted various types of features from each image or task. We applied SVM and RF classifiers to classify adult and child and calculated the classification accuracy. We used 30 hiragana words and extracted 30 features which are clearly explained in Table 3. Table 5 shows the performance scores of SVM and RF for better features combination of handwritten text dataset. It was observed that SVM with RBF kernel produced the classification accuracy of 87.7% for the combination of 15 selected features out of 30 features. Moreover, SVM also produced 92.4% recall, 85.9% precision, 89.1% f1-score, and 0.919 AUC for the selected 15 features, whereas RF classifier achieved an excellent classification accuracy of 93.5% along with 95.7% recall, 92.2% precision, 93.9% f1-score, and 0.983 AUC, respectively, for the combination of 18 selected features. Therefore, RF achieved more outstanding performance than SVM.

Table 5. Performance scores of SVM and RF for handwritten text dataset.

CT	# of Features	ACC	Rec	Prec	f1-Score	AUC
SVM	15	87.7	92.4	85.9	89.1	0.919
RF	18	93.5	95.7	92.2	93.9	0.983

CT: Classifier types.

We observed that 15 and 18 features were selected by SFFS with SVM and RF classifiers. A total of 11 common features was extracted from those two methods, which are shown in Figure 4, and the listed selected features are presented in Table 6. These 11 common features were used as input features and then we applied SVM and RF classifiers to distinguish adults from children.

The performance scores of SVM and RF classifiers for 11 common features are shown in Table 7. It was observed that SVM with RBF provided 87.4% accuracy, 90.8% recall, 86.6% precision, 88.7% f1-score, and 0.947 AUC, respectively, whereas RF gave 91.5% accuracy, 93.0% recall, 91.5% precision, 92.3% f1-score, and 0.967 AUC, receptively. Finally, we may conclude that RF had more outstanding performance scores than SVM for the prediction of the adult and child for handwritten text dataset.

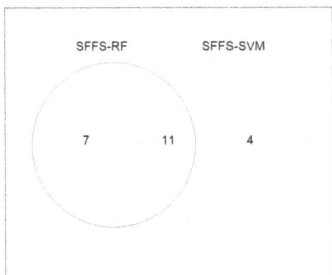

Figure 4. Identification of common features from SFFS-RF and SFFS-SVM for handwritten text dataset.

Table 6. List of common features selected for handwritten text dataset.

SN	Feature Names	SN	Feature Names
1	Height	7	GripAngleSDW
2	Velocity	8	GripAngleSDL
3	PIVL	9	PressureMean
4	PIAL	10	PressureSD
5	GripAngleMeanW	11	PCMin
6	GripAngleMeanL	–	–

Table 7. Performance scores of SVM and RF for common features for handwritten text dataset.

CT	ACC	Rec	Prec	f1-Score	AUC
SVM	87.4	90.8	86.6	88.7	0.947
RF	91.5	93.0	91.5	92.3	0.967

3.5. Experiment-2: Evaluation for Handwritten Pattern Dataset

To evaluate our proposed model, we used a handwritten pattern dataset and obtained a classification accuracy of up to 89.8%. In this section, we performed two experiments. Firstly, the best combination of the features set was identified using SFFS-based RF and SVM classifiers. We chose the feature combination at which the classification model provides the highest classification accuracy. The classification accuracy of RF and SVM for the handwritten pattern dataset is presented in Table 8. For the trace of zigzag lines, RF produced 83.3% classification accuracy for the combination of 19 selected features, whereas SVM produced 71.4% accuracy for the combination of 26 selected features. For the prediction of the zigzag, the RF classifier obtained the highest classification accuracy of 85.7% for 13 combinations of feature sets and the prediction of the zigzag line, whereas SVM provided 75.5% classification accuracy for 3 selected features. For the prediction and trace of the PL line, RF achieved 73.5% classification accuracy for the combination of 24 selected features, whereas SVM achieved 79.6% accuracy for 7 selected features and 87.7% accuracy for 12 selected features. RF classifier provided a good classification accuracy of 85.6% for the combination of all handwritten patterns, 25 features, whereas 82.1% classification accuracy was provided by SVM for the combination of all 28 features. Therefore, RF achieved better classification accuracy (89.8%) than SVM for the prediction of PL line.

Table 8. Classification accuracy (in %) of RF and SVM for handwritten pattern dataset.

Line Types	No. of Features	RF	No. of Features	SVM
Zigzag trace	19	83.7	26	71.4
Zigzag predict	13	85.7	3	75.5
PL trace	24	73.5	7	79.6
PL predict	9	89.8	12	87.7
All patterns	25	85.6	28	82.1

The second experiment was to take the common features from the two best combinations of feature sets and apply two classifiers for the prediction of adult and child. The number of selected common features was 18 features from the trace of zigzag line, 2 features from the prediction of zigzag line, 7 features from the trace of PL line, 9 features from the prediction of PL lines, and 23 features from all handwritten patterns (zigzag and PL lines), which are shown in Figure 5, and the corresponding list of selected common features is presented in Table 9.

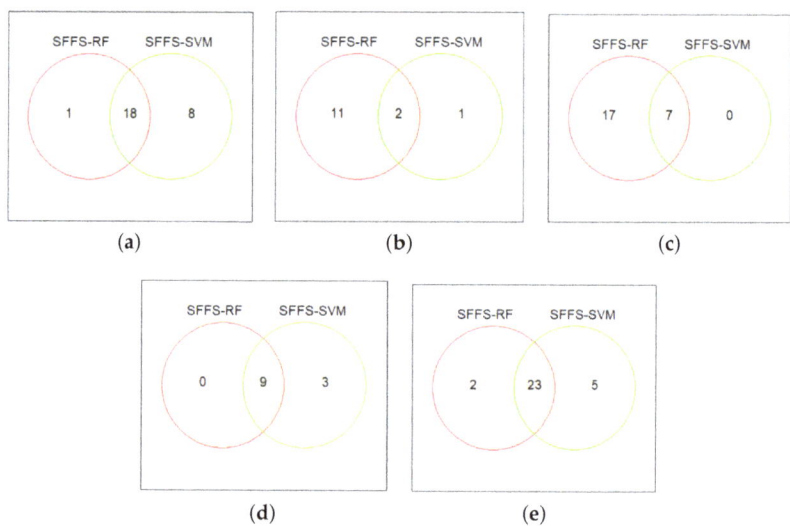

Figure 5. Identification of common features from SFFS-RF and SFFS-SVM for different lines of handwritten pattern dataset: (**a**) Feature from trace of zigzag line, (**b**) feature from the prediction of zigzag line, (**c**) feature from trace of PL line, (**d**) feature from prediction of PL line, and (**e**) feature from all handwritten patterns.

Table 9. List of common features of handwritten pattern dataset.

SN	ZigZag Trace	Zigzag Predict	PL Trace	PL Predict	All Patterns
1	Height	PIAH	Length	Velocity	Height
2	Length	AngleMean	PIVL	PIAH	Length
3	Velocity	–	PIAH	PIAL	Velocity
4	PIVH	–	PIAL	PCSDNeg	PIAH
5	PIAH	–	PeakpresMean	PeakpresMean	PIAL
6	PIAL	–	ReglineSlope	AngleMean	GripAngleMeanW
7	GripAngleMeanW	–	ReglineIntercept	AngleVar	GripAngleMeanL
8	GripAngleMeanL	–	–	ReglineIntercept	GripAngleSDW
9	PressureMean	–	–	ErrorRate	PressureMean
10	PCMax	–	–	–	PressureSD
11	Error	–	–	–	PCAvgPos
12	PeakpresMean	–	–	–	PCSDPos
13	ErrorStopTime	–	–	–	PCMax
14	AngleMean	–	–	–	PCAvgNeg
15	ReglineSlope	–	–	–	PCSDNeg
16	ReglineIntercept	–	–	–	PCMin
17	AngleSpeed	–	–	–	Error
18	ErrorRate	–	–	–	PeakpresMean
19	–	–	–	–	ErrorStopTime
20	–	–	–	–	AngleVar
21	–	–	–	–	ReglineSlope
22	–	–	–	–	ReglineIntercept
23	–	–	–	–	LoopCount

The classification accuracies of RF and SVM for these common features are presented in Table 10. It was observed that the RF classifier provided a higher classification accuracy of 79.5%, 73.4%, 83.6%, and 89.8% for the trace and prediction of the zigzag line than SVM for the trace and prediction of the PL line, respectively. On the other hand, the SVM classifier provided 85.1% accuracy for all handwritten patterns, whereas RF classifier gave 84.1% accuracy.

Table 10. Classification accuracy (in %) of RF and SVM for common features of handwritten pattern dataset.

Line Types	CF	RF	SVM
Zigzag trace	18	79.5	71.4
Zigzag predict	2	73.4	69.3
PL trace	7	83.6	79.5
PL predict	9	89.8	79.5
All patterns	25	84.1	85.1

CF: Common features.

The recall, precision, f1-score, and AUC of RF and SVM for common features of the handwritten dataset are presented in Table 11. It was observed that the RF classifier achieved comparatively better performance for all types of lines than SVM. RF classifier provided a higher recall of 87.0%, precision of 93.1%, f1-score of 90.0%, and AUC of 0.903 for the prediction of the PL line dataset, whereas SVM gave 83.8% recall, and 0.811 AUC, respectively. Table 11 shows that the highest performance scores are achieved by RF for four types of lines with all handwritten patterns. Finally, we can say that in our experiment, RF performed better than SVM.

Table 11. Performance scores of RF and SVM for common features of handwritten pattern dataset.

Line Types	CF	RF				SVM			
		Rec	Prec	f1-Score	AUC	Rec	Prec	f1-Score	AUC
Zigzag trace	18	80.6	86.2	83.3	0.870	67.7	84.0	75.0	0.820
Zigzag predict	2	83.8	76.4	80.0	0.784	77.4	75.0	76.1	0.732
PL trace	7	83.8	89.6	86.6	0.926	74.1	92.0	82.1	0.872
PL predict	3	87.0	93.1	90.0	0.903	83.8	83.8	83.8	0.811
All patterns	23	85.1	84.3	84.7	0.923	83.1	87.5	85.2	0.919

CF: Common features.

3.6. Comparison of Our Proposed Method with the Existing Method

The comparison of the classification accuracy of our proposed method with the existing method in the literature is presented in Table 12. Guimaraes et al. (2017) [4] applied different ML algorithms such as multilayer perception (MLP), deep convolutional neural network (DCNN), decision tree (DT), RF, and SVM for the classification of adult and teenager age groups based on sentences. They collected 7000 sentences for the classification of age groups (teenager vs. adult). They showed that DCNN had a better performance and obtained 95.0% precision. Rizwan et al. (2021) [31] proposed a novel method for the classification of human age. They extracted features using interior angle formulation, anthropometric model, carnio facial development, wrinkle detection, and heat maps. The best combination of feature sets was selected using SFS. They adopted CNN to classify human age and achieved 94.6% classification accuracy. Özkan and Turan (2018) [32] proposed a deep learning algorithm for the classification of people based on their age. They divided the people into 12 classes using age groups and collected 18,000 images. They took 10% of the images for testing and the rest of the images for training. They showed that the DL model can correctly classify people into different groups of age and achieved 78.5% classification accuracy.

Table 12. Accuracy comparison with the methods in the literature.

Authors	Year	Data Types	ACC (%)
Guimaraes et al. [4]	2017	Sentences	Prec: 95.0
Rizwan et al. [31]	2021	FG-NET	94.6
Özkan and Turan [32]	2018	FG-NET	78.5
Goshvarpour [33]	2019	ECG	94.6
Ilyas [34]	2020	Auditory perception	92.0
Reade [35]	2015	FG-Net	82.0
Tin [36]	2012	FG-Net	92.5
Our proposed	2022	Handwritten text	93.5
		Handwritten pattern	89.8

Goshvarpour (2019) [33] proposed a novel Poincare feature set to classify age and gender based on ECG. They collected ECG data from 79 respondents. Among them, 37 were males aged 31.24 years and 42 were females aged 25.8 years. They applied SVM for the classification of age and gender and obtained the highest classification accuracy of 94.6%. Ilyas et al. (2020) [34] investigated a novel biometric method for the classification of human age. For classification, RF, SVM, linear regression (LR), ridge regression (RR), polynomial regression (PR), and ANN were used. They collected a total of 837 subjects aged 6–60 years to evaluate the proposed biometric system. They showed that RF produced the highest classification accuracy of 92.0%. Voice is also used for user authentication and identification. Voiceprints were used in various forensic approaches to classify age, gender, and language. Reade et al. (2015) [35] conducted a study for the classification of adult, child, and senior using face images dataset. They extracted features using HOG, local binary pattern, and active appearance model. They adopted k-NN, SVM, and GB algorithms for the classification of adult, child, and senior and achieved 82.0% classification accuracy. Tin (2012) [36] applied PCA for the classification of age using face image and produced the highest classification accuracy of 92.5%. Our proposed SFFS with RF (SFFS-RF) model produced higher accuracy compared to SFFS with SVM (SFFS-SVM) to classify adult and child based on their handwritten text and handwritten pattern.

4. Conclusions

The purpose of this study was to clarify changes in the development of handwritten text and pattern between adult and child. Online handwritten text and pattern datasets were collected using a pen tablet system. We utilized SFFS for feature selection and adopted two classification algorithms, RF and SVM, for the classification of adult and child. We selected the common features from SFFS-RF and SFFS-SVM classifiers and then also applied RF and SVM classifiers for the classification of adult and child. For the handwritten text dataset, our proposed system SFFS with RF classifier produced 93.5% accuracy for 18 features, and 89.8% accuracy for 9 features in the handwritten pattern dataset. After identifying the common features, SFFS-RF also produced 91.5% and 87.7% classification accuracy for handwritten text and handwritten pattern datasets. We hope that this study will provide evidence of the possibility of classifying adults and child based on their handwritten text and handwritten pattern data. If we can find out the age range between adult and child, that will help our model to produce an estimated performance accuracy.

Author Contributions: Conceptualization, J.S., M.A.M.H. and A.Y.; methodology, M.M. and M.A.M.H.; software, M.M., Y.U. and M.A.M.H.; validation, J.S. and M.A.M.H.; formal analysis, M.M., Y.U. and M.A.M.H.; investigation, J.S. and M.A.M.H.; resources, J.S., A.M., A.S. and A.Y.; data curation and collection, A.M., A.S. and A.Y.; writing—original draft preparation, J.S., M.M and M.A.M.H.; writing—review and editing, M.M. and M.A.M.H.; visualization, M.M., Y.U. and M.A.M.H.; supervision, J.S. and M.A.M.H.; project administration, J.S.; funding acquisition, J.S. and A.Y. All authors have read and agreed to the published version of the manuscript.

Funding: This work was supported by the Japan Society for the Promotion of Science Grants-in-Aids for Scientific Research (KAKENHI), Japan (Grant Numbers JP20K11892, which was awarded to Jungpil Shin and JP21H00891, which was awarded to Akira Yasumura).

Institutional Review Board Statement: The study was conducted in accordance with the Declaration of Helsinki, and approved by The Ethics Committee of the Graduate School of Social and Cultural Sciences of Kumamoto University (approval number: 45, approved on: 25 May 2021).

Informed Consent Statement: Informed consent was obtained from all subjects involved in the study.

Data Availability Statement: Not applicable.

Conflicts of Interest: The authors declare no conflict of interest for this research.

References

1. Faundez-Zanuy, M.; Fierrez, J.; Ferrer, M.A.; Diaz, M.; Tolosana, R.; Plamondon, R. Handwriting biometrics: Applications and future trends in e-security and e-health. *Cognit. Comput.* **2020**, *12*, 940–953. [CrossRef]
2. Poon, C.; Gorji, N.; Latt, M.; Tsoi, K.; Choi, B.; Loy, C.; Poon, S. Derivation and analysis of dynamic handwriting features as clinical markers of Parkinson's disease. In Proceedings of the 52nd Hawaii International Conference on System Sciences, Maui, HI, USA, 8–11 January 2019; pp. 3721–3730.
3. Singhal, P.; Srivastava, P.K.; Tiwari, A.K.; Shukla, R.K. A Survey: Approaches to facial detection and recognition with machine learning techniques. In Proceedings of the Second Doctoral Symposium on Computational Intelligence, Lucknow, India, 6 March 2021; Springer: Singapore, 2022; pp. 103–125.
4. Guimaraes, R.G.; Rosa, R.L.; De Gaetano, D.; Rodriguez, D.Z.; Bressan, G. Age groups classification in social network using deep learning. *IEEE Access* **2017**, *5*, 10805–10816. [CrossRef]
5. Al Maadeed, S.; Hassaine, A. Automatic prediction of age, gender, and nationality in offline handwriting. *EURASIP J. Image Video Process.* **2014**, *2014*, 10. [CrossRef]
6. Illouz, E.; Netanyahu, N.S. Handwriting-based gender classification using end-to-end deep neural networks. In Proceedings of the International Conference on Artificial Neural Networks, Rhodes, Greece, 4–7 October 2018; pp. 613–621.
7. Maken, P.; Gupta, A. A method for automatic classification of gender based on text-independent handwriting. *Multimed. Tools Appl.* **2021**, *80*, 24573–24602. [CrossRef]
8. Morera, Á.; Sánchez, Á.; Vélez, J.F.; Moreno, A.B. Gender and handedness prediction from offline handwriting using convolutional neural networks. *Complexity* **2018**, *2018*, 3891624. [CrossRef]
9. Shin, J.; Okuyama, T. Detection of alcohol intoxication via online handwritten signature verification. *Pattern Recognit. Lett.* **2014**, *35*, 101–104. [CrossRef]
10. Gupta, U.; Bansal, H.; Joshi, D. An improved sex-specific and age-dependent classification model for Parkinson's diagnosis using handwriting measurement. *Comput. Methods Programs Biomed.* **2020**, *189*, 105305–105327. [CrossRef]
11. Wong, L.; Loh, W. Segregating Offline and Online handwriting for conditional classification analysis. *Proc. IOP Conf. Ser. Mater. Sci. Eng.* **2019**, *530*, 012058–012065. [CrossRef]
12. Lewis, J.A. *Forensic Document Examination: Fundamentals and Current Trends*; Elsevier: Amsterdam, The Netherlands, 2014.
13. Papaodysseus, C.; Rousopoulos, P.; Giannopoulos, F.; Zannos, S.; Arabadjis, D.; Panagopoulos, M.; Kalfa, E.; Blackwell, C.; Tracy, S. Identifying the writer of ancient inscriptions and Byzantine codices. A novel approach. *Comput. Vis. Image. Underst.* **2014**, *121*, 57–73. [CrossRef]
14. Ahmed, Z.E. Forensic documents examination. *Asian J. Forensic Sci.* **2022**, *1*, 2–3.
15. Ahmad, A.R.; Khalia, M.; Viard-Gaudin, C.; Poisson, E. Online handwriting recognition using support vector machine. In Proceedings of the 2004 IEEE Region 10 Conference TENCON 2004, Chiang Mai, Thailand, 24 November 2004; pp. 311–314.
16. Babu, U.R.; Venkateswarlu, Y.; Chintha, A.K. Handwritten digit recognition using K-nearest neighbour classifier. In Proceedings of the 2014 World Congress on Computing and Communication Technologies, Trichirappalli, India, 27 February–1 March 2014; pp. 60–65.
17. Ramzan, M.; Khan, H.U.; Akhtar, W.; Zamir, A.; Awan, S.M.; Ilyas, M.; Mahmood, A. A survey on using neural network based algorithms for hand written digit recognition. *Environment* **2018**, *9*, 519–528. [CrossRef]
18. Baldominos, A.; Saez, Y.; Isasi, P. A survey of handwritten character recognition with mnist and emnist. *Appl. Sci.* **2019**, *9*, 3169. [CrossRef]
19. Ashiquzzaman, A.; Tushar, A.K.; Rahman, A.; Mohsin, F. An efficient recognition method for handwritten arabic numerals using CNN with data augmentation and dropout. In *Data Management, Analytics and Innovation*; Springer: Berlin/Heidelberg, Germany, 2019; pp. 299–309.
20. Altwaijry, N.; Al-Turaiki, I. Arabic handwriting recognition system using convolutional neural network. *Neural. Comput. Appl.* **2021**, *33*, 2249–2261. [CrossRef]
21. Nishimura, R.; Lee, A.; Saruwatari, H.; Shikano, K. Public speech-oriented guidance system with adult and child discrimination capability. In Proceedings of the 2004 IEEE International Conference on Acoustics, Speech, and Signal Processing, Montreal, QC, Canada, 17–21 May 2004; Volume 1.
22. Reynolds, D.; Rose, R. RC: Robust Text-independent speaker identification using Gaussian mixture speaker models. *IEEE Trans. Speech Audio Process.* **1995**, *3*, 72–83. [CrossRef]

23. Makihara, Y.; Mannami, H.; Yagi, Y. Gait analysis of gender and age using a large-scale multi-view gait database. In Proceedings of the 10th Asian Conference on Computer Vision, Queenstown, New Zealand, 8–12 November 2010; pp. 440–451.
24. Faghel-Soubeyrand, S.; Kloess, J.A.; Gosselin, F.; Charest, I.; Woodhams, J. Diagnostic features for human categorisation of adult and child faces. *Front. Psychol.* **2021**, *12*, 775338. [CrossRef] [PubMed]
25. Pudil, P.; Novovičová, J.; Kittler, J. Floating search methods in feature selection. *Pattern Recognit. Lett.* **1994**, *15*, 1119–1125. [CrossRef]
26. Raschka, S. MLxtend: Providing machine learning and data science utilities and extensions to Python's scientific computing stack. *J. Open Source Softw.* **2018**, *3*, 638–640. [CrossRef]
27. Jan, S.U.; Lee, Y.D.; Shin, J.; Koo, I. Sensor fault classification based on support vector machine and statistical time-domain features. *IEEE Access* **2017**, *5*, 8682–8690. [CrossRef]
28. Hasan, M.A.M.; Nasser, M.; Pal, B.; Ahmad, S. Support vector machine and random forest modeling for intrusion detection system (IDS). *J. Intell. Learn. Syst. Appl.* **2014**, *2014*, 45–52. [CrossRef]
29. Nelli, F. Machine Learning with scikit-learn. In *Python Data Analytics*; Springer: Berlin/Heidelberg, Germany, 2018; pp. 313–347.
30. Hasan, M.A.M.; Nasser, M.; Ahmad, S.; Molla, K.I. Feature selection for intrusion detection using random forest. *J. Inform. Secur.* **2016**, *7*, 129–140. [CrossRef]
31. Rizwan, S.A.; Jalal, A.; Gochoo, M.; Kim, K. Robust active shape model via hierarchical feature extraction with SFS-optimized convolution neural network for invariant human age classification. *Electronics* **2021**, *10*, 465. [CrossRef]
32. Özkan, İ.; Turan, B. Classification of different age groups of people by using deep learning. *J. New Res. Sci.* **2018**, *7*, 9–16.
33. Goshvarpour, A.; Goshvarpour, A. Gender and age classification using a new Poincare section-based feature set of ECG. *Signal Image Video Process.* **2019**, *13*, 531–539. [CrossRef]
34. Ilyas, M.; Othmani, A.; Naït-Ali, A. Auditory perception based system for age classification and estimation using dynamic frequency sound. *Multimed. Tools Appl.* **2020**, *79*, 21603–21626. [CrossRef]
35. Reade, S.; Viriri, S. Hybrid age estimation using facial images. In Proceedings of the 12th International Conference, ICIAR 2015, Niagara Falls, ON, Canada, 22–24 July 2015; pp. 239–246.
36. Tin, H.H.K. Subjective age prediction of face images using PCA. *Int. J. Eng. Res. Appl.* **2012**, *2*, 296–299. [CrossRef]

Article

Provider Fairness for Diversity and Coverage in Multi-Stakeholder Recommender Systems

Evangelos Karakolis *, Panagiotis Kokkinakos and Dimitrios Askounis

School of Electrical & Computer Engineering, National Technical University of Athens, Iroon Polytechniou 9, 15780 Zografou, Greece; pkokkinakos@epu.ntua.gr (P.K.); askous@epu.ntua.gr (D.A.)
* Correspondence: vkarakolis@epu.ntua.gr

Abstract: Nowadays, recommender systems (RS) are no longer evaluated only for the accuracy of their recommendations. Instead, there is a requirement for other metrics (e.g., coverage, diversity, serendipity) to be taken into account as well. In this context, the multi-stakeholder RS paradigm (MSRS) has gained significant popularity, as it takes into consideration all beneficiaries involved, from item providers to simple users. In this paper, the goal is to provide fair recommendations across item providers in terms of diversity and coverage for users to whom each provider's items are recommended. This is achieved by following the methodology provided by the literature for solving the recommendation problem as an optimization problem under constraints for coverage and diversity. As the constraints for diversity are quadratic and cannot be solved in sufficient time (NP-Hard problem), we propose a heuristic approach that provides solutions very close to the optimal one, as the proposed approach in the literature for solving diversity constraints was too generic. As a next step, we evaluate the results and identify several weaknesses in the problem formulation as provided in the literature. To this end, we introduce new formulations for diversity and provide a new heuristic approach for the solution of the new optimization problem.

Keywords: multi-stakeholder recommender systems; diversity; fairness; coverage; optimization

Citation: Karakolis, E.; Kokkinakos, P.; Askounis, D. Provider Fairness for Diversity and Coverage in Multi-Stakeholder Recommender Systems. *Appl. Sci.* **2022**, *12*, 4984. https://doi.org/10.3390/app12104984

Academic Editors: Alessandro Micarelli, Giuseppe Sansonetti and Giuseppe D'Aniello

Received: 21 March 2022
Accepted: 13 May 2022
Published: 14 May 2022

Publisher's Note: MDPI stays neutral with regard to jurisdictional claims in published maps and institutional affiliations.

Copyright: © 2022 by the authors. Licensee MDPI, Basel, Switzerland. This article is an open access article distributed under the terms and conditions of the Creative Commons Attribution (CC BY) license (https://creativecommons.org/licenses/by/4.0/).

1. Introduction

Item recommendations from traditional recommender systems are in most cases based on the highest predicted ratings for each user. Of course, this criterion is crucial for the success of a recommender system, as it proposes items that the users seem to be interested in. However, it may not be sufficient when it is the only metric taken into consideration. This is because it fails to enable the users to explore new content, and the recommendations for each user tend to be very similar to each other. As a result, relatively unknown and innovative items may never be discovered by the users. Moreover, most users are recommended items of a certain type. This problem is magnified for content-based recommender systems.

These weaknesses have been realized early by scientists and researchers, and several alternative metrics for the evaluation of recommender systems have been defined, including coverage, diversity, fairness, serendipity, and novelty. Regarding coverage, according to [1] an item is covered if it is recommended to at least a certain number of users. Diversity is the average dissimilarity between all pairs of users or items in the result set (per item and per user diversity) [2]. Serendipity in a recommender system is the experience of receiving an unexpected and fortuitous item recommendation [3]. Fairness is mostly used for group recommendations and its intuitive meaning is that a set of items needs to be recommended to a group of users, so that each group member is satisfied in a fair manner [1]. Novelty of any information is defined as the proportion of unknown and known information with respect to the user [4].

Furthermore, it is worth mentioning that, in the last few years, the focus has been shifted from users that receive recommendations to all stakeholders involved in a recommender system. Specifically, the main beneficiaries are item providers, users, and the recommendation system operator. This holistic paradigm in recommender systems, in which all stakeholders need to be jointly taken into consideration, is widely known as the Multi-Stakeholder Recommender Systems (MSRS) paradigm [5].

1.1. Objectives

The paper at hand focuses on item providers that pose requirements for fair item recommendations in terms of the coverage and diversity of users to whom each provider's items are recommended. The notion of fairness here implies that the items of each provider are recommended to a similar number of users as items of other providers, as well as that the users to whom a provider's items are recommended are as diverse as the users that items of other providers are recommended. More specifically, provider fairness (p-fairness) [6] requirements in terms of coverage and diversity are introduced as constraints that need to be fulfilled for each item provider. Hence, average user coverage and diversity per item for each provider should reach at least a certain target, while the expected satisfaction from the provided recommendations should remain as high as possible.

To this end, the presented recommender system aims at maximizing the total predicted rating of the recommendation lists, as well as satisfying the requirements for fair user coverage and diversity per item provider. Specifically, the problem is formulated as a mathematical optimization problem as proposed in the literature [7]. Moreover, the proposed methodology for the solution of the problem is followed. As the problem is not linear and hence it is not computationally feasible to be solved, and due to the fact that in the literature the solution for diversity is not adequately described, a heuristic algorithm which finds solutions that are very close to the optimal one and also fulfill the diversity constraints is introduced. As a next step, the results and the limitations of the proposed approach are presented in brief, and several weaknesses of the proposed formulation are identified. Moreover, two new mathematical definitions for provider diversity are proposed to address the identified weaknesses of the planned approach by the literature, and a high-level methodology is proposed for the solution of the new optimization problem. Finally, several limitations of the newly proposed approach are concisely presented, along with future extensions for the proposed framework.

1.2. Research Questions

The paper at hand has as its main objective to answer the following research questions:

RQ1. How provider fairness in terms of user coverage and diversity can be achieved in recommender systems?

RQ2. Are the results of the proposed methodology satisfactory?

RQ3. What are the limitations of the proposed methodologies?

RQ4. How can the aforementioned limitations be abated?

1.3. Contribution

The purpose of the publication at hand is to build a recommender system that will provide fair recommendations across the different item providers in terms of the coverage and diversity of users to whom their items are proposed. In this context, its contribution is multifold and includes:

- Review of the recent literature about provider fairness, coverage, and diversity.
- Formalization of the problem at hand, as provided in the literature.
- Application and extension of the proposed methodology to a public and well-known dataset.
- Because the proposed solution in the literature is vague in terms of solving the diversity constraints, in this publication, we present a concrete heuristic approach for solving the latter.
- Evaluation of the results and discussion on the proposed solution.

- Introduction of new definitions for quantifying diversity, as the proposed approach in the literature posed significant limitations.
- Introduction of new high-level heuristic approaches for solving the newly-defined diversity constraints as the problem for diversity remains NP-Hard.

2. Background and Related Work

There is a plethora of publications that acknowledge the need to go beyond the accuracy of recommendations in recommender systems. However, there is no commonly accepted approach for quantification of the alternative evaluation metrics of the latter. To this end, in this section we provide a literature review of the recent work in the MSRS paradigm, and the evaluation metrics of diversity, coverage, and fairness, in terms of their definitions and usage. As a next step, we present a publication that proposes a framework for providing multi-stakeholder recommendations with provider coverage and diversity constraints and the proposed solution. Finally, we discuss how the approach proposed in the publication at hand goes beyond the state-of-the-art.

2.1. Multi-Stakeholder Recommender Systems (MSRS)

The MSRS paradigm has gained significant popularity in the last few years. This paradigm, in contrast to the traditional one that puts the user (recommendation receiver) in the center of attention, also takes into consideration other stakeholders of recommender systems [5,8] such as item providers, the recommender system operator, and society in general. In this context, several recent studies focus on this paradigm. Specifically, Refs. [5,8] identify the need for examining recommender systems from the different stakeholders' perspective and introduce the multi-stakeholder recommender systems paradigm. On the other hand, Abdollahpouri et al. [9] investigated multi-sided fairness in multi-stakeholder recommendations and how different fairness concerns can be introduced in such systems. Moreover, Milano et al. [10] analyzed the ethical aspects of MSRS. Furthermore, Refs. [11,12] review the existing case studies, methods, and challenges, and propose new research directions for MSRS. Last but not least, Refs. [13,14] present applications of the MSRS paradigm. The first introduces provider constraints to the multi-stakeholder recommendation problem and formulates it as an integer programming optimization model that is solved using an approximation and can achieve satisfactory results in real use cases. On the other hand, in the second article the authors developed a multi-objective binary integer programming model to allocate sponsored recommendations. As a next step, they present an algorithm to solve the problem in a computationally efficient way. The proposed approach was applied to a real use case with good results and is easily applicable to existing recommender systems as it is applied as a form of postprocessing.

Of course, in the publication at hand, we are not examining MSRS in general; instead, we focus on the metrics of coverage and diversity across provider items and address the need for fairness across the latter. Therefore, it is worth presenting the different definitions of coverage, diversity, and fairness.

2.2. Coverage

Ge et al. [15] defined as item coverage, the proportion of items for which the system is able to generate recommendations (prediction coverage), or the proportion of available items that are recommended to a user (catalog coverage).

It needs to be noted that the previous definitions refer to coverage in items. However, a less popular term is user coverage and is defined as the proportion of users to whom specific items have been recommended [16]. User coverage poses significant interest for lists of items, meaning that it is important for an item provider to be aware to what proportion of the users his items have been recommended. In this publication, one of the goals of the recommender system is to fulfill user coverage constraints for the different providers.

Although the coverage has been mentioned in a variety of publications, in most cases, it is calculated after the recommendations have been provided and is used only for system evaluation, without being taken into consideration during the recommendation process [15–18]. In [1], Koutsopoulos et al. include constraints for user coverage to the recommendation problem definition. Specifically, they formulate the system recommendation problem as a mathematical optimization problem with constraints for coverage. This approach is close to the approach followed in the context of this publication.

2.3. Diversity

Concerning diversity, a variety of definitions is available in the literature [16]. More specifically, it has been defined as the average pairwise distance among the proposed items, or the total pairwise distance among the recommended items [19]. The first definition is the most popular in the literature. Another interesting aspect of diversity is how the distance (or dissimilarity) is calculated. For instance, when items are modeled as content descriptors, the dissimilarity is calculated through taxonomies [19], while when they are represented as vectors of terms the dissimilarity is calculated through the complement of Jaccard [20] or cosine similarity [21]. On the other hand, when items are represented as vectors of rankings, the most suitable metrics are Hamming distance, the complement of Pearson similarity [22], or the complement of cosine similarity [21]. In our case, we use the complement of Pearson similarity, and the items are represented as vectors of rankings. Of course, there are more approaches for the calculation of distance among items, but these will not be examined in the scope of this publication [20,23].

Similar to diversity for items, diversity for users can also be defined. The idea is that an item would be worth recommending to as many differentiated users as possible. In this context, the diversity for users can be defined as the average pairwise distance among the users to whom an item is recommended. This definition may not be important for individual items or users, but can be of utmost importance for item providers who intend to maximize the diversity of users to whom their items are recommended. This type of diversity is thoroughly analyzed under the context of this paper.

2.4. Fairness

Regarding fairness, in most cases it refers to omitting or reducing prejudice from a machine learning model or a recommender system [6]. However, recommender systems in most cases provide personalized recommendations, and as a result it is difficult to omit prejudice. Moreover, in most cases recommender systems are used by different stakeholders, and fairness issues may be raised for different stakeholder groups. Therefore, a recommender system, as well as for securing fairness for customers, should also be fair for item providers. For this reason, Burke et al. [6] defined fairness for different stakeholders. Specifically, fairness is divided to customer fairness (C-fairness) and provider fairness (P-fairness). In the current publication, provider fairness implies that the items of each provider are recommended to a similar number of users as items of other providers, and that the users to whom a provider's items are recommended are as diverse as the users that items of other providers are recommended to, and this is the ultimate goal of this study.

There are a number of practical applications for addressing provider fairness in recommender systems. For instance, Borrato et al. [24] evaluate provider fairness in terms of disparities in relevance, visibility, and exposure for minority groups and propose a treatment that combines observation up-sampling and loss regularization for user-item relevance scores, with satisfactory experimental results. In comparison to the paper at hand, the aforementioned work focuses on fairness across minorities, while the current study approaches all item providers as equal and provides a recommendation strategy that achieves at least a certain amount of average coverage and diversity per item for each provider, if such a strategy exists. On the other hand, the previous study enables the association of an item with more than one provider, which is not the case in the paper at hand, because the problem would be much more complex.

Furthermore, Sonboli et al. [25] acknowledged that individual preferences may limit the ability of an RS to produce fair recommendations. Moreover, they introduced a re-ranking approach for fairness-aware recommendations that learns users' preferences across multiple fairness dimensions instead of a single sensitive feature such as race. This approach achieved better experimental results than other approaches in the literature. Unlike our study, the aforementioned approach also focuses on re-ranking of recommended items, which is not the case in this study. Moreover, the aforementioned work pays attention to the diversity of items a user is recommended, while our approach focuses on user diversity for provider items, as well as user coverage of items for each provider.

Moreover, Gomez et al. [26] acknowledge the provider fairness in terms of geographic imbalance in educational recommender systems. Their study was based on real-world data and observe that data are highly imbalanced in favor of the United States, in terms of open courses and interactions. As traditional RSs tend to reinforce the most represented countries (or providers in general), this study proposes an approach that regulates the share of recommendations for each country and their position in the recommendation list. The definition of fairness in the aforementioned publication is the closest to the definition of the paper at hand, as it recognizes the lack of equity in recommendations across different countries in terms of visibility and exposure, while the paper at hand aims to address the imbalance (lack of equity) in the coverage and diversity of users across the different item providers.

Another notable work about fairness in recommender systems is presented in [27]. In this study, Beutel et al. evaluate algorithmic fairness in a real-world recommender system and showcase that measuring fairness based on pairwise comparisons from randomized experiments is a tractable means for reasoning fairness in rankings and propose a new regularizer that helps to significantly enhance pairwise fairness to a large-scale production RS. The proposed approach can identify and improve systematic mis-ranks or underranks of items of a particular group. On the other hand, the current study produces recommendations based on predefined thresholds for average diversity and coverage in users for the items of each provider, and as already mentioned does not take into consideration the rank of each item in the recommendation list.

Other practical applications for addressing fairness in recommender systems are presented in [28,29].

2.5. Problem Formalization

After presenting MSRS, as well as the main evaluation metrics that are examined in the context of this publication, alongside several applications of the latter in RS, it is worth presenting the work conducted by Koutsopoulos et al. [7]. In this publication, authors provide a mathematical formalization for the problem of fairness across item providers in terms of user coverage and diversity. Therefore, the problem at hand, along with the proposed solution as defined by Koutsopoulos et al. [7], are presented in brief.

According to Koutsopoulos et al. [7], a set of items, I, and a set of users, U, are considered along with some baseline recommendation system (in our case, item-based Collaborative Filtering) that generates recommendation lists, L_u, for each user, u. Additionally, that C is the number of providers, and each item belongs to exactly one provider, while I_c is the set of items of provider c, $c = 1, \ldots, C$ for each user, u, and item, i, of a provider. Finally, r_{iu} denotes the predicted rating with the baseline RS algorithm.

It is worth mentioning that the proposed methodology can be generalized for categories or classes of items. In particular, C can also denote the number of categories or classes and not necessarily the different providers. Hence, from now on, provider and category will be used interchangeably in the problem at hand.

The output of the recommendation algorithm is a list of recommended items for each user. Each recommendation list has size L. Additionally, L_u denotes the set of items recommended to the user, u. L_u' denotes the new list of recommended items that should satisfy coverage and diversity constraints.

Moreover, $x = (x_{iu}: i \in I, u \in U)$ denotes the new Boolean recommendation policy (if $x_{iu} = 0$, item i is not recommended to user u, if $x_{iu} = 1$, item i is recommended to user u)

that should be found. The total deviation between the baseline recommendation lists and the new recommended ones is:

$$Cost(x) = \frac{1}{L|U|}\left(\sum_{u \in U}\sum_{i \in Lu} riu - \sum_{i \in I}\sum_{u \in U} riu * xiu\right), \quad (1)$$

Regarding provider coverage, it is calculated as the sum of the item coverage values for each item of a provider or category. Thus, the average per item user coverage for items of a given provider, c, is:

$$Cov(c, x) \frac{1}{|Ic|} \sum_{i \in Ic}\sum_{u \in U} xiu, \quad (2)$$

Concerning per item diversity of users of provider, c, it is calculated for each provider, c, as follows:

$$Div(c, x) \frac{1}{|Ic|} \sum_{i \in Ic}\sum_{u \in U}\sum_{v \in U, v \neq u} duv * xiu * xiv, \quad (3)$$

where duv is the dissimilarity between users u and v.

Instead of the absolute category diversity, the average normalized diversity (per item, per user pair) that will be used is defined as:

$$\overline{Div(c, x)} = \frac{2}{|Ic|} \frac{\sum_{i \in Ic}\sum_{u \in U}\sum_{v \in V: u \neq v} duv * xiu * xiv}{(\sum_{i \in Ic}\sum_{u \in U} xiu) * (\sum_{i \in Ic}\sum_{u \in U} xiu - 1)}, \quad (4)$$

or simpler:

$$\overline{Div(c, x)} = \frac{2}{|Ic|} \frac{\sum_{i \in Ic}\sum_{u \in U}\sum_{v \in V: u \neq v} duv * xiu * xiv}{Kc|Ic| * (Kc|Ic| - 1)}, \quad (5)$$

and the constraints for provider coverage become:

$$Cov(c, x) >= Kc, \quad (6)$$

and for diversity:

$$Divc(c, x) >= Dc, \quad (7)$$

where Kc is the minimum user coverage for the items of a provider, c, and Dc is the minimum threshold for average per item and per user pair diversity for provider, c.

Along with the optimization problem in the same publication, a solution for the coverage constraints in polynomial time is provided, and a low complexity heuristic approach on top of the coverage solution is proposed in order to also solve the diversity constraints. The results of the proposed approach show satisfactory performance, according to the authors.

In the context of this publication, we follow the proposed methodology and introduce a heuristic algorithm for the solution of the diversity constraints. This is because the approach proposed in the aforementioned publication was too generic, suggesting the recommendation of items to users that increase diversity of a category the most and substituting item recommendations with low ranking on the condition that they do not violate coverage constraints. Because it is not easy or obvious to calculate which item recommendations should be substituted, we provide a more concrete approach, and evaluate the results. As a next step we identify several weaknesses with the current definition for diversity and provide two new definitions for it that address the identified weaknesses. Finally, we propose a new heuristic approach for the solution of the newly defined problem.

3. Heuristic Solution Approach

In this section, we present the overall methodology that has been used in order to solve the recommendation problem. It consists of three steps: (a) find a solution for the unconstrained problem, according to the best predicted ratings, (b) find a solution that deviates the least from the unconstrained solution, subject to the constraints for coverage, and (c) based on the solution that fulfills the constraints for coverage and with the minimum possible deviation from the previous solution, change the recommendation lists in a way that increases the most the diversity for the providers that do not fulfill the aforementioned constraints, on condition that the constraints for coverage and diversity that are fulfilled still hold.

3.1. Baseline Solution (Unconstrained Problem)

The first step of the methodology is to find a solution for the unconstrained recommendation problem. For this problem, item-based collaborative filtering is selected as the most proper technique. Specifically, as a first step we create the user-item matrix and predict the ratings of the users for items that they have not rated yet. To do this, we use the K nearest neighbors (KNN) technique [30]. After some experimentation, the best predictions were occurring for $K = 20$. As a similarity metric we used the Pearson similarity [22].

The result of this methodology is a list of items with the best predicted ratings for each user. The results of this approach are used to form the optimization problem of the next steps.

3.2. Addressing Coverage Constraints

Having available the solution from the unconstrained problem, we can solve the problem subject only to the coverage constraints. The optimization problem as described in [7] can be described from the following equations:

$$\max(\frac{1}{L|U|} \sum_{i \in I} \sum_{u \in U} r_{iu} * x_{iu}), \tag{8}$$

subject to:

$$Cov(c, x) = \frac{1}{|Ic|} \sum_{i \in Ic} \sum_{u \in U} x_{iu} \geq Kc, \; x_{iu} \in \{0,1\}, \tag{9}$$

$$\sum_{i \in I} x_{iu} = L, \; \text{for all } u \in U \tag{10}$$

where all the symbols have been described in the previous section. Specifically, the goal is to maximize the total predicted rating of the new recommendation list, while average coverage for the items of each provider should be at least Kc (items of provider, c, should be recommended on average to at least Kc users. The last constraint denotes that each user receives exactly L recommendations.

As described also in [7], the optimization problem with coverage constraints is linear and hence solvable and can be solved in polynomial time. The results of the problem subject to coverage constraints are a list of recommended items for each user; these have the highest predicted rating by each user, that allow coverage constraints to be fulfilled. The aforementioned recommendation lists will be used as a starting point, in order to find a solution that also fulfills the constraints for diversity.

3.3. Addressing Diversity Constraints

Regarding the objective of diversity, the following constraints should be added to the previous model:

$$Div(c, x) = \frac{2}{|Ic|} \frac{\sum_{i \in Ic} \sum_{u \in U} \sum_{v \in U, v \neq u} d_{uv} * x_{iu} * x_{iv}}{Kc|Ic| * (Kc|Ic| - 1)} \geq D, \tag{11}$$

where all the symbols have been described in the previous section. Specifically, the constraints for diversity denote that the average user diversity per item for items of each provider should be at least D.

As the formula for computing the Diversity is in quadratic form, the problem becomes much more complex.

Specifically, the problem at hand, as also identified by Koutsopoulos in [7], is a quadratic constraint integer programming problem, which is NP-Hard. As such, instead of trying to find the optimal solution to the problem, it was decided to develop a heuristic approach that finds a solution close to the optimal, which is also the approach proposed by Koutsopoulos et al. [7]. In the aforementioned publication, the approach proposed for fulfilling diversity constraints is applying item substitutions across providers, through switching of recommended items between users with priority to switches that mostly improve diversity, after making sure that coverage constraints are still satisfied. However, this approach is too generic, as calculating provider diversity and coverage at every step is not an easy task, while it is even more complex to identify the items that increase the diversity the most and the ones that should be omitted from the recommendation list. For this reason, we investigated the definition of diversity more deeply and propose a more concrete approach.

Specifically, for the development of a heuristic solution, a thorough exploratory analysis has been performed on the data to identify how diversity is influenced by other metrics, such as dissimilarity. At first, the intuition was that diversity would increase if an item was recommended to the most dissimilar set of users. However, the initial intuition has been proven wrong.

What was observed instead is that the most influential factor for increasing or decreasing the diversity was the number of users to whom an item has been recommended. Specifically, the higher the number of users to whom the most recommended item of a provider was recommended, the higher is the diversity for the specific provider.

Therefore, a high-level algorithm has been designed in order to address diversity constraints. It uses as a starting point the solution that fulfills the coverage constraints as described in the previous subsection. As a next step, it calculates the diversity for all categories (providers) and finds the ones that do not fulfill the diversity constraints. For all the aforementioned categories, it finds their most recommended item and the users to whom it was not recommended, ordered by their predicted rating for that item, and if they have been recommended with items from categories for which diversity constraint is fulfilled, the item with the worst predicted rating that belongs to a category which fulfills the diversity constraint is no longer recommended to them. Instead, it is substituted by the most recommended item of the category that needs to fulfill the diversity constraint, on condition that diversity constraint for the category of the swapped item still holds. This procedure is applied until the diversity constraint of the specific category is fulfilled.

A high-level but more detailed overview of the developed algorithm of the heuristic solution that was applied is presented in the following algorithm (Algorithm 1):

For more information, the code of the experiment is available in Github [31].

The algorithm that has been developed follows a greedy approach and avoids backtracking to find a better solution, in case a good enough solution has been found at a previous step. Therefore, the problem becomes efficiently solvable in terms of time and memory. The complexity of the algorithm in the worst-case scenario is $O(L*U*Ctc*Ccc)$, where Ctc are the providers for which the diversity constraint does not hold and Ccc are the providers for which the diversity constraint holds. More simply, the complexity in the worst-case scenario is $O(L*U*C^2)$. This is because, in the worst case, all items need to be accessed for all users, and the latter have to be accessed for all changeable categories and for all categories or providers that need to change. However, the algorithm completes much earlier in most cases because once a category's diversity constraint is fulfilled it proceeds with a new category of items, without accessing all the users and items.

Algorithm 1 Heuristic algorithm for addressing diversity constraints

1: Xcov = calculate_solution_for_Coverage(A_pred, Kc) // as a linear programming problem through cvxpy
2: Xnew = Xcov   // copy Coverage solution as a starting point for the final solution
3: **foreach** category c:
4:   Div[c] = calculate_Diversity_for_category(Xnew, c) // as described in Formula (11)
5: categories_ordered_by_Diversity = argsort(Div)
6: categories_ordered_by_Diversity_desc = reverse(categories_ordered_by_Diversity)
7: categories to change, changeable_categories = [], categories_ordered_by_Diversity
8: **foreach** category c in categories_ordered_by_Diversity:
9:   **If** Div[c] < D:
10:     Categories_to_change.add(c)
11:     changeable_categories.remove(c)
12: **foreach** category c in categories_to_change:
13:   Most_recommended_item[c] = find_most_recommended_item_of_category(c)
14:   Users_to_recommend = argsort(A_pred[most_recommended_item[c]])
15:   Users_to_recommend_except_recommended=users not recommended with the most recommended item
16:   **foreach** user u in users_to_recommend_except_recommended:
17:     **foreach** category c2 in changeable_categories:
18:       rec_items_of_categ_to_usr = find_rec_items_of_category_to_user(c2, u)
19:       **foreach** item i in rec_items_of_categ_to_usr:
20:         Xnew[u,i] = 0
21:         Xnew[u, most_recommeded_item[c]] = 1
           // Calculate new Coverage for category c2 according to formula (9)
22:         Cov_xnew[c2] = calculate_Coverage_for_category(X_new, c2)
           // Calculate new Diversity for categories c2 and c according to Formula (11)
23:         Div_xnew[c2] = calculate_Diversity_for_category(Xnew, c2)
24:         Div_xnew[c] = calculate_Diversity_for_category(Xnew, c)
           // If Coverage or Diversity constraints for c2 are violated rollback
25:         **If** Cov_xnew[c2]< K[c2] or Div_xnew[c2]<D:
26:           Xnew[u,i] = 1
27:           Xnew[u, most_recommended_item[c2]] = 0
28:         **Else if** Div_xnew[c] > D:
29:           Break (line 12)

4. Experiment Details and Results

In this section, we present the dataset that was used and the preprocessing that took place for the solution of the problem. Afterwards, we present the experiment and the results of the heuristic solution.

4.1. Dataset Overview and Preprocessing

The dataset that was used in the context of this publication is a dataset for movie recommendations, and it was taken from MovieLens [32]. The initial dataset of ratings consists of 9724 items (movies) and 610 users. As the dataset is very sparse and there were limited computational resources, the experiments were executed for a subset of items. More specifically, movies that had 10 or less ratings were filtered out, and as a result only 2121 movies were kept.

4.2. Baseline (Unconstrained) Solution

After the preprocessing phase, the user-item ratings matrix was created, based on which the analysis and modeling was performed. Regarding the baseline solution and the coverage constraints, the methodology described in [7] has been followed.

Specifically, after applying five-fold cross validation to the user item (80–20% training and test sets, accordingly), the Pearson similarity between items in the training set is computed, and after some experimentation only the 20 nearest neighbors are kept for each item. More specifically, 20 nearest neighbors led to the lowest Mean Squared Error in predicted ratings on the test set of the initial dataset. Afterwards, based on this similarity metric, ratings for items of the test set were estimated. As a next step, the performance of the item-based Collaborative

Filtering (CF) based on mean squared error, which measures the squared difference between predicted and actual ratings, was evaluated. The resulting mean squared error was 0.8605 for the test set. This means that the model had a relatively satisfactory performance, as the predicted ratings deviate 0.86 stars from actual ones in the test set.

Therefore, by adopting this baseline recommendation system, the ratings that users would give to items, which they have not rated so far, are predicted (null values in the initial user-item ratings matrix). The final output is a list of 10 items for each user with the highest predicted ratings that the algorithm produces with no further constraint. The resulted average rating for recommended items for all users was 4.938.

4.3. Item Providers or Categories

The result of the item-based CF method maximizes the sum of ratings that the recommended movies provide; however, the total rating is not the only metric that should be taken into consideration as the objective of the problem. In particular, the goal of this problem is to make sure that the coverage and the diversity metrics for each item provider will also be taken into consideration. As described by Koutopoulos et al. [7], this is achieved by setting thresholds for the value of both Coverage and Diversity for each item provider.

Because coverage and diversity are defined as provider-related metrics, different providers have been created, and each item is connected to a specific one. The experiment has been conducted for different numbers of item providers with similar results. Therefore, only the results for twenty item providers are presented. Specifically, twenty item providers were created, and only one provider was assigned to each item. The provider assignment was performed randomly, with uniform probability, and as a result the distribution of providers is almost uniform, as shown in the Table 1:

Table 1. Different providers or categories and their number of items.

Category	1	2	3	4	5	6	7	8	9	10	11	12	13	14	15	16	17	18	19	20
Items	101	91	105	115	105	107	90	121	101	101	90	114	94	120	115	104	114	108	107	118

4.4. Coverage Solution

The coverage constrained problem was solved for discrete values of x (specifically 0 and 1) because an item is either recommended or not recommended to a user. For the solution of the problem, the cvxpy [33] python library was utilized. Although the new lists for different values of K are different from the baseline lists, Lu, the results of the total rating for different coverage thresholds found with the discrete solution are the same for different values of coverage. In particular, the total rating is the same for $K = 1, 1.5, 2, 2.5$, while for $K \geq 3$ there is no solution. This fact occurs because there are many excellent ratings (rating = 5) in the predicted rating matrix, and the items of the new lists Lu' are different from the ones of the baseline recommendation list Lu, but the new recommended items are rated perfectly, the same as the old ones (5 stars). However, the distribution of coverage changes per item provider (category), as shown in Figures 1 and 2:

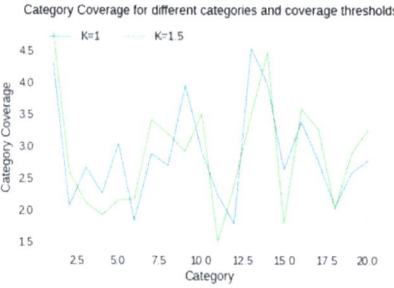

Figure 1. Coverage per provider for $K = 1$ and $K = 1.5$.

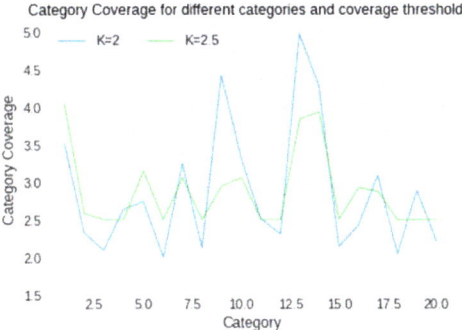

Figure 2. Coverage per provider for $K = 2$ and $K = 2.5$.

As observed from the figures above, as K increases, the coverage tends to be more uniform.

4.5. Final Solution

As already mentioned, the solution for coverage is used as a starting point in order to find a feasible solution that also fulfills the diversity constraints. As a next step, the heuristic approach of Algorithm 1 has been followed. The result of the proposed solution is shown in Figure 3:

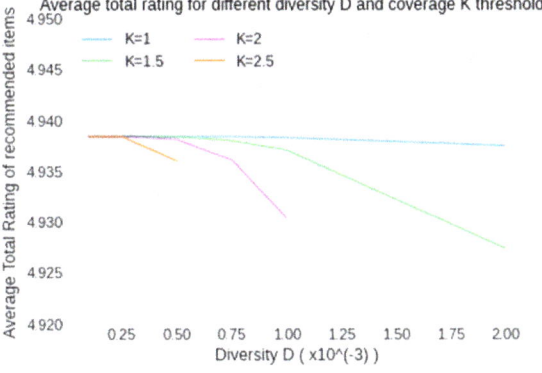

Figure 3. Average rating for different values of diversity and coverage.

As observed in the figure above, for every value of coverage, as the diversity threshold increases, the average rating of recommended items is similar or drops. Furthermore, for greater coverage thresholds, the problem becomes more difficult to solve and the average rating drops more rapidly with the increase of diversity threshold. More specifically for coverage threshold $Kc = 2.5$, the problem has no solution for a Diversity threshold greater than 0.0005. In Table 2, the total and average rating of the recommended items are presented for different values of Diversity threshold and for Coverage threshold $Kc = 1$ and $Kc = 2.5$.

As observed from the table and the figure above, the total rating of all the recommended items for all users is very high and, in most cases, it is the same as the coverage solution. This shows that the heuristic solution chosen may not be the optimal one but produces a result that is very close to the optimal.

Table 2. Average and total rating for different values of diversity and coverage.

Diversity Threshold D	Total Rating (Kc = 1)	Average Rating (Kc = 1)	Total Rating (Kc = 2.5)	Average Rating (Kc = 2.5)
0.0001	30,124.411	4.9384	30,124.411	4.9384
0.0002	30,124.411	4.9384	30,124.411	4.9384
0.00025	30,124.411	4.9384	30,124.411	4.9384
0.0005	30,124.411	4.9384	30,110.984	4.9362
0.00075	30,124.411	4.9384	No solution	No solution
0.001	30,124.120	4.9383	No solution	No solution
0.002	30,119.159	4.9375	No solution	No solution

In order to provide a better view on how diversity influences the total rating, in the following plots the total rating for coverage threshold $K = 1.5$ and $K = 2$ for different thresholds of diversity is presented (Figures 4 and 5):

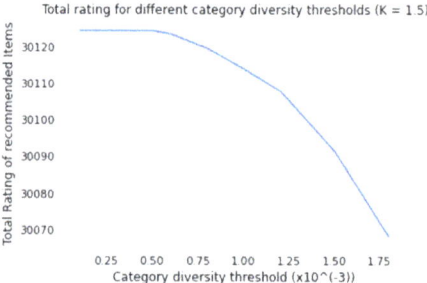

Figure 4. Total rating for different category Diversity thresholds for $K = 1.5$.

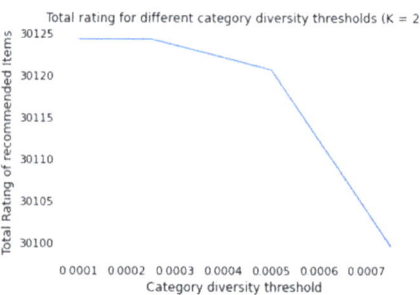

Figure 5. Total rating for different category Diversity thresholds for $K = 2$.

From the plots above (Figures 4 and 5), it is observed that, as the diversity threshold increases, the total rating decreases, while the decrease of the total rating is more important for greater diversity thresholds and not proportional. Concerning the relation between diversity and coverage, it is observed that, for a bigger coverage threshold, the total rating decreases faster with the increase of diversity. This is also what was expected, because, for a higher coverage threshold, on the one hand a stricter constraint is introduced, and on the other, less exchanges with items from other categories are allowed.

Moreover, as observed in Table 2, the results in terms of total or average rating are very close to the ones of the unconstrained solution. The total rating of the latter serves as the upper bound for the optimal solution, because the constrained problem cannot provide more relevant recommendations than the unconstrained one.

As stated before, the approach that has been followed is heuristic and does not always provide the optimal solution to the optimization problem. Instead, it always provides a solution that is close to the optimal one. Specifically, the algorithm that has been proposed follows a greedy approach and avoids backtracking to find a better solution, in case a good enough solution has been found at a previous step.

The cases where the algorithm fails to identify the optimal solution stem from the fact that when the diversity constraint is fulfilled for a certain category, the recommendations of this category can only change on the condition that the diversity and coverage constraints for this category are not violated with a certain swap. Thus, it is possible to find a different recommendation plan in which the total predicted rating would be slightly higher than the total rating of the heuristic solution. For example, the approach at hand, fails to identify cases where the most recommended item's rating of a category is not as good as the second ones, while by recommending the second most recommended item of the category fulfills the diversity constraint. Moreover, if in a certain swap a category's diversity constraint is violated, the heuristic solution will not revisit the latter category to fix its diversity, even if there are options to do that. However, at least for the specific dataset the results are very close, not only to the optimal solution but also to the baseline solution, which is the upper bound for the total rating.

5. Discussion

5.1. Evaluation

The approach that has been followed to solve the optimization problem for provider fairness in terms of coverage and diversity finds a heuristic solution to the problem that, according to the results, is very close to the optimal solution as it is very close to the results of the baseline recommender system that does not pose any constraint for coverage or diversity and hence acts as an upper bound for the proposed solution.

Concerning the comparison of our results to other studies, only Koutsopoulos et al. [7] dealt with the same problem of provider fairness in terms of user coverage and diversity. The results of the proposed algorithm are similar to the experimental results in the aforementioned work, meaning that as diversity and coverage requirements increase, the total predicted rating of the recommended items tends to drop, while diversity constraints have a larger impact to the total rating. Moreover, our approach shows smaller deviation from the baseline solution, which poses the upper bound on the total predicted rating. Of course, this is not a safe inference as the solution of [7] does not clearly describes neither the preprocessing that took place nor the hyperparameters that have been used.

In general, the advantage of treating the recommendation problem as a constraint optimization problem is that it ensures those certain criteria (the constraints) will be fulfilled. Of course, the constraints should be defined, also taking into consideration the tradeoff between the constraint requirements and the user preferences. This means that if the constraints are too stringent the relevance of the recommended items with the user may be critically low. Therefore, recommender system operators should always evaluate and update their requirements.

Consequently, the proposed approach or a similar optimization approach with constraints for equal representation of providers or other stakeholder groups can be used in order to soften "The winner takes it all phenomenon" [26], in which the most significant providers (or groups of providers) have huge advantage over the others. This is because the optimization algorithm would secure at least a certain target of exposure (coverage, diversity, or any other metric) for all groups, if such a solution existed. Moreover, the proposed approach can also ensure that minority item providers will be protected as they would be guaranteed at least a certain amount of coverage and diversity. Additionally, with a similar approach, exposure and visibility for minority groups can also be achieved.

However, the literature-based definition of the problem as an optimization one [7], poses several limitations. Firstly, it does not take into account the position of an item in the recommendation list. This is problematic because in most cases the position of the items in the user's list of recommendations plays a vital role in users' behavior, especially when the recommended items are numerous. For instance, a user may pay attention to the top few recommendations but not to the bottom ones.

Another limitation of the problem definition at hand is that the definition of diversity fails to capture the dissimilarity among users to whom an item is recommended. Instead, it is observed that the most recommended items of a provider influence the diversity significantly more than the dissimilarity among the users to whom an item is recommended, and this fact is exploited by the heuristic approach that has been developed.

To make this weakness more evident, we tried to approach the item's diversity as defined in the literature, by supposing that the dissimilarity of two different users is fixed and equals the average dissimilarity between two different users. With this approximation, if d is considered as the average dissimilarity between any two different users, the formula for computing the diversity becomes:

$$Div(c,x) = \frac{2}{|Ic|} \frac{\sum_{i \in Ic} \sum_{u \in U} \sum_{v \in U, v \neq u} d * x_{iu} * x_{iv}}{Kc|Ic| * (Kc|Ic| - 1)} \geq D, \quad (12)$$

However, if the item's total Diversity is analyzed further, Formula (12) results in:
If the item was recommended to one user, it is:
$Div(item) = 0$
If the item was recommended to two users 1 and 2 it is:
$Div(item) = d(1,2) = d$
If the item was recommended to three users 1, 2, and 3 it is:
$Div(item) = d(1,2) + d(1,3) + d(2,3) = 2d + d = 3d$
With the same approach for n users:
$Div(item) = (1 + 2 + 3 + \ldots + (n-1))*d$
Given that $(1 + 2 + 3 + \ldots + n - 1)$ is a well-known sum and is equal to $\frac{1}{2}n(n-1)$, finally:

$$Div(item) = \frac{1}{2}n(n-1)d, \quad (13)$$

The item diversity for all user pairs was approximated according to the formula above, and the result is illustrated in the following figures:

Specifically, in the Figures 6 and 7, the red line illustrates the diversity approximation for different values of n (number of users to whom an item is recommended), while the blue points illustrate the actual item diversity calculated for different items and their correlation with n. Figure 6 shows the results after removing outliers, while Figure 7 illustrates all the item diversities for every different occurred value of n.

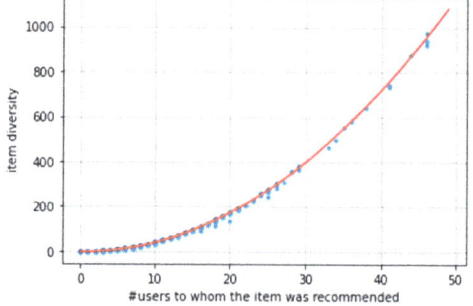

Figure 6. Relation between item diversity and number of users to whom the items were recommended (outliers removed).

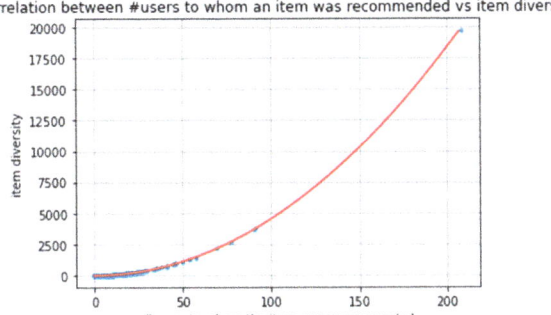

Figure 7. Relation between item diversity and number of users to whom the items were recommended.

As observed, the approximation seems to fit almost perfectly, at least for the specific dataset. In particular, the actual values of item Diversity are very close to the predicted ones, and this is the case for every item in the dataset.

At this point it is worth mentioning that several tests (e.g., the Kolmogorov–Smirnov for both uniformity and normality) were applied to check if the dissimilarity is either fixed or follows normal distribution around d. However, in all cases the null hypothesis was rejected and statistically the approximation is not valid for this dataset. However, the approximation is more than satisfactory and illustrates that instead of paying attention to user dissimilarity, the proposed definition for item diversity favors the most recommended items.

Of course, item providers would not be happy if their most recommended items were recommended to more users but their least recommended ones would be recommended to less users, even if the average diversity of their items would be significantly increased.

Hence, the definition of diversity needs to be reexamined. Moreover, new solutions should be designed to solve the problem efficiently.

5.2. Redefining User Diversity and a New Heuristic Solution Approach

In order to identify the problems of the proposed definition of provider diversity, we will revisit its definition as presented in the literature [7]:

First of all, user diversity for an item is defined as the sum of dissimilarities among users to whom the specific item is recommended:

$$Div(i) = \sum_{u \in U} \sum_{v \in U,\ v \neq u} d_{uv} * x_{iu} * x_{iv}, \tag{14}$$

while the diversity of an item provider is defined as the sum of item diversities for items that belong to the specific provider.

As a next step the category diversity is normalized by dividing with the number of pairs of users to whom items of a specific provider are recommended. As a result, provider diversity per item and user pair is defined in Formula (4) or simplified (using the coverage constraint) in Formula (5).

Conceptually it can be defined as:

$$\overline{Div(c)} = \frac{2}{|Ic|} \frac{sum\ of\ all\ diversities\ for\ items\ in\ the\ specific\ category}{all\ pairs\ of\ users\ to\ whom\ items\ of\ class\ c\ are\ assigned}, \tag{15}$$

Although this definition normalizes the category diversity according to the pairs of users its items are recommended to, it fails to capture the dynamics of most recommended items, as illustrated in the previous section. Consequently, if a provider owns x items that are recommended to y users, the provider diversity can be significantly increased if all users are recommended with the most recommended item, and this should not be happening.

As a result, the definition of provider diversity per item and user pair should either normalize the diversities per item or define a new diversity for the entire set of items for the specific provider. In the first case, the definition of provider diversity per item and user pair becomes:

$$\overline{Div\prime(c)} = \frac{2}{|I_c|} \sum_{i \in I_c} \frac{\sum_{u \in U} \sum_{v \in V: u \neq v} d_{uv} * x_{iu} * x_{iv}}{(\sum_{u \in U} x_{iu}) * (\sum_{u \in U} x_{iu} - 1)}, \quad (16)$$

Of course, with the previous definition for diversity, the optimization problem is not solvable, and of course the heuristic solution that was presented in Section 4 is not valid anymore.

In the second case, the provider diversity, instead of the sum of item diversities, can be calculated for each provider by considering all the items of diversity as one super item. In this case, we should create a new matrix $Y(c,u)$ ($c \in C$, $u \in U$) that denotes the Boolean recommendation policy (if $y(c,u) = 0$ no item for category c has been recommended to user u, and if $y(c,u) = 1$ one or more items of category c have been recommended to user u).

As a result, category diversity can be defined as:

$$\overline{Div\prime\prime(c)} = \frac{2}{|I_c|} \frac{\sum_{u \in U} \sum_{v \in V: u \neq v} d_{uv} * y_{cu} * y_{cv}}{(\sum_{u \in U} y_{cu}) * (\sum_{u \in U} y_{cu} - 1)}, \quad (17)$$

In the previous definition, the denominator can be simplified as in [7]:

$$\overline{Div\prime\prime(c)} = \frac{2}{|I_c|} \frac{\sum_{u \in U} \sum_{v \in V: u \neq v} d_{uv} * y_{cu} * y_{cv}}{K_c |I_c| * (K_c |I_c| - 1)}, \quad (18)$$

The last definition is much simpler (although it still places quadratic constraints on the problem, which is NP-Hard), as the number of categories is much smaller than the number of items. However, it is not suitable to be used for small systems, for which the users have rated a significant amount of items. Instead, for large-scale instances of the recommendation problem and large user-item matrices that are sparse, this definition can prove very helpful.

Regarding the solution of the newly defined diversity per provider, in both cases a heuristic approach can also solve the problem. For example, in the first case, by recommending the best rated item to users that have the greatest distance from the centroid of the users to whom the item was recommended, substituting the worst rated item (from another category) that was recommended to them, on condition that the coverage and the diversity constraints for the category of the swapped item are still fulfilled. The same approach can also be applied in the second solution, by recommending the best rated items to users that are the most distant from the centroid of the users who they had recommendations from the specific provider. Moreover, in case the number of users is very large, clustering techniques can be used to cluster the users according to the similarity between them, and the item swaps can be performed only for users from the most distant clusters.

5.3. Answers to Research Questions

After presenting the literature, the methodology, and the results and acknowledging the limitations of the proposed approach, it is time to answer the research questions that were presented in Section 1.

RQ1. How can provider fairness in terms of user coverage and diversity be achieved in recommender systems?

The literature provides a formalization of the problem at hand, alongside a solution. The literature models the problem as a quadratic constraint mathematical programming problem, which is NP-Hard and hence finding the optimal solution is not feasible. As a result they propose a heuristic solution for solving diversity constraints. However, as the latter is not adequately described, we introduce a new heuristic approach for the solution of diversity constraints.

RQ2. Are the results of the proposed methodology satisfactory?

The results of the proposed heuristic approach not only achieve the targets of sufficient user coverage and diversity per item provider, but are also very close in terms of predicted ratings with the recommendation lists provided by the solution of the unconstrained problem, which acts as an upper bound for the heuristic solution.

RQ3. What are the limitations of the proposed methodologies?

There are several limitations with the proposed approach that stem from the formalization of the problem. Specifically, the definition of diversity fails to capture the dissimilarity among users to whom an item is recommended. Instead, it is observed that the most recommended items of a provider influence the diversity significantly more than the dissimilarity among the users to whom an item is recommended, and this fact is exploited by the heuristic approach that has been developed. Hence, the problem definition and in particular the quantification of diversity should be redefined. As a result, new solutions to the problem should be developed.

Moreover, the proposed approach does not take into consideration the position of an item in the recommendation list, which is of utmost importance in synchronous recommender systems.

RQ4. How can the aforementioned limitations be abated?

In order to address the problematic definition of diversity, we redefine it under the context of Section 5.2. Specifically, two new definitions are introduced, discussing their pros and cons. Moreover, new high-level heuristic approaches are proposed for the solution of the newly-defined diversity constraints.

6. Conclusions and Future Extensions

The contribution of this publication is multifold. Specifically, the methodology proposed by the literature is applied to solve the problem of provider fairness for user diversity and coverage in Multi-stakeholder recommender systems. However, as the problem at hand is NP-Hard and the methodology proposed in the literature is too generic, a new heuristic algorithm for solving diversity constraints is proposed, along with its complexity and limitations. The results of the proposed solution are presented, and they are very close to the optimal solution for the specific dataset.

As a next step, the results were evaluated and discussed thoroughly, and several weaknesses to the proposed problem definition were identified. Specifically, the proposed approach is not taking into consideration the ranking of items in the recommendation list. Furthermore, after digging further to the definition of diversity to explore how the latter is influenced by dissimilarity and other metrics, it was discovered that, with the current definition, the most influential factor for provider diversity is the number of users to whom the most diverse item of a provider has been recommended instead of the dissimilarity among the users to whom it was recommended. To this end, in this publication, we analyzed the proposed definition for diversity and identified why this weakness appears. Finally, we proposed two new definitions for provider diversity as well as a new high-level heuristic solution approach.

The next step forward regarding the proposed approach is to take into consideration the ranking of recommendations in the recommendation problem, as the position of an item in a recommendation list is very important for the recommender system users. This will be carried out using more evaluation metrics such as the Normalized Discounted Cumulative Gain (NDCG), which measures the quality of ranking of the recommended items for each user.

Furthermore, minority representation will be examined in terms of user coverage and diversity, against other well-known approaches (e.g., [24]). Moreover, different metrics will be examined in order to be formalized as constraints in the optimization problem.

Finally, in the near future, we intend to apply the proposed approach to several test datasets, as well as to a real-life recommender system. Finally, the results will be evaluated in terms of coverage, diversity, and satisfaction by real users including item providers and simple users of the system.

Author Contributions: Conceptualization, E.K.; methodology, E.K.; software, E.K.; validation, P.K. and D.A.; formal analysis, E.K.; investigation, E.K.; data curation, P.K.; writing—original draft preparation, E.K.; writing—review and editing, E.K. and P.K.; visualization, P.K. and E.K.; supervision, D.A.; project administration, D.A. All authors have read and agreed to the published version of the manuscript.

Funding: This research was funded by Institute of Communication and Computer Systems, ICCS.

Informed Consent Statement: Not applicable.

Data Availability Statement: The dataset that has been used for this publication has been taken from MovieLens and is available at [32].

Conflicts of Interest: The authors declare no conflict of interest. The funders had no role in the design of the study; in the collection, analyses, or interpretation of data; in the writing of the manuscript; or in the decision to publish the results.

References

1. Koutsopoulos, I.; Halkidi, M. Efficient and Fair Item Coverage in Recommender Systems. In Proceedings of the IEEE 16th Intl Conf on Dependable, Autonomic and Secure Computing, Athens, Greece, 12–15 August 2018; pp. 912–918.
2. Kunaver, M.; Požrl, T. Diversity in recommender systems—A survey. *Knowl.-Based Syst.* **2017**, *123*, 154–162. [CrossRef]
3. McNee, S.M.; Riedl, J.; Konstan, J.A. Being accurate is not enough: How accuracy metrics have hurt recommender systems. In *CHI'06 Extended Abstracts on Human Factors in Computing Systems*; Association for Computing Machinery: New York, NY, USA, 2006. [CrossRef]
4. Saranya, K.G.; Sudha Sadasivam, G. Personalized News Article Recommendation with Novelty Using Collaborative Filtering Based Rough Set Theory. *Mob. Netw. Appl.* **2017**, *22*, 719–729. [CrossRef]
5. Abdollahpouri, H.; Burke, R.; Mobasher, B. Recommender systems as multistakeholder environments. In Proceedings of the 25th Conference on User Modeling, Adaptation and Personalization, Bratislava, Slovakia, 9 July 2017. [CrossRef]
6. Burke, R.; Sonboli, N.; Ordoñez-Gauger, A. Balanced Neighborhoods for Multi-sided Fairness in Recommendation. In Proceedings of the Conference on Fairness, Accountability and Transparency, New York, NY, USA, 23–24 February 2018.
7. Koutsopoulos, I.; Halkidi, M. Optimization of Multi-stakeholder Recommender Systems for Diversity and Coverage. In Proceedings of the IFIP Advances in Information and Communication Technology, Crete, Greece, 25–27 June 2021; Volume 627. [CrossRef]
8. Burke, R.; Abdollahpouri, H.; Malthouse, E.C.; Thai, K.P.; Zhang, Y. Recommendation in multistakeholder environments. In Proceedings of the 13th ACM Conference on Recommender Systems, Copenhagen, Denmark, 16–19 September 2019. [CrossRef]
9. Abdollahpouri, H.; Burke, R. Multi-stakeholder recommendation and its connection to multi-sided fairness. In Proceedings of the CEUR Workshop, Lviv, Ukraine, 16–17 May 2019; Volume 2440.
10. Milano, S.; Taddeo, M.; Floridi, L. Ethical aspects of multi-stakeholder recommendation systems. *Inf. Soc.* **2021**, *37*, 35–45. [CrossRef]
11. Abdollahpouri, H. Multistakeholder recommendation: Survey and research directions. *User Model. User-Adapt. Interact.* **2020**, *30*, 127–158. [CrossRef]
12. Zheng, Y. Multi-stakeholder recommendations: Case studies, methods and challenges. In Proceedings of the 13th ACM Conference on Recommender Systems, Copenhagen, Denmark, 16–20 September 2019. [CrossRef]
13. Sürer, Ö.; Burke, R.; Malthouse, E.C. Multistakeholder recommendation with provider constraints. In Proceedings of the 12th ACM Conference on Recommender Systems, Vancouver, BC, Canada, 2 October 2018. [CrossRef]
14. Malthouse, E.C.; Vakeel, K.A.; Hessary, Y.K.; Burke, R.; Fudurić, M. A multistakeholder recommender systems algorithm for allocating sponsored recommendations. In Proceedings of the CEUR Workshop, Lviv, Ukraine, 16–17 May 2019; Volume 2440.
15. Ge, M.; Delgado-Battenfeld, C.; Jannach, D. Beyond accuracy: Evaluating recommender systems by coverage and serendipity. In Proceedings of the Fourth ACM conference on Recommender Systems, Barcelona, Spain, 26–30 September 2010. [CrossRef]
16. Kaminskas, M.; Bridge, D. Diversity, serendipity, novelty, and coverage: A survey and empirical analysis of beyond-Accuracy objectives in recommender systems. ACM Transactions on Interactive Intelligent Systems. *ACM Trans. Interact. Intell. Syst.* **2016**, *7*, 1–42. [CrossRef]
17. Rahman, M.; Oh, J.C. Graph bandit for diverse user coverage in online recommendation. *Appl. Intell.* **2018**, *48*, 1979–1995. [CrossRef]
18. Hammar, M.; Karlsson, R.; Nilsson, B.J. Using maximum coverage to optimize recommendation systems in E-commerce. In Proceedings of the 7th ACM Conference on Recommender Systems, Hong Kong, China, 12–16 October 2013. [CrossRef]
19. Ziegler, C.-N.; McNee, S.M.; Konstan, J.A.; Lausen, G. Improving recommendation lists through topic diversification. In Proceedings of the 14th International Conference on World Wide Web, Chiba, Japan, 10–14 May 2005. [CrossRef]
20. Vargas, S.; Castells, P.; Vallet, D. Intent-oriented diversity in recommender systems. In Proceedings of the 34th International ACM SIGIR Conference on Research and Development in Information Retrieval, Beijing, China, 24–28 July 2011. [CrossRef]
21. Ekstrand, M.D. Collaborative Filtering Recommender Systems. In *The Adaptive Web*; Springer: Berlin/Heidelberg, Germany, 2011.

22. Kelly, J.P.; Bridge, D. Enhancing the diversity of conversational collaborative recommendations: A comparison. *Artif. Intell. Rev.* **2006**, *25*, 79–95. [CrossRef]
23. Yu, C.; Lakshmanan, L.V.S.; Amer-Yahia, S. Recommendation diversification using explanations. In Proceedings of the 2009 IEEE 25th International Conference on Data Engineering, Shanghai, China, 29 March–2 April 2009. [CrossRef]
24. Boratto, L.; Fenu, G.; Marras, M. Interplay between upsampling and regularization for provider fairness in recommender systems. *User Modeling User-Adapt. Interact.* **2021**, *31*, 421–455. [CrossRef]
25. Sonboli, N.; Eskandanian, F.; Burke, R.; Liu, W.; Mobasher, B. Opportunistic Multi-aspect Fairness through Personalized Re-ranking. In Proceedings of the 28th ACM Conference on User Modeling, Adaptation and Personalization, Genoa, Italy, 12–18 July 2020. [CrossRef]
26. Gómez, E.; Shui Zhang, C.; Boratto, L.; Salamó, M.; Marras, M. The Winner Takes it All: Geographic Imbalance and Provider (Un)fairness in Educational Recommender Systems. In Proceedings of the 44th International ACM SIGIR Conference on Research and Development in Information Retrieval, Online, 11–15 July 2021. [CrossRef]
27. Beutel, A.; Chen, J.; Doshi, T.; Qian, H.; Wei, L.; Wu, Y.; Goodrow, C. Fairness in recommendation ranking through pairwise comparisons. In Proceedings of the 25th ACM SIGKDD International Conference on Knowledge Discovery & Data Mining, Anchorage, AK, USA, 4–8 August 2019. [CrossRef]
28. Lin, X.; Zhang, M.; Zhang, Y.; Gu, Z.; Liu, Y.; Ma, S. Fairness-aware group recommendation with pareto-efficiency. In Proceedings of the Eleventh ACM Conference on Recommender Systems, Como, Italy, 27–31 August 2017. [CrossRef]
29. Serbos, D.; Qi, S.; Mamoulis, N.; Pitoura, E.; Tsaparas, P. Fairness in package-to-group recommendations. In Proceedings of the 26th International Conference on World Wide Web, Perth, Australia, 3–7 April 2017. [CrossRef]
30. Sarwar, B.; Karypis, G.; Konstan, J.; Riedl, J. Item-based collaborative filtering recommendation algorithms. In Proceedings of the 10th International Conference on World Wide Web, Hong Kong, China, 1–5 May 2001. [CrossRef]
31. Provider Fairness for Coverage and Diversity Experiments Github Repository. Available online: https://github.com/vkarakolis-epu/recsys_provider_fairness_optimization (accessed on 20 March 2022).
32. Movielens Datasets. Available online: http://files.grouplens.org/datasets/movielens/ml-latest-small.zip (accessed on 20 March 2022).
33. Cvxpy Python Library, Convex Optimization, for Everyone. Available online: https://www.cvxpy.org/ (accessed on 20 March 2022).

Article

Outcome Prediction for SARS-CoV-2 Patients Using Machine Learning Modeling of Clinical, Radiological, and Radiomic Features Derived from Chest CT Images

Lorenzo Spagnoli [1,2], Maria Francesca Morrone [1,2], Enrico Giampieri [3,4], Giulia Paolani [1,2], Miriam Santoro [1,2], Nico Curti [4], Francesca Coppola [5], Federica Ciccarese [5], Giulio Vara [5], Nicolò Brandi [5], Rita Golfieri [5], Michele Bartoletti [6], Pierluigi Viale [6] and Lidia Strigari [1,*]

[1] Medical Physics Unit, IRCCS Azienda Ospedaliero-Universitaria di Bologna, 40138 Bologna, Italy; lorenzo.spagnoli@studio.unibo.it (L.S.); maria.morrone@studio.unibo.it (M.F.M.); giulia.paolani@studio.unibo.it (G.P.); miriam.santoro@studio.unibo.it (M.S.)
[2] Medical Physics Specialization School, Alma Mater Studiorum, University of Bologna, 40126 Bologna, Italy
[3] National Institute of Nuclear Physics (INFN), 40127 Bologna, Italy; enrico.giampieri@unibo.it
[4] Experimental, Diagnostic and Specialty Medicine-DIMES, University of Bologna, 40126 Bologna, Italy; nico.curti2@unibo.it
[5] Department of Radiology, IRCCS Azienda Ospedaliero-Universitaria di Bologna, 40138 Bologna, Italy; francesca_coppola@hotmail.it (F.C.); federica.ciccarese@aosp.bo.it (F.C.); giulio.vara@studio.unibo.it (G.V.); nicolo.brandi@studio.unibo.it (N.B.); rita.golfieri@aosp.bo.it (R.G.)
[6] Infectious Diseases Unit, Department of Medical and Surgical Sciences, IRCCS Azienda Ospedaliero-Universitaria di Bologna, 40138 Bologna, Italy; m.bartoletti@unibo.it (M.B.); pierluigi.viale@aosp.bo.it (P.V.)
* Correspondence: lidia.strigari@aosp.bo.it

Featured Application: The present study demonstrates that semi-automatic segmentation enables the identification of regions of interest affected by SARS-CoV-2 infection for the extraction of prognostic features from chest CT scans without suffering from the inter-operator variability typical of segmentation, hence offering a valuable and informative second opinion. Machine Learning methods allow identification of the prognostic features potentially reusable for the early detection and management of other similar diseases.

Abstract: (1) Background: Chest Computed Tomography (CT) has been proposed as a non-invasive method for confirming the diagnosis of SARS-CoV-2 patients using radiomic features (RFs) and baseline clinical data. The performance of Machine Learning (ML) methods using RFs derived from semi-automatically segmented lungs in chest CT images was investigated regarding the ability to predict the mortality of SARS-CoV-2 patients. (2) Methods: A total of 179 RFs extracted from 436 chest CT images of SARS-CoV-2 patients, and 8 clinical and 6 radiological variables, were used to train and evaluate three ML methods (Least Absolute Shrinkage and Selection Operator [LASSO] regularized regression, Random Forest Classifier [RFC], and the Fully connected Neural Network [FcNN]) for their ability to predict mortality using the Area Under the Curve (AUC) of Receiver Operator characteristic (ROC) Curves. These three groups of variables were used separately and together as input for constructing and comparing the final performance of ML models. (3) Results: All the ML models using only RFs achieved an informative level regarding predictive ability, outperforming radiological assessment, without however reaching the performance obtained with ML based on clinical variables. The LASSO regularized regression and the FcNN performed equally, both being superior to the RFC. (4) Conclusions: Radiomic features based on semi-automatically segmented CT images and ML approaches can aid in identifying patients with a high risk of mortality, allowing a fast, objective, and generalizable method for improving prognostic assessment by providing a second expert opinion that outperforms human evaluation.

Keywords: radiomics; CT images; Machine Learning; SARS-CoV-2; mortality

Citation: Spagnoli, L.; Morrone, M.F.; Giampieri, E.; Paolani, G.; Santoro, M.; Curti, N.; Coppola, F.; Ciccarese, F.; Vara, G.; Brandi, N.; et al. Outcome Prediction for SARS-CoV-2 Patients Using Machine Learning Modeling of Clinical, Radiological, and Radiomic Features Derived from Chest CT Images. *Appl. Sci.* 2022, *12*, 4493. https://doi.org/10.3390/app12094493

Academic Editors: Alessandro Micarelli, Giuseppe Sansonetti and Giuseppe D'Aniello

Received: 11 March 2022
Accepted: 24 April 2022
Published: 28 April 2022

Publisher's Note: MDPI stays neutral with regard to jurisdictional claims in published maps and institutional affiliations.

Copyright: © 2022 by the authors. Licensee MDPI, Basel, Switzerland. This article is an open access article distributed under the terms and conditions of the Creative Commons Attribution (CC BY) license (https://creativecommons.org/licenses/by/4.0/).

1. Introduction

Since the beginning of 2020, the SARS-CoV-2 virus (Severe Acute Respiratory Syndrome-Corona Viruses-2) has triggered the outbreak of a world-wide pandemic, leading to restrictive measures of isolation and closure. To face the health emergency, hospitals increased the number of beds in intensive care units (ICUs) and introduced novel indicators for prioritizing patient admission and predicting patient outcome.

A reverse-transcriptase polymerase chain reaction (RT-PCR) assay from nasopharyngeal swabs or bronchoalveolar lavage is the reference test for diagnosing SARS-CoV-2 infection [1]. Chest Computed Tomography (CT) has recently been considered to be a potential non-invasive method for independently confirming the diagnosis of suspected COVID-19 patients with a sensitivity of 97%, specificity of 25%, and accuracy of 68% [2]. Consequently, many COVID-19 patients underwent CT scans to evaluate the extent of the damage and improve prognosis estimation thus increasing the possibility of an overdiagnosis.

In addition, CT-based radiological findings (e.g., Ground Glass Opacity [GGO], Crazy Paving, Lung Consolidation) can detect SARS-CoV-2 virus based on 2D/3D imaging techniques in one or both lungs and can be used as a surrogate of disease severity. These findings were reached by means of a consensus in the European Society of Radiology (ESR) [3].

Furthermore, images can convey a large amount of information which the human eye cannot objectively quantify, providing other potential predictive or prognostic factors related to the COVID-19 disease. For this reason, the field of radiomics uses rigorous mathematical definitions and well-defined approaches [4] to quantitatively describe the image-based properties contained within radiological images, such as texture and shape/volumetric information.

Semi-automatic segmentation has recently been suggested as a tool for quickly sectioning the lungs or the COVID-19 lesions, enabling the extraction of the radiomic features in order to improve the prediction of several clinical endpoints, including ICU admission, need for ventilators [5–10], and severe vs. critical conditions [9]. However, only a limited number of papers have investigated patient mortality as an outcome, often having only a relatively limited patient cohort or a short patient follow-up [5–10].

The former limitation is likely related to the manual nature of the segmentation methods used in the papers published, which represents a very time-consuming task. During the pandemic, various semi-automatic segmentation COVID-19-dedicated tools became available; therefore, the performance of Machine Learning models built on the radiomic features extracted was investigated, using one of these tools for predicting mortality in a high-risk COVID-19-positive group.

2. Materials and Methods

2.1. Study Design

The study, regarding the prognostic value of radiomic features, was conducted and included all patients suitable for analysis, according to the guidelines of the Declaration of Helsinki. The study was approved by the Institutional Review Board (or Ethics Committee) of the IRCCS University Hospital of Bologna (protocol code no. EM949-2020_507/2020/Oss/AOUBo, approved on 16 September 2020).

All patients identified according to the inclusion/exclusion criteria before the Ethics committee approval were included retrospectively, while the remaining population (after 16 September 2020) was included prospectively; informed consent forms were obtained. All the clinical data were retrieved from an ad hoc clinical database for SARS-CoV-2 patient management, while the radiological data and CT chest images were retrieved from structured reports and Digital Image Communication in Medicine (DICOM) files available from the Radiology Information System (RIS) and Picture Archiving and Communication System (PACS), respectively.

2.2. Patient Cohort

The patient cohort was made up of a subset of patients, confirmed positive for COVID-19 using RT-PCR, admitted to the IRCCS University Hospital of Bologna–Polyclinic Sant'Orsola-Malpighi (IRCCS AOSP), redirected from neighboring hospitals from February 2020 to March 2021 since the authors' Institute is a regional emergency hub capable of managing patients at high risk of SARS-CoV-2 infection [11]. Consequently, the present cohort of hospitalized patients was considered at high risk irrespective of the referring hospital.

Chest CT scan findings (radiological and radiomics) and the clinical data available at patient admission were used to develop a predictive model of patient mortality.

The inclusion criteria were the following: having a chest CT scan with slice thicknesses of between 1 and 1.25 mm without contrast medium acquired after patient admission and recorded on the RIS-PACS of the IRCCS AOSP associated with radiological findings, and a complete set of clinical baseline information including RT-PCR positivity to COVID-19. When multiple CT scans were available, only that closest to the date of admission was analyzed.

The duration of hospitalization is reported in Supplementary Materials Table S1 according to patient outcome as well as period of first diagnosis (first or second wave). Moreover, the days elapsed between the CT scan and the hospitalization date were not statistically significantly different (p-value = 0.29) using a standard t-test comparing patients by outcome. The average days of survival were 21 and 14 in the patients hospitalized in the first and second wave, respectively; this difference showed a trend (p-value = 0.074), indicating that patients with severe disease were better selected during the second wave, albeit with expected improvement in the treatment strategies available over time.

The inclusion criteria were fulfilled by 436 patients, i.e., 286 males (65.6%) and 150 females (34.4%). The main patient characteristics and baseline comorbidities are reported in Table 1. The median age was 68.5 (21–99) years; a hypertension status was recorded in 241 patients. Two-hundred and fifty-one had a fever (Temperature $\geq 38°$) at hospital admission. The choice of using this cutoff for fever was based on the variability of body temperature occurring on the day of admission. Information regarding fever and hypertensive state were included in the routine admission procedure and, hence, were available for all patients; however, no additional details were recorded at admission.

Table 1. Clinical characteristics of the patients as well as the radiological findings obtained upon radiologist inspection of the CT scans.

Variables	Median (Min–Max)
Age (years)	68.5 (21–99)
Respiratory rate (Breaths/m)	20 (10–98)
Days of hospitalization	13 (0.25–99)
	Yes N (%)—No N (%)
Hypertension	241 (55.3%)—195 (44.7%)
History of smoking	347 (79.5%)—89 (20.5%)
Obesity	363 (83.3%)—73 (16.7%)
Sex	Male 286 (65.6%)—Female 150 (34.4%)
Fever	251 (57.6%)—185 (42.4%)
Lung Consolidation	225 (51.6%)—211 (48.4%)
Ground Glass Opacity (GGO)	382 (87.6%)—54 (12.4%)
Crazy Paving	336 (77.1%)—100 (22.9%)
Bilateral involvement	403 (92.4%)—33 (7.6%)

It is also worth noting that the present cohort presented a large prevalence of obese individuals (83%). This could have been a bias as since an estimation of the visceral fat surface and muscular surface obtained with the segmentation software by segmenting a slice at the height of vertebra T12 of the thoracic region was available, the authors expected some dependency on body composition to arise from the respective radiomic feature, which allowed much more nuance in patient characterization.

The CT scans were obtained using an Ingenuity CT (Philips Medical Systems, Cleveland, OH, USA) in 56% of patients, a Lightspeed VCT (General Electric Healthcare, Chicago, IL, USA) in 41% of patients, and an ICT SP (Philips Medical Systems, Cleveland, OH, USA) in 3% of patients.

The scanners can be considered equivalent as the CT chest acquisition protocols were set to produce comparable image quality as verified during the Quality Assurance (QA) controls. In addition, the acquisition protocols remained unchanged during the entire data collection period.

For the most part, the kilovolt peaks (kVps) were set to 120 kV (91.5% scans), with a few exceptions which were set to 100 kV (5.0% scans) or 140 kV (3.5% scans), according to patient characteristics.

2.3. Image Segmentation

Sophia DDM for Radiomics [12] is a CE/FDA-marked software for SARS-CoV-2 patients which offers a CT-based automated workflow for whole-lung segmentation and disease quantification. It was used for both lung and disease volume of interest (VOI) segmentation as well as for radiomic feature extraction [12].

The segmentation was based on region growing techniques, and used gradient detection and volume stability to regulate the convergence of the process. The majority of the radiomic features were defined and extracted following the workflow as per Image Biomarker Standardization Initiative (IBSI) [4] regulations.

Sophia Radiomics also uses two thresholds which correspond qualitatively to the portion of GGO (from -740 HU to -400 HU) and the range of pixel values representing the vascular tree (from -400 HU to about 1000 HU).

These voxels are counted and kept as a measure of damage volume (in mL). In particular, these ranges are generally appropriate for differentiating GGO from the vascular tree; they may affect the quality of the radiomic features extracted and can be manually modified by the user upon visual inspection, if required.

The software produces one-hundred and seventy-seven features relative to both lungs as a single VOI. In addition, quantification of the visceral fat and abdominal mass surface, as a surrogate of Sarcopenia, was computed using manual segmentation of the abdominal cavity on a single slice at the height of vertebra T12. These surfaces identified via the thresholding method were computed by counting the pixels identified and were expressed in cm^2. In all cases in which the segmentation obtained semi-automatically was incomplete, the patients were eliminated from the study both in cases of partial imaging scans as well as in cases of widespread infection affecting software segmentation capability. All patients were checked manually after the segmentation process for a final approval of inclusion.

2.4. Patient and Image Characteristics

The dataset was composed of 436 patients, each with a set of assigned features. For convenience, the features were categorized into three subsets: Clinical, Radiomic, and Radiological.

The clinical features available at hospital admission were divided into (a) continuous: age at the time of the CT exam and respiratory rate in breaths/min, and (b) binary: Sex of the patient, obesity status, fever at the hospital admission, hypertension condition, and smoking history.

One-hundred and seventy-nine radiomic features were supported by the segmentation software, the majority of which were described in [4], with the addition of visceral fat and abdominal mass.

The six radiological features included the acquisition parameters (kVp, current, and slice thickness) extracted from the DICOM header and Boolean features (such as the bilaterality of the lung damage, the presence of GGO, lung consolidations, and crazy paving) assessed by expert radiologists and extracted from the structured medical report.

Different models were built using each of the feature groups to compare performance in a single and/or combined fashion and evaluate the potential benefits in terms of prognostic value. The structure of the training and testing, reported below in this section, was the same in all subsets regardless of the input features; the models were named using the same name as the family given in the input features. The outcome investigated was the mortality observed in 78/436 patients.

2.5. Predictive Models

All the analyses were conducted using Python-3 [13], utilizing the scikit-learn libraries [14], imblearn [15], pandas [16], numpy [17], scipy [18], and ELI5 [19], while the plotting was carried out using matplotlib [20] and seaborn [21].

The data were analyzed using Machine Learning (ML) methods, including regression regularized via Least Absolute Shrinkage and Selection Operator (LASSO) [22], the Random Forest classifier [23], or the Fully connected Neural Network (FcNN) [24].

Details regarding the implementation of all the algorithms can be found in the scikit-learn [14] documentation of LASSO cross-validation (LassoCV), Random Forest classifier, and Multi-layer Perceptron Classifier (FcNN Classifier) functions. Lasso CV has been utilized with all the default parameters since they are automatically optimized by means of a built-in cross-validation procedure. The random forest was built using 200 decision trees with balanced class weights, and the FcNN classifier was utilized with alpha = 10, a single hidden layer with five nodes, max number of iterations = 1000, activation function ReLu, and "lbfgs" solver. The RF hyper-parameters were chosen using a parametric scan to explore the main possible combination of values including number of estimators, max depth, max number of features, and oob score. In addition, the impact of dataset dimensionality reduction was also investigated for the RF approach.

Different pre-processing procedures were followed for the different algorithms. Since the present dataset was heavily unbalanced (18% mortality), the Random Forest, which was the most sensitive to imbalances in the dataset, was preceded by a Synthetic Minority Oversampling Technique [25] which created new instances of the minority class using the convex combination of a set of samples in the minority class. The Standard Scaler was used to carry out z-score scaling on all the features before Random Forest implementation.

In the case of LASSO and FcNN, normalization and scaling of the features was achieved using the Box–Cox transformation and the Standard Scaler, respectively. The number of features was reduced by using a threshold of 0.6 in the Spearman correlation. In addition, the single feature which was best correlated with the patient outcome using the Spearman correlation test was re-included in the set of remaining features.

For all the algorithms, evaluation of the models was carried out using a 10-fold cross-validation approach, with stratification with respect to the outcome, to obtain a more realistic evaluation of the model performance, using the "cross-val-predict" scikit-learn function. The data analysis pipeline is represented schematically in Figure 1.

The hyperparameter search for the Lasso was carried out automatically, using an additional stratified 10-fold cross-validation in the training phase. To avoid data leakage, the entire cross-validation procedure was managed using the scikit-learn library.

In all cases, performance was evaluated using the Area Under the Curve (AUC) of the respective Receiver Operator characteristic (ROC) curves as well as sensitivity and specificity.

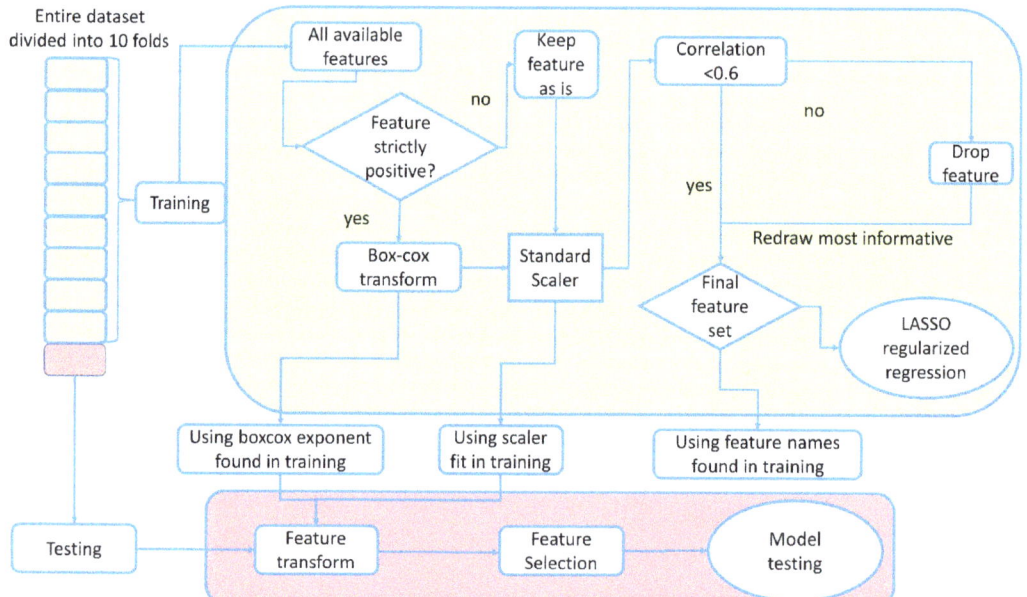

Figure 1. Representation of the 10-fold cross-validation approach used for the training and testing of a single feature group classifier. The input family was selected before entering the green box.

3. Results

The plots in Figures 2, 3 and S1 are ROC curves relative to the LASSO, FcNN, and RFC methods, respectively. In all cases, the fainter lines represent the 10 curves relative to the 10 testing phases. The bold blue line is the average performance, the turquoise bands represent the standard deviation around the mean, and the red line is the performance of a random guesser blindly predicting mortality. The performances of all the developed models are reported in Table 2. To investigate the capabilities of the LASSO model based on all the available features to describe our cohort irrespective of admission rate, the present cohort was divided into two groups according to hospitalization date (before or after 20 July 2020). The AUCs were determined, resulting in 0.73 and 0.76, which were found not to be statistically significantly different in demonstrating the capability of the model to describe the present dataset, irrespective of the wave of belonging. Similar results have been reported in [26] using a semi-quantitative score based on a database including only radiological information.

DeLong's tests were used to compare the ROCs. Without considering the radiological models, only the Lasso clinical and the Lasso radiomic models were statistically different, with a p-value of 0.044.

The relevant features in the Lasso models are reported in Table 3; a graphical representation of the importance of the features in each model is reported in Figure 4.

For the Lasso regularized regression, the importance is expressed by the coefficient of the feature in the linear combination. For the RFC, the importance is the Gini importance built into the implementation of the sklearn function, and for the FcNN, the importance is obtained using a Permutation Importance approach implemented in the ELI5 library (25). It should be noted that the performances, as well as the values of the importance produced by the models, are directly affected by the kind of regularization, or lack thereof, employed in the training. This can also be seen in the performance evaluation of the train dataset, which is obtained as the average over the different folds used for the cross-validation. Regularized models (i.e., LASSO and FcNN) tend to perform better while non-regularized models (i.e., RF) have slightly worse performances. It is also worth mentioning that the lack of balance in the training labels particularly affects the performance of the RFs, despite the attempts made to reduce these effects.

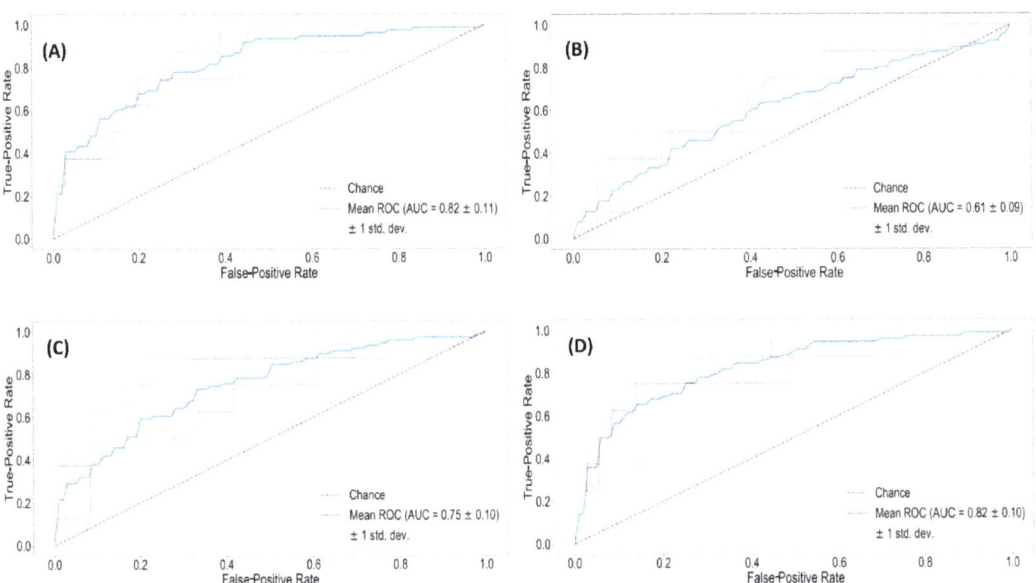

Figure 2. Performance of the Lasso regularized classifier on different input features: (**A**) Clinical, (**B**) Radiological, (**C**) Radiomic, and (**D**) All available features.

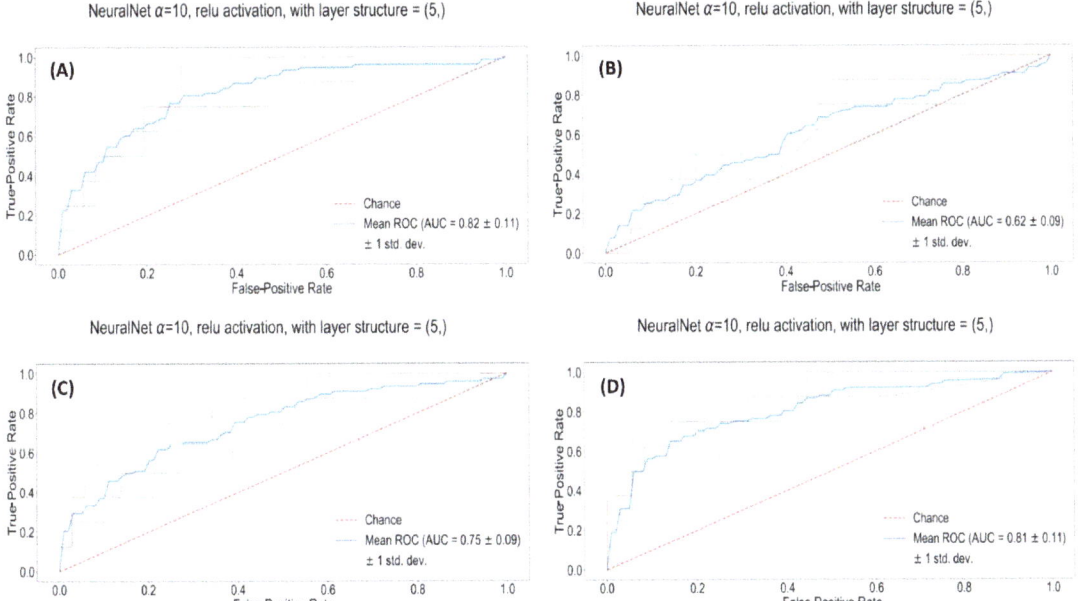

Figure 3. Performance of the FcNN classifier on different input features: (**A**) Clinical, (**B**) Radiological, (**C**) Radiomic, and (**D**) All available features.

Table 2. Table with all the AUCs obtained using the different models during both training and testing. Sensitivity (Sens) and Specificity (Spec) are relative to the testing phase.

	Model Name	Training AUC	Testing AUC	Sens	Spec
Random Forest classifier	Clinical	0.98 ± 0.01	0.63 ± 0.09	44%	83%
	Radiomic	1.00 ± 0.01	0.64 ± 0.08	41%	86%
	Radiological	0.93 ± 0.01	0.49 ± 0.07	19%	79%
	All	1.00 ± 0.01	0.67 ± 0.11	44%	88%
Fully connected Neural Network	Clinical	0.82 ± 0.11	0.82 ± 0.01	76%	75%
	Radiomic	0.83 ± 0.09	0.75 ± 0.01	77%	64%
	Radiological	0.69 ± 0.09	0.62 ± 0.02	63%	56%
	All	0.91 ± 0.11	0.81 ± 0.01	69%	83%
Lasso regularized classifier	Clinical	0.84 ± 0.01	0.82 ± 0.11	69%	83%
	Radiomic	0.81 ± 0.01	0.75 ± 0.10	64%	78%
	Radiological	0.67 ± 0.01	0.61 ± 0.09	70%	51%
	All	0.88 ± 0.01	0.82 ± 0.10	85%	68%

Table 3. Set of variables chosen via Lasso regression with respective weights of the linear combination in the tested version of the model. Arranged in descending order of absolute value; in all cases, the Intercept = 0.178899.

Model Name	Relevant Features (Coefficient)
Clinical	Age (years) (0.116771), Respiratory Rate (0.082292), Sex (−0.037591), Fever (−0.022923)
Radiomic	10th intensity percentile (−0.125094), Intensity-based interquartile range (0.103349), Complexity (−0.102924), Cluster prominence (−0.064690), Area density-aligned bounding box (−0.039374), Entropy (0.033002), Number of compartments (GMM) (−0.032441), Asphericity (0.028517), Local intensity peak (0.028478), Global intensity peak (−0.024832), Intensity range (0.012509), Fat surface (0.007267)
Radiological	Ground-glass opacity (−0.043875), Lung consolidation (0.038143), X-ray Tube Current (−0.017264), kVp (0.004995)
All	Age (years) (0.092963), Intensity-based interquartile range (0.057260), Respiratory Rate (0.049603), Ground-glass opacity (−0.031423), Sex (−0.028895), Complexity (−0.028606), Lung consolidation (0.017272), Fever (−0.016933), X-ray Tube Current (−0.016908), Area density—aligned bounding box (−0.009676), Cluster prominence (−0.006663), Fat surface (0.004984), Number of compartments (GMM) (−0.001448), Local intensity peak (0.000195)

To clarify the impact of regularization on performance of the RF approach, a dataset of reduced dimensionality obtained using the LASSO approach was implemented. However, the resulting performance in terms of AUC of this second attempt (data not shown) remained very similar to the authors' previous attempt. Thus, the sub-optimal result was likely due to the application of this classifier on a strongly imbalanced dataset [27]. In addition, the RF hyper-parameters were chosen using a parametric scan to explore the main possible combination of values, including number of estimators, max depth, max number of features, and oob score. None of these parameter combinations produced any relevant improvement in the RF models when applied to the dataset being tested.

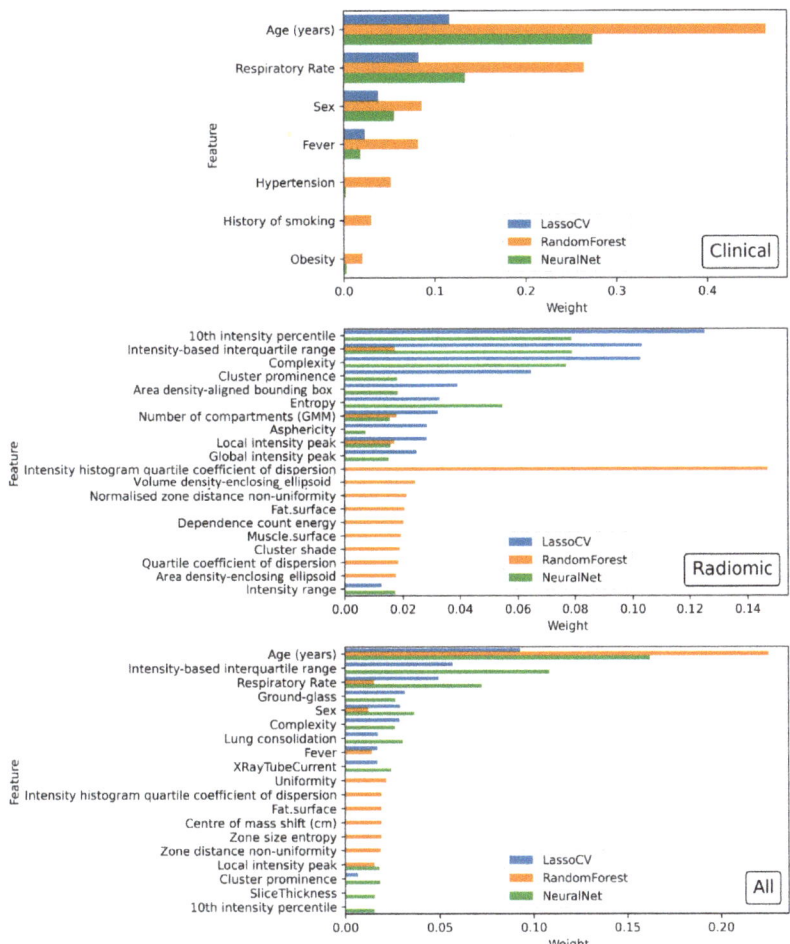

Figure 4. Barplot with the most important features of each model and their respective importance within each model. The barplots for the radiological models are not reported since, in all cases, these did not have any predictive ability at all.

Figure S2 shows an example of how age and SARS-CoV-2 disease affect CT image appearance and grey level inhomogeneities, consequently impacting the values of the radiomic features. One such example is entropy, which did not remain in the final model, being related to patient age. In particular, the entropy values obtained from the images were 8.29, 9.96, 8.22, and 9.97 for the patients illustrated in panels A, B, C, and D, respectively. A and C were both under 70 years of age while B and D were both older. A and B were successfully discharged from the hospital while C and D died from SARS-CoV-2 disease.

These findings suggested investigating the impact of ageing on several relevant radiomic features, as shown in Supplementary Materials Figure S3 (entropy/complexity).

Figure 5 shows the misclassification distribution with respect to patient age (which was, in all cases, the most relevant feature included in the Lasso model). Moreover, from Figure 5, it can be noted that the radiomic and clinical models seem to have different weaknesses while having a slight overlap in patient misclassification.

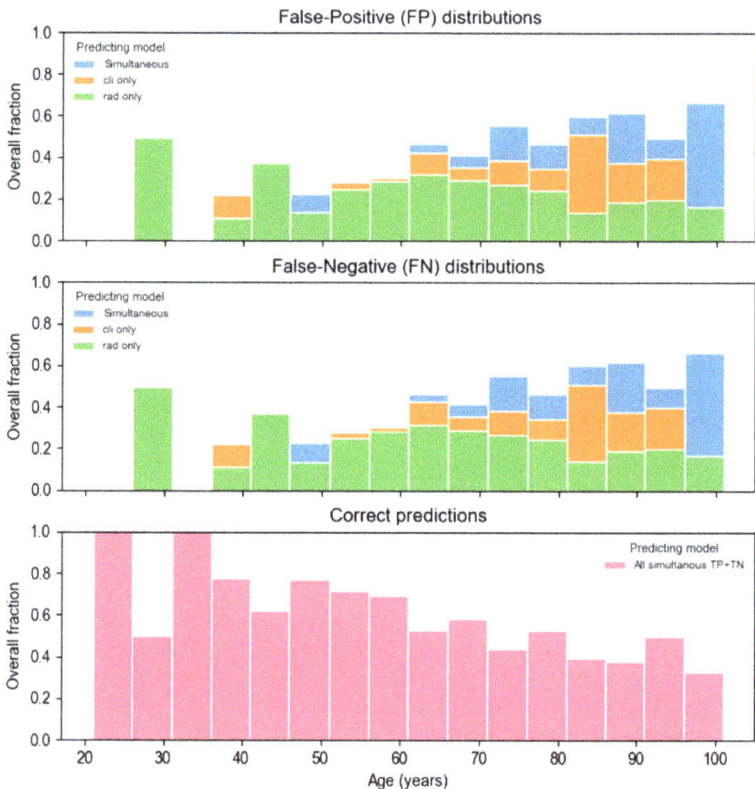

Figure 5. Histogram of different misclassifications individually made by the clinical (Orange), the radiomic (Green), or both simultaneously (Blue) LASSO models. The pink-colored histogram represents the distribution of the simultaneously correct predictions for both the clinical and the radiomic models. False positives are cases in which the model predicts patient mortality and the patient survives; false negatives are cases in which the model predicts that the patient survives, but the patient dies. In all cases, all the bins are normalized with the size of the population in the corresponding age range.

This peculiar behavior suggested that, in the clinical model, the risk for older patients was overestimated (more False Positives) and was somewhat underestimated (more False Negatives) in the younger population, while the opposite was true for the radiomic model. Similar behavior was found for the FcNN classifier, as reported in Supplementary Materials Figure S4.

4. Discussion

The radiological findings extracted from the clinical database and assessed by expert radiologists have, in no instance, proven to be informative regarding the outcome investigated. Correspondingly, in all cases, the radiological model was statistically different from all others as well as the worst performing.

At no point in the analysis did the history as a smoker seem to be relevant within the models in this study, despite what was shown by [8]. This could be due to the present dataset having a high percentage of patients with a smoking history. Although this variable could be indirectly associated with hypertension, respiratory rate, or other clinical variables (i.e., age and sex) in the present dataset, history as a smoker was found not to be correlated above the correlation threshold set at 0.6 before the preprocessing phases.

The set of clinical variables in the present study contained fewer features which attempted to predict prognosis than the majority of those used in the available literature [8,28]. Of note, the clinical model used in the present study had a performance (AUC = 0.82) comparable to that obtained by [28] and slightly worse than that of Shiri et al. [8].

In concordance with what was found in [29], the present model outperformed the radiological assessment obtained by the expert radiologists who took part in the study cited.

As one would have expected from the World Health Organization (WHO) guidelines [30], when included in the present dataset, age was the most relevant variable in the model, followed by respiratory rate and sex.

In this study, three different ML models were investigated in terms of ability to predict the relevant clinical outcome, i.e., death. The combination of the segmentation method with predictive models was chosen with the intent of identifying the most important predictive features while keeping the interpretation of the results as simple as possible and facilitating their application in clinical practice. The authors recognized that Convolutional Neural Networks could be applied to image analysis and segmentation [31]. However, these approaches require large computational power as well as large training datasets and can be of difficult interpretability, often resulting as black boxes [32].

Looking at Figure 4, it can be noted that the features contributing the most to the compared models are the same irrespective of the algorithm adopted which included age (years), respiratory rate, ground glass (GGO), and intensity-based interquartile range.

It should also be noted that the relative importance (weight) of the features in each model was similar in the two models (i.e., Lasso and FcNN) which better described the present cohort. Their performance, as well as the magnitude of the importance estimated, can be attributed to the regularized nature of the methods utilized.

Furthermore, these findings supported the presence of an association between patient outcome, clinical parameters (e.g., age and respiratory rate), and radiological (e.g., GGO) and hidden image properties not noticeable by the human eye but requiring ad hoc computation (e.g., intensity-based interquartile range). All of the above features enabled taking into consideration the deterioration of lung tissues related to SARS-CoV-2 disease as well as the ageing process.

The most relevant radiomic variables in the model used in the present study were related to the Gray-Level distribution and disorder/inhomogeneity in the image (i.e., entropy, complexity, 10th intensity percentile). Some of these features were found in models developed by [28] and were also informative in a univariate analysis carried out by [33].

As expected, looking at the same univariate analysis as in the study of [33], the performance of a more complex model is consistently better than that of a single radiomic variable.

The dimensionality (2D vs. 3D) of the images probably affected performance; in fact, the present models consistently outperformed those obtained using radiographic chest images as in the studies of [6,28,33].

The authors hypothesized that ageing of pulmonary tissue may affect several of the relevant radiomic features left after the LASSO feature reduction, as shown in Supplementary Materials Figure S3. Unfortunately, the current dataset used in the study did not allow discriminating the impact of lung tissue ageing, even when using the Neural Network approach.

Figure S2 highlights how disorder and inhomogeneity in the grey levels are related to damage in the lungs as well as to the age of the patient. To the best of the authors' knowledge, this has not previously been highlighted.

As a final consideration, it is important to note that the semi-automatic segmentation tool significantly reduced human costs in terms of manpower and time with respect to a manual approach. Moreover, the segmentation of a single patient may require from 10 to 60 min when performed manually against the 2–6 min necessary with an automatic tool, depending on computer and software specifics. It is noteworthy that manual segmentation, which is feasible only with small patient cohorts, may achieve a slightly better predictive performance [6,8]. On the other hand, the time utilized by trained radiologists to manually segment all the chest CT images may be unavailable in a busy department, especially during pandemic events.

Some of the limitations of the present study include the imbalanced nature of the majority of the clinical variables available as well as the reduced number of clinical features available.

However, this may also represent one of the strong points of the study, since it showed that, even with a basic amount of information, it was still possible to obtain acceptable results.

Another similar point is that of the length of time from the date of the CT scan to the outcome. It is a clear limitation since only the first CT was considered, hence concealing all the disease progression after the first scan. However, it showed that it was possible to have a quick and reliable evaluation of patients at admission, allowing better allocation of hospital resources.

Some future prospects in this regard may include an additional analysis of the dataset in a delta-radiomics setting in which disease progression is also included in the patient evaluation by looking at the changes in radiomic features in successive CT scans.

Another interesting prospect would be to additionally investigate the relationship between patient characteristics, such as age, and radiomic variables extracted from various organs.

5. Conclusions

The present study pointed out that semi-automatic segmentation tools allowed the extraction of the radiomic features, which allowed the construction of predictive ML models, having a performance not reaching those obtained using clinical variables but more accurate than the models based on radiological findings. The models developed could provide valuable support to clinicians and radiologists in discerning CT-based RFs representative of the extension and severity of areas affected by SARS-CoV-2.

Supplementary Materials: The following supporting information can be downloaded at: https://www.mdpi.com/article/10.3390/app12094493/s1, Figure S1: Performance of the RFC on different input features: (A) Clinical, (B) Radiological, (C) Radiomic, or (D) All available features; Figure S2: CT axial images from different patients. In first column (A,C), the patients are both under 70 years of age while, in the second column (B,D), both are older. Similarly, patients in the same row (A,B and C,D) were alive and deceased at the end of the follow-up, respectively; Figure S3: Behavior of relevant radiomic features (A) Entropy and (B) Complexity as a function of Age. The patients are also represented as a Misclassification group, using the marker in the plot, and by outcome, by color, as Deceased (Red) or Alive (Blue); Figure S4. Histogram of different misclassifications individually made by the clinical model (Orange), the radiomic model (Green), or both simultaneously (Blue) according to the FcNN. The pink-colored histogram represents the distribution of the simultaneously correct predictions for both the clinical and the radiomic models. False positives are cases in which the model predicts patient mortality and the patient survives; false negatives are cases in which the model predicts that the patient survives, but the patient dies. In all cases, all the bins are normalized with the size of the population in the corresponding age range; Table S1: Description of follow-up lengths for various groups of patients in the study.

Author Contributions: Conceptualization, L.S. (Lorenzo Spagnoli), F.C. (Francesca Coppola), P.V., R.G. and L.S. (Lidia Strigari); methodology and formal analysis, L.S. (Lorenzo Spagnoli), M.S., G.P., E.G., N.C. and L.S. (Lidia Strigari); investigation, L.S. (Lorenzo Spagnoli), F.C. (Francesca Coppola), P.V., R.G., E.G., N.C. and L.S. (Lidia Strigari); resources, R.G.; data curation, F.C. (Federica Ciccarese), M.B., G.V., N.B., M.F.M. and F.C. (Francesca Coppola); writing—original draft preparation, L.S. (Lorenzo Spagnoli), E.G., M.F.M., G.P., M.S. and L.S. (Lidia Strigari); writing—review and editing, all authors; supervision, F.C. (Francesca Coppola), E.G. and L.S. (Lidia Strigari); funding acquisition, L.S. (Lidia Strigari) and R.G. All authors have read and agreed to the published version of the manuscript.

Funding: This research was funded by the S. Orsola Polyclinic Foundation (Fondazione Policlinico S. Orsola) of Bologna, Italy. The APC was funded by the Alma Mater Studiorum University of Bologna, Specialization in Radiology Bologna, Italy.

Institutional Review Board Statement: This study regarding the prognostic value of radiomic features was conducted and included all the patients suitable for analysis, according to the guidelines of the Declaration of Helsinki. The study was approved by the Institutional Review Board (or Ethics Committee) of IRCCS University Hospital of Bologna (protocol code No. EM949-2020_507/2020/Oss/AOUBo, approved on: 16 September 2020).

Informed Consent Statement: Written informed consent was obtained from all patients before publishing this paper.

Data Availability Statement: Data will be available after reasonable request to the corresponding author.

Acknowledgments: We would like acknowledge Eng. Stefano Vezzani from the S. Orsola Polyclinic Foundation of Bologna, Italy.

Conflicts of Interest: The authors declare no conflict of interest.

References

1. Corman, V.M.; Landt, O.; Kaiser, M.; Molenkamp, R.; Meijer, A.; Chu, D.K.; Bleicker, T.; Brünink, S.; Schneider, J.; Schmidt, M.L.; et al. Detection of 2019 novel coronavirus (2019-nCoV) by real-time RT-PCR. *Eurosurveillance* **2020**, *25*, 2000045. [CrossRef] [PubMed]
2. Ai, T.; Yang, Z.; Hou, H.; Zhan, C.; Chen, C.; Lv, W.; Tao, Q.; Sun, Z.; Xia, L. Correlation of Chest CT and RT-PCR Testing for Coronavirus Disease 2019 (COVID-19) in China: A Report of 1014 Cases. *Radiology* **2020**, *296*, E32–E40. [CrossRef] [PubMed]
3. Revel, M.P.; Parkar, A.P.; Prosch, H.; Silva, M.; Sverzellati, N.; Gleeson, F.; Brady, A. COVID-19 patients and the radiology department—Advice from the European Society of Radiology (ESR) and the European Society of Thoracic Imaging (ESTI). *Eur. Radiol.* **2020**, *30*, 4903–4909. [CrossRef] [PubMed]
4. Zwanenburg, A.; Vallières, M.; Abdalah, M.A.; Aerts, H.; Andrearczyk, V.; Apte, A.; Ashrafinia, S.; Bakas, S.; Beukinga, R.J.; Boellaard, R.; et al. The Image Biomarker Standardization Initiative: Standardized Quantitative Radiomics for High-Throughput Image-based Phenotyping. *Radiology* **2020**, *295*, 328–338. [CrossRef]
5. Wang, S.; Dong, D.; Li, L.; Li, H.; Bai, Y.; Hu, Y.; Huang, Y.; Yu, X.; Liu, S.; Qiu, X.; et al. A Deep Learning Radiomics Model to Identify Poor Outcome in COVID-19 Patients With Underlying Health Conditions: A Multicenter Study. *IEEE J. Biomed. Health Inform.* **2021**, *25*, 2353–2362. [CrossRef]
6. Ke, Z.; Li, L.; Wang, L.; Liu, H.; Lu, X.; Zeng, F.; Zha, Y. Radiomics analysis enables fatal outcome prediction for hospitalized patients with coronavirus disease 2019 (COVID-19). *Acta Radiol.* **2021**, *63*, 319–327. [CrossRef]
7. Xiao, F.; Sun, R.; Sun, W.; Xu, D.; Lan, L.; Li, H.; Liu, H.; Xu, H. Radiomics analysis of chest CT to predict the overall survival for the severe patients of COVID-19 pneumonia. *Phys. Med. Biol.* **2021**, *66*, 10. [CrossRef]
8. Shiri, I.; Sorouri, M.; Geramifar, P.; Nazari, M.; Abdollahi, M.; Salimi, Y.; Khosravi, B.; Askari, D.; Aghaghazvini, L.; Hajianfar, G.; et al. Machine learning-based prognostic modeling using clinical data and quantitative radiomic features from chest CT images in COVID-19 patients. *Comput. Biol. Med.* **2021**, *132*, 104304. [CrossRef]
9. Wang, D.; Huang, C.; Bao, S.; Fan, T.; Sun, Z.; Wang, Y.; Jiang, H.; Wang, S. Study on the prognosis predictive model of COVID-19 patients based on CT radiomics. *Sci. Rep.* **2021**, *11*, 11591. [CrossRef]
10. Li, C.; Dong, D.; Li, L.; Gong, W.; Li, X.; Bai, Y.; Wang, M.; Hu, Z.; Zha, Y.; Tian, J. Classification of Severe and Critical COVID-19 Using Deep Learning and Radiomics. *IEEE J. Biomed. Health Inform.* **2020**, *24*, 3585–3594. [CrossRef]
11. Gamberini, L.; Coniglio, C.; Cilloni, N.; Semeraro, F.; Moro, F.; Tartaglione, M.; Chiarini, V.; Lupi, C.; Bua, V.; Gordini, G. Remodelling of a regional emergency hub in response to the COVID-19 outbreak in Emilia Romagna. *Emerg. Med. J.* **2021**, *38*, 308. [CrossRef] [PubMed]
12. Bettinelli, A.; Marturano, F.; Avanzo, M.; Loi, E.; Menghi, E.; Mezzenga, E.; Pirrone, G.; Sarnelli, A.; Strigari, L.; Strolin, S.; et al. A Novel Benchmarking Approach to Assess the Agreement among Radiomic Tools. *Radiology* **2022**, 211604. [CrossRef] [PubMed]
13. Van Rossum, G.A.D.; Fred, L. *Python 3 Reference Manual*; CreateSpace: Scotts Valley, CA, USA, 2009.
14. Pedregosa, F.; Varoquaux, G.; Gramfort, A.; Michel, V.; Thirion, B.; Grisel, O.; Blondel, M.; Prettenhofer, P.; Weiss, R.; Dubourg, V.; et al. Scikit-learn: Machine learning in Python. *J. Mach. Learn. Res.* **2011**, *12*, 2825–2830.
15. Lemaître, G.; Nogueira, F.; Aridas, C.K. Imbalanced-learn: A Python Toolbox to Tackle the Curse of Imbalanced Datasets in Machine Learning. *J. Mach. Learn. Res.* **2017**, *18*, 1–5.
16. McKinney, W.A.O. Data Structures for Statistical Computing in Python. In Proceedings of the 9th Python in Science Conference, Austin, TX, USA, 28 June–3 July 2010; Volume 445, pp. 51–56.
17. Harris, C.R.; Millman, K.J.; van der Walt, S.J.; Gommers, R.; Virtanen, P.; Cournapeau, D.; Wieser, E.; Taylor, J.; Berg, S.; Smith, N.J.; et al. Array programming with NumPy. *Nature* **2020**, *585*, 357–362. [CrossRef] [PubMed]
18. Virtanen, P.; Gommers, R.; Oliphant, T.E.; Haberland, M.; Reddy, T.; Cournapeau, D.; Burovski, E.; Peterson, P.; Weckesser, W.; Bright, J.; et al. SciPy 1.0: Fundamental algorithms for scientific computing in Python. *Nat. Methods* **2020**, *17*, 261–272. [CrossRef]
19. Fan, A.; Jernite, Y.; Perez, E.; Grangier, D.; Weston, J.; Auli, M. ELI5: Long Form Question Answering. *arXiv* **2019**, arXiv:1907.09190.
20. Hunter, J.D. Matplotlib: A 2D Graphics Environment. *Comput. Sci. Eng.* **2007**, *9*, 90–95. [CrossRef]

21. Waskom, M.; Botvinnik, O.; O'Kane, D.; Hobson, P.; Lukauskas, S.; Gemperline, D.C.; Augspurger, T.; Halchenko, Y.; Cole, J.B.; Warmenhoven, J.; et al. *Mwaskom/Seaborn: v0.8.1 (September 2017)*; Zenodo: Geneva, Switzerland, 2017.
22. Tibshirani, R. Regression Shrinkage and Selection via the Lasso. *J. R. Stat. Society. Ser. B* **1996**, *58*, 267–288. [CrossRef]
23. Tin Kam, H. Random decision forests. In Proceedings of the 3rd International Conference on Document Analysis and Recognition, Montreal, QC, Canada, 14–16 August 1995; Volume 271, pp. 278–282.
24. Haykin, S. *Neural Networks: A Comprehensive Foundation*; Prentice Hall PTR: Lebanon, IN, USA, 1994.
25. Chawla, N.; Bowyer, K.; Hall, L.; Kegelmeyer, W. SMOTE: Synthetic Minority Over-sampling Technique. *J. Artif. Intell. Res. (JAIR)* **2002**, *16*, 321–357. [CrossRef]
26. Balacchi, C.; Brandi, N.; Ciccarese, F.; Coppola, F.; Lucidi, V.; Bartalena, L.; Parmeggiani, A.; Paccapelo, A.; Golfieri, R. Comparing the first and the second waves of COVID-19 in Italy: Differences in epidemiological features and CT findings using a semi-quantitative score. *Emerg. Radiol.* **2021**, *28*, 1055–1061. [CrossRef] [PubMed]
27. Boulesteix, A.L.; Janitza, S.; Kruppa, J.; König, I.R. Overview of random forest methodology and practical guidance with emphasis on computational biology and bioinformatics. *Wiley Interdiscip. Rev. Data Min. Knowl. Discov.* **2012**, *2*, 493–507. [CrossRef]
28. Bae, J.; Kapse, S.; Singh, G.; Gattu, R.; Ali, S.; Shah, N.; Marshall, C.; Pierce, J.; Phatak, T.; Gupta, A.; et al. Predicting Mechanical Ventilation and Mortality in COVID-19 Using Radiomics and Deep Learning on Chest Radiographs: A Multi-Institutional Study. *Diagnostics* **2021**, *11*, 1812. [CrossRef] [PubMed]
29. Homayounieh, F.; Ebrahimian, S.; Babaei, R.; Mobin, H.K.; Zhang, E.; Bizzo, B.C.; Mohseni, I.; Digumarthy, S.R.; Kalra, M.K. CT Radiomics, Radiologists, and Clinical Information in Predicting Outcome of Patients with COVID-19 Pneumonia. *Radiol. Cardiothorac. Imaging* **2020**, *2*, e200322. [CrossRef]
30. World Health Organization, Coronavirus Disease (COVID-19) Advice for the Public, Mythbusters. Available online: https://www.who.int/emergencies/diseases/novel-coronavirus-2019/advice-for-public/myth-busters (accessed on 10 March 2022).
31. Manco, L.; Maffei, N.; Strolin, S.; Vichi, S.; Bottazzi, L.; Strigari, L. Basic of machine learning and deep learning in imaging for medical physicists. *Phys. Med.* **2021**, *83*, 194–205. [CrossRef]
32. Santoro, M.; Strolin, S.; Paolani, G.; Della Gala, G.; Bartoloni, A.; Giacometti, C.; Ammendolia, I.; Morganti, A.G.; Strigari, L. Recent Applications of Artificial Intelligence in Radiotherapy: Where We Are and Beyond. *Appl. Sci.* **2022**, *12*, 3223. [CrossRef]
33. Varghese, B.A.; Shin, H.; Desai, B.; Gholamrezanezhad, A.; Lei, X.; Perkins, M.; Oberai, A.; Nanda, N.; Cen, S.; Duddalwar, V. Predicting clinical outcomes in COVID-19 using radiomics on chest radiographs. *Br. J. Radiol.* **2021**, *94*, 20210221. [CrossRef]

Article

Deep Variational Embedding Representation on Neural Collaborative Filtering Recommender Systems

Jesús Bobadilla, Jorge Dueñas, Abraham Gutiérrez * and Fernando Ortega

Departamento de Sistemas Informáticos, Escuela Técnica Superior de Ingeniería de Sistemas Informáticos, Universidad Politécnica de Madrid, 28031 Madrid, Spain; jesus.bobadilla@upm.es (J.B.); jorge.duenas.lerin@gmail.com (J.D.); fernando.ortega@upm.es (F.O.)
* Correspondence: abraham.gutierrez@upm.es

Featured Application: This paper proposes a new deep learning design to obtain accurate plots of RS information. The innovative model incorporates embedding layers of small (representable) sizes, variational layers to improve the latent space and to spread samples, and a Euclidean similarity measure to place samples according to the intuitive human interpretation of distances.

Abstract: Visual representation of user and item relations is an important issue in recommender systems. This is a big data task that helps to understand the underlying structure of the information, and it can be used by company managers and technical staff. Current collaborative filtering machine learning models are designed to improve prediction accuracy, not to provide suitable visual representations of data. This paper proposes a deep learning model specifically designed to display the existing relations among users, items, and both users and items. Making use of representative datasets, we show that by setting small embedding sizes of users and items, the recommender system accuracy remains nearly unchanged; it opens the door to the use of bidimensional and three-dimensional representations of users and items. The proposed neural model incorporates variational embedding stages to "unpack" (extend) embedding representations, which facilitates identifying individual samples. It also replaces the join layers in current models with a Lambda Euclidean layer that better catches the space representation of samples. The results show numerical and visual improvements when the proposed model is used compared to the baselines. The proposed model can be used to explain recommendations and to represent demographic features (gender, age, etc.) of samples.

Keywords: embedding; collaborative filtering; variational method; deep learning; recommender systems; recommendation explanations; data visual interpretation

1. Introduction

Recommender Systems (RS) [1] are machine learning-based personalization applications. They facilitate human/machine integration by providing accurate recommendations of items to users; mainly, items are products or services recommended to collaborative clients. Remarkable commercial companies that incorporate RS are Spotify, Netflix, TripAdvisor, and Amazon. RS can be classified according to their filtering strategy: demographic [2], content-based [3], context-aware [4], social [5], collaborative [6] and different ensembles [7]. Of the mentioned filtering approaches, the Collaborative Filtering (CF) is the most relevant since it returns the most accurate predictions and recommendations. The first CF implementations made use of the memory-based K-Nearest Neighbors (KNN) algorithm [8] due to its simplicity and because it conceptually fits with the recommendation task. Nevertheless, the KNN algorithm has some drawbacks when applied to CF RS: it is not accurate enough and it is not efficient, since successive executions are necessary to make successive recommendations. For these reasons, the KNN approach was replaced by model-based methods, such as Matrix Factorization (MF) [9], non-Negative Matrix Factorization (NMF) [10] and Bayesian NMF (BNMF) [11].

MF models generate hidden factors for both users and items. Hidden factors can be considered as embedding representations. The rating prediction of each item to each user is obtained by just making the dot product of both (user and item) embeddings. MF models are not specifically designed to plot visual representations, but their embedding values can be processed to provide recommendation explanations [12] and to draw relationships [13]. Currently, MF models are being replaced by neural networks (NN) approaches, and consequently, hidden factors are replaced by neural embedding layers. The most relevant NN models in the CF area are DeepMF [14] and Neural Collaborative Filtering (NCF) [15]. DeepMF makes a deep learning implementation of the MF model; thus, it contains a user embedding layer, an item embedding layer, and a 'Dot' layer to join the preceding embeddings. NCF replaces the DeepMF 'Dot' layer with an MLP and eventually combines deep and shallow learning.

Embeddings are abstract, low-dimensional representations of information. There are a large number of fields where embeddings are used to encode data structures; network embeddings are largely applied to graphs [16], where they embed entities and relationships in low-dimensional spaces [17]. Gene sequences have been predicted from embeddings [18], and biomedical networks [19] have been evaluated as social graphs. Social communities are detected in the embedding-based Silhouette [20], via clustering of network node embeddings. Image applications are also a recurrent target for embedding processing, such as the person identification based on pose invariant embedding [21], image tag refinement through deep collaborative embedding [22], and handcrafted image retrieval [23] using supervised deep feature embedding. Natural language processing is also an area where tokens must be coded through different types of embedding models. Beyond the most known word2vec model, there are specific models, such as the convolution-deconvolution fusion word embedding [24], which makes a fusion of context and task information. Finally, fairness in RS is improved by means of an embedding-based combination of MF and deep learning models [25].

In the CF area, embedding layers have been used to implement autoencoders, such as the probabilistic autoencoder in [26] fed with the user-item data, the combination of stacked convolutional autoencoders, and stacked denoising autoencoders [27] to extract knowledge in RS. Context-aware information is coded using a deep learning autoencoder [28] that predicts scores and extracts features. However, the usual embedding-based architecture in the RS area exploits collaborative relationships, such as in [29] where they use embeddings to code user-item bipartite graphs for recommendation and representation learning. Relationships are also managed by means of low-dimensional dense embeddings learned from the sparse features in a wide and deep RS architecture [30]. A k-partite graph is used in [31] to characterize several types of information in recommendation tasks, and embeddings for different kinds of information are projected in the same latent space. A collaborative user network embedding has been proposed for social RS, where the cold start problem is addressed by combining MF and Bayesian personalized ranking [32]. A method to automatically set embedding sizes in RS [33] is based on the use of a reinforcement learning agent that adaptatively selects adequate sizes. Finally, internal embedding information is combined in [34] to obtain prediction and recommendation reliabilities.

Using deep learning variational approaches, we obtain wider, more representative, and more robust latent spaces and embedding representations. Neural autoencoders are, in certain cases, reinforced with a variational stage, building Variational Auto-Encoder (VAE) models. Mainly, VAEs are particularly applied to the image processing area, e.g., reconstructing images [35], creating super-resolution images by encoding low-resolution images in a dense latent space vector [36], and reducing blurring by adding a conditional sampling mechanism [37]. This paper proposes adding variational layers to the neural model suggested to improve the latent space where the embedding samples are located. It mimics the underlying VAE operative to obtain super-resolution images, reducing blurring, and handling low-resolution samples: using VAE, the latent space is enriched, and samples are spread. Enriched embeddings are used in image processing to decode high-resolution images, unblurred images, etc., whereas we propose the use of enriched embeddings to improve the visual representation of RS information.

From the explained research, this paper proposes an innovative deep learning model that incorporates two embedding layers: one for code users and the other for code items. Both embeddings will have small sizes to make it possible to draw bi- or three-dimensional graphs of user and item samples. The accuracy loss caused by the small embedding sizes (two or three neurons each embedding) will be tested in the paper. The proposed model also incorporates a variational stage, designed to spread the latent space where item and user embeddings are represented. Both user and item embeddings will be followed by their own Gaussian variational layers whose parameter values are learned in the whole neural model. The expected results are accurate low-dimensional item and user graphs, where samples are spread in a latent space area and not 'compressed' in a reduced space region, making it easier to discriminate between adjacent samples. Finally, a 'Lambda' join layer is added to the model to implement the Euclidean distance between the embeddings of the items and the embeddings of the users. This layer replaces the 'Dot' product layer of the traditional DeepMF model or the MLP stage of the NCF model. The Euclidean Lambda layer's purpose is to keep near to related user or item embeddings and to keep far from nonrelated user or item embeddings, such as humans intuitively understand distances.

In short, this paper proposes a new deep learning design to obtain accurate plots of RS information. The innovative model incorporates embedding layers of small (representable) sizes, variational layers to improve the latent space and to spread samples, and a Euclidean similarity measure to place samples according to the intuitive human interpretation of distances. Experiments have been conducted using representative CF data sets to test the proposed model. The rest of the paper has been structured as follows: In Section 2 the proposed model is explained, Section 3 shows the experiments' design and results, Section 4 the results are discussed. Finally, Section 5 contains the main conclusions of the paper and future work.

2. Models and Methods

The current deep CF state of the art includes two remarkable neural models: DeepMF (Figure 1a) and NCF (Figure 1b). As shown in Figure 1, both DeepMF (Figure 1a) and NCF (Figure 1b) models provide two embedding layers: the first codes users and the second codes items. These are the embeddings that this paper addresses. DeepMF (Figure 1a) uses only a dot product to combine user and item factors, as well as the MF machine learning method. It is simple and it provides accurate results; nevertheless, it does not catch the nonlinear complex relations existing among users and items embedding. To solve the drawback mentioned, the NCF model (Figure 1b) incorporates an MLP that non-linearly combines factors of the user and the item, returning scalar regression values (predictions). Previously, a concatenate layer joined the embedding values of the user and the item and provided a single tensor flow to the MLP.

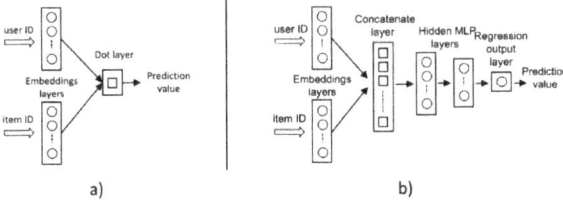

Figure 1. Deep matrix factorization (**a**) vs. neural collaborative filtering (**b**) recommender system models.

The embedding layers of the existing CF models are not designed for visual representations due to the following reasons: (1) they are vectors of excessive large sizes to be visualized, (2) their values tend to cluster in small representation areas, and (3) The neural learning process does not consider visually understandable similarity measures (such as the Euclidean distance). To tackle the aforementioned drawbacks, we will provide three different contributions: (1) Testing the accuracy impact of reducing the embedding sizes just to two or three dimensions, (2) Expanding the embedding values representation by using the variational approach, and (3) Incorporating the Euclidean similarity measure in the deep neural model.

Contribution 1: In the CF field, it is particularly useful to visually represent users and items in such a way that clients, company managers, and technical staff can understand the existing relations among users, items, and between both users and items. This leads us to code users and items using only two or three dimensions. The key question here is: is it affordable for the accuracy we will lose in the process? As we will show in the next section, the answer is yes, tested datasets show little significant accuracy decrease.

Contribution 2: We borrow the variational method from the variational autoencoder field; they expand the embedding representation of samples, making it possible to improve clustering and classification, and to return a progressive morphing when needed. Figure 2 explains the variational approach, where each sample embedding (white circle) can be probabilistically located (grey circles) nearby (green circle) to its nonvariational fixed location (white circle). Variational methods are usually implemented by setting a Gaussian distribution in each embedding dimension. The defined set of parameters of the Gaussian distributions (blue and orange distributions) establishes the probabilistic area where the samples lay out. Our neural model specifically learns both the mean and variance of each Gaussian distribution.

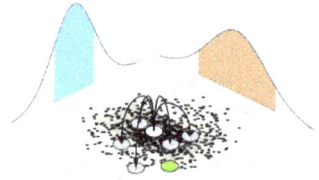

Figure 2. Representation of the variational method, where each sample embedding (white circle) is probabilistically placed (grey circles) at a nearby position (green circle).

As explained, the variational approach expands the area where the sample embeddings lay out. This is particularly adequate for our embedding representation goal since it will make it easier to visually catch our attention on the existing sample relations. As an example, in Figure 3 we show the variational result (embeddings) of the proposed model applied to the MNIST dataset, where samples have been stochastically spread to make the classification of the classes easier. Figure 3 left and right graphs show, respectively, the obtained latent space and its cumulative normal distribution. The cumulative normal distribution is frequently used to support generative tasks; in this case, it can be used to generate fake embeddings, and then to obtain fake samples (MNIST numbers). In the CF area, this opens the door to implementing augmentation data and to obtaining augmented RS datasets.

Contribution 3: Traditional DeepMF and NCF models implement, respectively, a dot layer and an MLP network (Figure 1). Both approaches (dot layer and MLP network) can be considered as similarity functions, and none of them are designed to arrange embedding representations in a visual disposition. Our proposed model replaces these functions with a visually convenient similarity measure: the Euclidean distance. It will set the embedding representation of samples in such a way that similar samples will be located at nearby locations. It is expected that what is gained in understanding the RS representation is not lost in the accuracy of the CF.

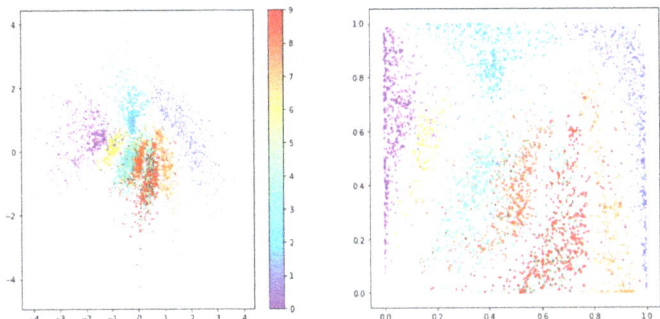

Figure 3. Representation of a VAE latent space for the MNIST dataset (left side) and its cumulative normal distribution (right side).

By combining the three mentioned contributions, we have designed the deep neural models shown in Figure 4. The user model (orange) and the item model (blue) are conceptually identical: their first stage "embedding layers" (bottom-left of Figure 4) is an embedding layer that maps user or item IDs to coded values. It is expected that users with similar behavior (similar casted votes) will be assigned similar embedding values. Same for items; items similarly voted will be coded in an equivalent way. Please note that an embedding size of two or three neurons is expected to adequately capture the diversity of the existing sets of users and items in the recommender system. In this case, we will be able to visually represent users and items by drawing graphs in two or three dimensions.

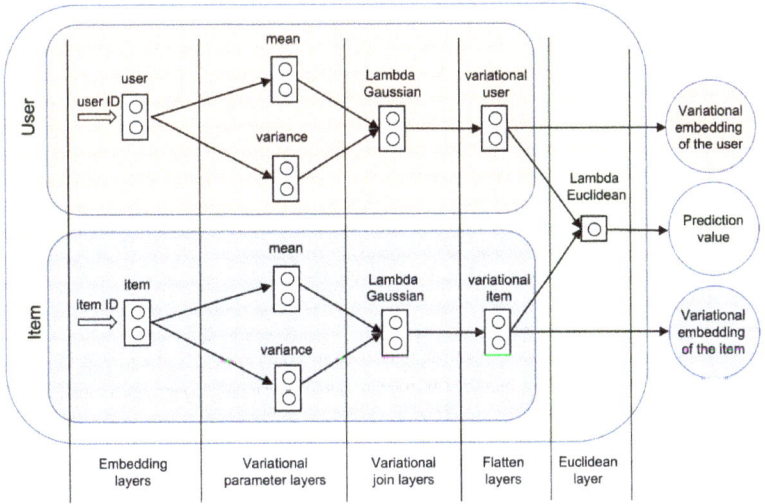

Figure 4. Proposed models: Green: collaborative filtering prediction; Orange: variational Euclidean model for users; Blue: variational Euclidean model for items.

The next stage of the proposed model: 'variational parameter layers', at the bottom of Figure 4, is responsible for learning the most adequate values of the Gaussian distributions that implement the variational behavior of our model (Figure 2). We split the user embedding into two separated tensor flows, implemented through both the 'mean' layer and the 'variance' layer, providing us with the mean and variance of each Gaussian distribution (two or three distributions, in our case). We also split the item embedding into two separated tensor flows. The user mean and the user variance layers must be combined to obtain the user variational embedding (same for the item to obtain the item variational embedding). To implement the Figure 2 operation a 'Lambda' layer is used; this layer makes the variational sample generation. Each sample is stochastically generated attending to the Gaussian distributions that the model has learned; in the Figure 2 example, the generated sample has more probability to be spread through the orange Gaussian distribution than the blue one, since the orange one has a higher variance.

Please note that the variational sample has the same vector size as the user (or item) embedding, its 'mean' layer and its 'variance' layer. Finally, "Flatten" layers are added to the model to reshape data to unidimensional users and item vectors (of size 2 or 3).

The parallel user and item flows (orange and blue ones) provide both the user variational vector and the item variational vector ("Flatten layers" stage in Figure 4). Traditionally, they would be merged using a dot product or an MLP model (Figure 1). As explained, instead, we will incorporate a 'Lambda' layer that implements the Euclidean similarity measure ("Euclidean layer' in the bottom right of Figure 4). It will force the main model (the green one) to arrange variational user embeddings and variational item embeddings in a joined spatial area susceptible to being visually represented and easily understandable to humans. We use the regression model (green) to make training; once the model is trained, we can easily predict user variational embeddings from user IDs (orange model), and item variational embeddings from item IDs (blue model). It is important to stress that the proposed model is not designed to improve prediction accuracy (green model). The model is designed to obtain visually understandable representations of the users and the items embeddings (orange and blue inner models).

To get a deeper understanding of the proposed model, Code 1 provides the Keras/Python implementation of the model kernel.

Code 1. Keras/Python kernel of the proposed model.

```
def sampling(args):
z_mean, z_var = args
epsilon = K.random_normal(shape=(1, latent_dim), mean=0., stddev=1)
return z_mean + K.exp(z_var) * epsilon

def euclidean(args):
movie_v, user_v = args
return K.sqrt(K.sum(K.square(movie_v - user_v), axis=−1))

def variational_Euclidean(latent_dim):
user_input = Input(shape=[1])
user_embedding = Embedding(num_users + 1, latent_dim)(user_input)
user_embedding_mean = Dense(latent_dim) (user_embedding)
user_embedding_var = Dense(latent_dim) (user_embedding)
user_embedding_z = Lambda(sampling) ([user_embedding_mean, user_embedding_var])
user_vec = Flatten()(user_embedding_z)
movie_input = Input(shape=[1])
movie_embedding = Embedding(num_movies + 1, latent_dim)(movie_input)
movie_embedding_mean = Dense(latent_dim) (movie_embedding)
movie_embedding_var = Dense(latent_dim) (movie_embedding)
movie_embedding_z = Lambda(sampling) ([movie_embedding_mean,
movie_embedding_var], latent_dim)
movie_vec = Flatten()(movie_embedding_z)
similar = Lambda(euclidean)([movie_vec, user_vec])
var_eucl_pred = Model([user_input, movie_input], similar)
var_eucl_user = Model(user_input, user_vec)
var_eucl_item = Model(movie_input, movie_vec)
return var_eucl_pred, var_eucl_user, var_eucl_item
```

3. Experiments and Results

To run the designed experiments, we have chosen a set of open and representative CF databases. Table 1 shows the main parameter values of the selected datasets: MovieLens 100K [38], MovieLens1M [38], and a subset (Netflix*) of the Netflix database [39]. Please note the high number of Netflix* users compared to the MovieLens datasets. The chosen datasets have a similar structure, where their kernel is the CF information of ratings stored in files containing tuples: <user_id, item_id, rating>. Basically, they differ from each other in their sizes: number of users, items, and ratings. Additionally, the combination of the previous values determines the sparsity of the CF data. Please note that MovieLens 100K and MovieLens 1M not only differ in their number of ratings, but also in the number of users and items, and consequently in their sparsity (Table 1). Since the MovieLens 1M version is richer than the MovieLens 100K, its accuracy will be also better, as we will see in Table 2.

Table 1. Values of the main parameters of the tested datasets.

Dataset	#Users	#Items	#Ratings	Scores	Sparsity
Movielens 100K	943	1682	99,831	1 to 5	93.71
MovieLens 1M	6040	3706	911,031	1 to 5	95.94
Netflix*	23,012	1750	535,421	1 to 5	98.68

Table 2. Mean absolute error results using the proposed variational Euclidean method (embedding sizes = 2, 3, 5, 10 and the comparative accuracy obtained by setting an embedding size 2 versus an embedding size 10. The lower the MAE values, the better the result.

Dataset\Embedding Size	Mean Absolute Error (MAE)				Achieved Accuracy
	2	3	5	10	
Movielens 100K	0.7355	0.7368	0.7297	0.7213	98.04%
MovieLens 1M	0.6927	0.6875	0.6839	0.6801	98.15%
Netflix*	0.7260	0.7250	0.7248	0.7243	99.76%

From the aforementioned contributions, we will provide three different experiments to substantiate the proposed neural model: (1) CF quality impact by setting different embedding sizes, (2) numerical improvement of the proposed model versus the DeepMF baseline, (3) visual improvement of the proposed model versus the DeepMF baseline.

Experiment 1: This experiment tests the 'Contribution 1' assessment stated in the preceding section. As explained in the preceding section, it is necessary to test the RS accuracy when a bottleneck is set to the embedding layers. Since we need to visually represent embedding samples, we use embedding sizes: 2 (two-dimensional representation) or 3 (three-dimensional representation), whereas the usual implementation sizes range from 5 to 10. Our first experiment tests the accuracy loss when small embedding sizes are set. For each tested dataset (Table 1), we obtain the Mean Absolute Error (MAE) by setting embedding sizes = {2, 3, 5, 10}. Table 2 shows the MAE results, as well as the achieved accuracy percentage comparing the embedding sizes 2 and 10. As can be seen, very little accuracy is lost setting visualizable embedding sizes (2 and 3) compared to the usual sizes (5 to 10). Notably, only 2% of accuracy is lost in the worst-case scenario. The results in Table 2 open the door to visually represent the sample embeddings of items and users, knowing that the embedding values are meaningful to provide accurate CF predictions.

Experiment 2: This experiment numerically shows the improvement obtained by combining the three contributions stated in the preceding section. Once we have validated the adequacy of using visualizable embedding sizes, it is time to test the obtained improvement using the proposed approach. We will test visual improvement using the standard intra-clustering quality measure equation that processes the distance of all the samples to their centroid. That is:

$$\text{intra-clustering} = \frac{1}{|S|} \sum_{x \in S} d(x, \bar{v}) \qquad (1)$$

where:

$$d(x, \bar{v}) = \sqrt{\sum_{i \in \{1,\dots,n\}} (x_i - \bar{v}_i)^2} \qquad (2)$$

S is the set of samples, $\bar{v}$ is the S centroid and 'n' is the dimension size. Please note that whereas in the clustering field we look for low intra-clustering values, our embedding visualization aim is to spread embedding representations and to avoid them too being packed together. In this way, we will be able to better catch relations among samples. So, the higher our 'intra-clustering' quality measure, the better the results. In the CF embedding visualization field, we could call this quality measure an 'unpacking measure'. Table 3 shows the comparative results that test the non-variational dot product DeepMF baseline versus the proposed variational Euclidean model. Table 3 provides quality results for both user embeddings and item embeddings. As can be seen, representative improvements are obtained when the proposed model is used.

Table 3. Unpacking quality measure results (intra-clustering results from the quality measure defined in the 'Experiment 2') for both user and item embeddings. The higher the quality value, the better the result. "proposed" and "baselines" are absolute values, whereas "improv." shows the improvement percentage of the proposed model versus the baseline one.

Embedding Dataset/Model	Users			Items		
	Proposed	Baseline	Improv.	Proposed	Baseline	Improv.
Movielens 100K	0.8317	0.4789	73.66%	0.8943	0.6346	40.92%
MovieLens 1M	0.8289	0.5552	49.29%	0.9762	0.7372	32.42%
Netflix*	0.8790	0.4572	92.25%	0.9160	0.8569	6.90%

Experiment 3: This experiment visually shows the improvement obtained by combining the three contributions stated in the preceding section. The visual results of the proposed variational Euclidean model have been compared to the proposed non-variational dot product baseline (DeepMF). Figure 5 shows the returned graphs for both models when applied to the datasets in Table 1. The top graphs in Figure 5 show the baseline results, whereas the bottom graphs plot the proposed model results. As can be seen, the MovieLens 100K dataset (left graphs) displays an unpacked (extended) vision of both user and item samples when the proposed model (bottom-left graph) is used, compared to the baseline (top-left graph) one. The proposed model makes it easier to compare the relationship between samples by visually inspecting the (Euclidean) distances in the graphs. It also decreases intersections between users and items embedding representations. What we are looking at here explains the 'unpacked' quality values shown in Table 3. MovieLens 1M (center graphs) and Netflix* (right graphs) show similar layouts to MovieLens 100K, suggesting that, on CF datasets, the proposed model performs as expected.

As an example of the proposed model application, Figure 6 shows some demographic information from MovieLens 100K. Both graphs in Figure 6 show the location of the users. The graph on the left plots gender information: female (red) versus male (blue). The right graph plots age information: over 40 years of age (red) versus younger users (blue). Please note that the user plot in the bottom left graph of Figure 5 (MovieLens 100K) is not the same as the shapes shown in Figure 6; this is because they belong to different model trainings. Figure 6 is just an example that shows some type of demographic information: male versus female, and younger versus older users. Similar graphs can be obtained from different demographic features of users and from the item's type: zip code, incomings, educational level, genre of movies, type of music, year of book publication, etc. Figure 6 shows that there is not a clear pattern to cluster users attending to their gender or age; that is, in the

MovieLens 100K dataset, males and females rate movies in a similar way, analogously to the younger and older user case. What is relevant here is that we can obtain representative two and three-dimension representative graphs showing the location of CF demographic features. This big data visual information can be useful to take commercial decisions, implement segmented marketing, understand business data, improve RS information, balance data, correct biased datasets, etc.

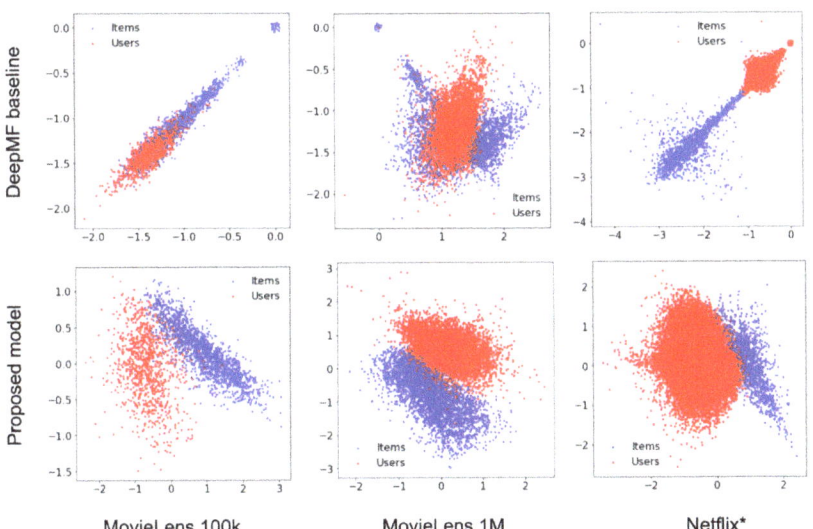

Figure 5. Embedding visualizations of users (red points) and items (blue points) for datasets in Table 1 when the DeepMF baseline is applied (top graphs) versus the proposed variational Euclidean method (bottom graphs).

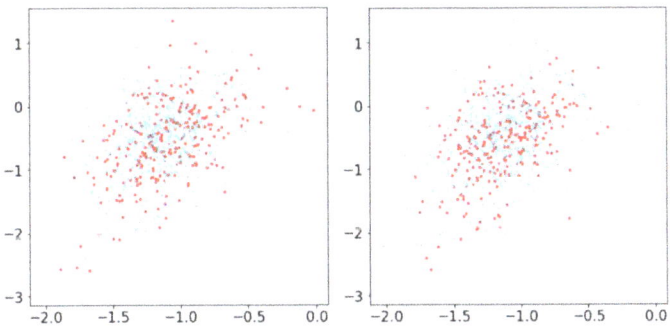

Figure 6. MovieLens 100K demographic information. The left graph shows the latent embedding location of female (red) versus male (blue) users, whereas the right graph shows over 40 years old users (red) versus younger users (blue).

To facilitate reproducibility, Table 4 shows the selected values of the involved parameters in the learning process of the proposed model.

Table 4. Parameter values chosen for the proposed model learning.

# Layers	5
# Neurons in each non-exit layer	Experiment #1: {2, 3, 5, 10}, Experiments #2 and #3: {2}
# Epochs	20
Batch size	16
Activation function of the non-exit layers	ReLu
Activation function of the exit layer	Linear
Loss function	Euclidean distance
Gaussian random distribution	Mean: 0, Variance: 1

4. Discussion

The variational proposed model has proven to adequately afford the visual representation of samples in the CF field. To fulfill this objective, we have tested the impact of limiting the embedding sizes to two neurons (obtaining two-dimensional graphs) or to three neurons (obtaining three-dimensional graphs). Results show a prediction quality of over 98% when the embedding sizes are two or three, compared to the usual five to ten embedding sizes (Table 2). Combining the variational approach and the Euclidean distance loss function, the intra-clustering quality measure improves in the proposed model compared to the DeepMF baseline (Table 3). This improvement can be visually observed by plotting each of the dataset embeddings (Figure 5). The proposed variational model has a better performance than the DeepMF baseline due to two different factors: (a) the designed variational stochasticity (which does not exist in the DeepMF model) spreads the embedding samples through the latent space (Figures 3 and 5), as the generative learning does to obtain fake images by interpolating embeddings; and (b) the proposed Euclidean function can arrange sample embeddings in the latent space in a comprehensible way to humans, compared to the non-Euclidean loss functions that usually implement the baseline model: mean squared differences, mean absolute error, etc.

5. Conclusions

Recommender Systems research is focused on accuracy, but there are some other relevant goals that should be achieved, such as the representative visualization of the collaborative filtering items and users. This can be considered a big data analytics tool that helps system managers. This paper provides an innovative neural model to make visual representations of user and item embeddings. First, we have shown that it is possible to reduce the model embedding sizes to just two or three neurons without any significant loss in prediction accuracy. Then, we have introduced Gaussian variational layers to the proposed model in order to spread the area where samples are located. Finally, a Lambda layer replaces the DeepMF Dot layer (or the NCF MLP); this layer implements the Euclidean distance. Both the Gaussian variational layers and the Lambda-Euclidean layer running together in the proposed model return suitable accuracy results and improved sample representations.

Experiment results show that the user and item embedding representations are conveniently spread through visual representation areas, making it possible to discriminate close samples and to relate between sample pairs. The centroid-based intra-cluster quality measure shows a significant improvement in the proposed neural model compared to the baseline. The plotted graphs also show better embedding representations when the proposed model is tested using the three selected representative collaborative filtering datasets. Results open the door to future works such as: (1) representing demographic features (gender, age, etc.) of samples; (2) explaining recommendations by providing a graph showing, in the same area, the active user, their recommendations and the nearest voted items to both the active user and the recommended items; and (3) incorporating three-dimensional embedding representations in 3D commercial environments.

Author Contributions: J.B. proposed the key idea for this paper and prepared to the manuscript. J.D. made the whole set of experiments and proposed several improvements. F.O. prepared the testing datasets from existing ones. A.G. helped with reviewing the manuscript and he is the corresponding author. All authors have read and agreed to the published version of the manuscript.

Funding: This work was partially supported by *Ministerio de Ciencia e Innovación* of Spain under the project PID2019-106493RB-I00 (DL-CEMG) and the *Comunidad de Madrid* under *Convenio Plurianual* with the *Universidad Politécnica de Madrid* in the actuation line of *Programa de Excelencia para el Profesorado Universitario*.

Institutional Review Board Statement: The study did not require ethical approval.

Informed Consent Statement: Not applicable.

Data Availability Statement: Movielens datasets: https://grouplens.org/datasets/movielens/, accessed on 19 April 2022; Netflix* dataset: http://cf4j.etsisi.upm.es/, accessed on 19 April 2022.

Conflicts of Interest: The authors declare no conflict of interest.

References

1. Batmaz, Z.; Yurekli, A.; Bilge, A.; Kaleli, C. A review on deep learning for recommender systems: Challenges and remedies. *Artif. Intell. Rev.* **2019**, *52*, 1–37. [CrossRef]
2. Bobadilla, J.; González-Prieto, Á.; Ortega, F.; Cabrera, L. Deep learning feature selection to unhide demographic recommender systems factors. *Neural Comput. Appl.* **2020**, *33*, 7291–7308. [CrossRef]
3. Deldjoo, Y.; Schedl, M.; Cremonesi, P.; Pasi, G. Recommender Systems Leveraging Multimedia Content. *ACM Comput. Surv.* **2021**, *53*, 1–38. [CrossRef]
4. Saurabh, K.; Sunil, F.R. Context Aware Recommendation Systems: A review of the state of the art techniques. *Comput. Sci. Rev.* **2020**, *37*, 100255. [CrossRef]
5. Shokeen, J.; Rana, C. A study on features of social recommender systems. *Artif. Intell. Rev.* **2020**, *53*, 965–988. [CrossRef]
6. Bobadilla, J.; Alonso, S.; Hernando, A. Deep Learning Architecture for Collaborative Filtering Recommender Systems. *Appl. Sci.* **2020**, *10*, 2441. [CrossRef]
7. Forouzandeh, S.; Berahmand, K.; Rostami, M. Presentation of a recommender system with ensemble learning and graph embedding: A case on MovieLens. *Multimed. Tools Appl.* **2021**, *80*, 7805–7832. [CrossRef]
8. Zhu, B.; Hurtado, R.; Bobadilla, J.; Ortega, F. An Efficient Recommender System Method Based on the Numerical Relevances and the Non-Numerical Structures of the Ratings. *IEEE Access* **2018**, *6*, 49935–49954. [CrossRef]
9. Mnih, A.; Salakhutdinov, R.R. Probabilistic matrix factorization. In *Advances in Neural Information Processing Systems 21*; Koller, D., Schuurmans, D., Bengio, Y., Bottou, L., Eds.; MIT Press: Cambridge, MA, USA, 2008; pp. 1257–1264.
10. Lee, D.D.; Seung, H.S. Algorithms for Non-negative Matrix Factorization. In *Advances in Neural Information Processing Systems 13*; Leen, T.K.; Dietterich, T.G., Tresp, V., Eds.; MIT Press: Cambridge, MA, USA, 2001; pp. 556–562.
11. Hernando, A.; Bobadilla, J.; Ortega, F. A non negative matrix factorization for Collaborative Filtering Recommender Systems based on a Bayesian probabilistic model. *Knowl.-Based Syst.* **2016**, *97*, 188–202. [CrossRef]
12. Hernando, A.; Moya, R.; Ortega, F.; Bobadilla, J. Hierarchical graph maps for visualization of collaborative recommender systems. *J. Inf. Sci.* **2014**, *40*, 97–106. [CrossRef]
13. Hernando, A.; Bobadilla, J.; Ortega, F.; Gutiérrez, A. Method to interactively visualize and navigate related information. *Expert Syst. Appl.* **2018**, *111*, 61–75. [CrossRef]
14. Hong-Jian, X.; Xinyu, D.; Jianbing, Z.; Shujian, H.; Jiajun, C. Deep Matrix Factorization Models for Recommender Systems. In Proceedings of the Twenty-Sixth International Joint Conference on Artificial Intelligence, Melbourne, Australia, 19–25 August 2017; pp. 3203–3209. [CrossRef]
15. Xiangnan, H.; Lizi, L.; Hanwang, Z. Neural Collaborative Filtering. In Proceedings of the International World Wide Web Conference, Perth, Australia, 3–7 April 2017; ACM: New York, NY, USA, 2017. ISBN 978-1-4503-4913-0/17/04. [CrossRef]
16. Cui, P.; Wang, X.; Pei, J.; Zhu, W. A Survey on Network Embedding. *IEEE Trans. Knowl. Data Eng.* **2019**, *31*, 833–852. [CrossRef]
17. Guan, N.; Song, D.; Liao, L. Knowledge graph embedding with concepts. *Knowl.-Based Syst.* **2019**, *164*, 38–44. [CrossRef]
18. Zou, Q.; Xing, P.; Wei, L.; Liu, B. Gene2vec: Gene Subsequence Embedding for Prediction of Mammalian N6-Methyladenosine Sites from mRNA. *RNA* **2018**, *25*, 205–218. [CrossRef] [PubMed]
19. Xiang, Y.; Zhen, W.; Jingong, H.; Srinivasan, P.; Moosavinasab, S.; Huang, Y.; Lin, S.M.; Zhang, W.; Zhang, P.; Sun, H. Graph embedding on biomedical networks: Methods, applications and evaluations. *Bioinformatics* **2020**, *36*, 1241–1251. [CrossRef]
20. Škrlj, B.; Kralj, J.; Lavrač, N. Embedding-based Silhouette community detection. *Mach. Learn.* **2020**, *109*, 2161–2193. [CrossRef]
21. Zheng, L.; Huang, Y.; Lu, H.; Yang, Y. Pose-Invariant Embedding for Deep Person Re-Identification. *IEEE Trans. Image Proces.* **2019**, *28*, 4500–4509. [CrossRef]
22. Li, Z.; Tang, J.; Mei, T. Deep Collaborative Embedding for Social Image Understanding. *IEEE Trans. Pattern Anal. Mach. Intell.* **2019**, *41*, 2070–2083. [CrossRef]

23. Kan, S.; Cen, Y.; He, Z.; Zhang, Z.; Zhang, L.; Wang, Y. Supervised Deep Feature Embedding with Handcrafted Feature. *IEEE Trans. Image Process.* **2019**, *28*, 5809–5823. [CrossRef]
24. Shuang, K.; Zhang, Z.; Loo, J.; Su, S. Convolution–deconvolution word embedding: An end-to-end multi-prototype fusion embedding method for natural language processing. *Inf. Fusion* **2020**, *53*, 112–122. [CrossRef]
25. Bobadilla, J.; Lara-Cabrera, R.; González-Prieto, A.; Ortega, F. DeepFair: Deep Learning for Improving Fairness in Recommender Systems. *Int. J. Interact. Multimed. Artif. Intell.* **2021**, *6*, 86–94. [CrossRef]
26. Huang, T.; Zhang, D.; Bi, L. Neural embedding collaborative filtering for recommender systems. *Neural Comput. Applic.* **2020**, *32*, 17043–17057. [CrossRef]
27. Fuzheng, Z.; Nicholas, J.Y.; Defu, L.; Xing, X.; Wei-Ying, M. Collaborative Knowledge Base Embedding for Recommender Systems. In Proceedings of the 22nd ACM SIGKDD International Conference on Knowledge Discovery and Data Mining (KDD '16), San Francisco, CA, USA, 13–17 August 2016; pp. 353–362. [CrossRef]
28. Jeong, S.-Y.; Kim, Y.-K. Deep Learning-Based Context-Aware Recommender System Considering Contextual Features. *Appl. Sci.* **2022**, *12*, 45. [CrossRef]
29. Chih-Ming, C.; Chuan-Ju, W.; Ming-Feng, T.; Yi-Hsuan, Y. Collaborative Similarity Embedding for Recommender Systems. In Proceedings of the The World Wide Web Conference (WWW '19), San Francisco, CA, USA, 13–17 May 2019; pp. 2637–2643. [CrossRef]
30. Heng-Tze, C.; Levent, K.; Jeremiah, H.; Tal, S.; Chandra, T.; Aradhye, H.; Anderson, G.; Corrado, G.; Chai, W.; Ispir, M.; et al. Wide & Deep Learning for Recommender Systems. In Proceedings of the 1st Workshop on Deep Learning for Recommender Systems (DLRS 2016), Boston, MA, USA, 15 September 2016; pp. 7–10. [CrossRef]
31. Zhao, W.X.; Huang, J.; Wen, J.R. Learning Distributed Representations for Recommender Systems with a Network Embedding Approach. In Proceedings of the Information Retrieval Technology (AIRS 2016), Beijing, China, 30 November–2 December 2016; Lecture Notes in Computer Science 9994. Springer: Berlin/Heidelberg, Germany, 2016; pp. 224–236. [CrossRef]
32. Zhang, C.; Yu, L.; Wang, Y.; Shah, C.; Zhang, X. Collaborative User Network Embedding for Social Recommender Systems. In Proceedings of the 2017 SIAM International Conference on Data Mining (SDM), Houston, TX, USA, 27–29 April 2017; pp. 381–389. [CrossRef]
33. Liu, H.; Zhao, X.; Wang, C.; Liu, X.; Tang, J. Automated Embedding Size Search in Deep Recommender Systems. In Proceedings of the 43rd International ACM SIGIR Conference on Research and Development in Information Retrieval, Online, 25–30 July 2020; pp. 2307–2316. [CrossRef]
34. Bobadilla, J.; Gutierrez, A.; Alonso, S.; González-Prieto, A. Neural Collaborative Filtering Classification Model to Obtain Prediction Reliabilities. *Int. J. Interact. Multimed. Artif. Intell.* **2021**. [CrossRef]
35. Liu, X.; Gherbi, A.; Wei, Z.; Li, W.; Cheriet, M. Multispectral Image Reconstruction from Color Images Using Enhanced Variational Autoencoder and Generative Adversarial Network. *IEEE Access* **2021**, *9*, 1666–1679. [CrossRef]
36. Liu, Z.-S.; Siu, W.-C.; Wang, L.-W.; Li, C.-T.; Cani, M.-P.; Chan, Y. Unsupervised Real Image Super-Resolution via Generative Variational AutoEncoder. In Proceedings of the EEE/CVF Conference on Computer Vision and Pattern Recognition Workshops (CVPRW), Seattle, WA, USA, 14–19 June 2020; pp. 1788–1797. [CrossRef]
37. Liu, Z.-S.; Siu, W.-C.; Chan, Y.-L. Photo-Realistic Image Super-Resolution via Variational Autoencoders. *IEEE Trans. Circuits Syst. Video Technol.* **2021**, *31*, 1351–1365. [CrossRef]
38. Harper, F.M.; Konstan, J.A. The movielens datasets: History and context. *ACM Trans. Interact. Intell. Syst.* **2015**, *5*, 1–19. [CrossRef]
39. Ortega, F.; Zhu, B.; Bobadilla, J.; Hernando, A. CF4J: Collaborative filtering for Java. *Knowl.-Based Syst.* **2018**, *152*, 94–99. [CrossRef]

Article

Using Deep Learning for Collecting Data about Museum Visitor Behavior

Alessio Ferrato [1], Carla Limongelli [1], Mauro Mezzini [2] and Giuseppe Sansonetti [1,*]

1 Department of Engineering, Roma Tre University, 00146 Rome, Italy; ale.ferrato@stud.uniroma3.it (A.F.); limongel@dia.uniroma3.it (C.L.)
2 Department of Education, Roma Tre University, 00185 Rome, Italy; mauro.mezzini@uniroma3.it
* Correspondence: gsansone@dia.uniroma3.it; Tel.: +39-06-5733-3220

Abstract: Nowadays, technology makes it possible to admire objects and artworks exhibited all over the world remotely. We have been able to appreciate this convenience even more in the last period, in which the pandemic has forced us into our homes for a long time. However, visiting art sites in person remains a truly unique experience. Even during on-site visits, technology can help make them much more satisfactory, by assisting visitors during the fruition of cultural and artistic resources. To this aim, it is necessary to monitor the active user for acquiring information about their behavior. We, therefore, need systems able to monitor and analyze visitor behavior. The literature proposes several techniques for the timing and tracking of museum visitors. In this article, we propose a novel approach to indoor tracking that can represent a promising and non-expensive solution for some of the critical issues that remain. In particular, the system we propose relies on low-cost equipment (i.e., simple badges and off-the-shelf RGB cameras) and harnesses one of the most recent deep neural networks (i.e., Faster R-CNN) for detecting specific objects in an image or a video sequence with high accuracy. An experimental evaluation performed in a real scenario, namely, the "Exhibition of Fake Art" at Roma Tre University, allowed us to test our system on site. The collected data has proven to be accurate and helpful for gathering insightful information on visitor behavior.

Keywords: cultural heritage fruition; human factors in artificial intelligence; museum visitors analysis; computer vision; machine learning; deep neural networks

1. Introduction

The fruition modalities of cultural heritage sites can benefit from advanced technologies and methodologies of data analysis that propose solutions aimed at visitor engagement. These proposals must cleverly balance the different needs of a large and diverse set of visitors and the peculiarities of the specific site. It is necessary to understand how visitors use the different spaces within a museum and how their behavior can help identify the strengths and weaknesses of the cultural offerings and, consequently, possible engagement strategies for museum institutions. An in-depth analysis of visitor behavior would stimulate new ways of promoting artworks. It would also serve as a spur for implementing more appropriate measures for museum security and visitor care. Moreover, collecting data about visitor behavior would allow museum curators and staff members to offer stakeholders a better settlement, both in displaying and narrating the artworks and in terms of marketing-related services. There are many studies on audience engagement. Some of them promote the integration of visitor tracking technology with mobile devices that users carry with them [1,2]. Other studies analyze user tracking to examine the flow of visits through complex and expensive tracking systems [3–8].

In this paper, we propose to collect visitor data through accurate, non-intrusive, cheap, and anonymity-preserving tools. The approach is based on computer vision techniques and leverages off-the-shelf RGB cameras and badges such as those provided free to attendees by event and conference organizers. Therefore, the overall cost of the entire instrumentation is reduced, which is certainly a significant advantage over other state-of-the-art technologies. The methodology is based on deep learning, more specifically, methods and techniques for image detection and classification through Convolutional Neural Networks (CNNs) capable of providing excellent performance in terms of accuracy. Our approach can represent the solution to some of the criticalities shown by the other visitor localization technologies in the museum environment. With the proposed setting, the estimate of the visitor position is extremely accurate (on the order of 10^{-2} m). Moreover, the intrusiveness of the proposed approach is minimal. The user is not required to wear additional devices such as active and passive sensors, Personal Digital Assistants (PDAs), smartphones, or portable Graphics Processing Units (GPUs), but a simple badge. Another advantage of our solution is that the coverage is guaranteed at a low cost, through simple commercial RGB cameras or any video surveillance system present in most national and international museums and exhibitions. Lastly, the collected videos can be processed using a free platform such as the Google Colaboratory environment, thus allowing museum curators and staff to save the cost of hardware and system usage. The contributions of this paper are as follows:

- The analysis of the state of the art to identify the main problems in the timing and tracking of users in indoor environments (i.e., high intrusiveness, low accuracy, high cost, and high consumption);
- The design of a novel, low-cost, and highly accurate system to overcome the aforementioned problems;
- The development of a user timing and tracking system capable of providing data useful both for museum curators and staff (e.g., the possibility to analyze and monitor how visitors enjoy museum objects) and for the visitors themselves (e.g., the possibility to receive personalized suggestions during the visit).

We propose a deep learning-based approach to comprehensively and accurately collect visitor experience data. Specifically, we describe in detail the characteristics of its architecture and the experimental results obtained. We also illustrate a case study in a real environment and finally show how the collected data can be stored and used to provide valuable information relating to the behavior of each visitor.

This paper is structured as follows. In Section 2, we briefly review current technologies used to monitor visitors, focusing on computer vision approaches. In Section 3, we present the proposed system for detecting the exact visitor location anytime. In Section 4, we report the experimental results of the proposed system and the analysis of the data collected in the "Exhibition of Fake Art" at Roma Tre University. In Section 5, we discuss the obtained findings. Finally, in Section 6, we draw our conclusions and identify some of the possible uses of the data collected through the proposed system.

2. State of the Art

Nowadays, technology is increasingly exploited to improve users' quality of life anywhere and anytime, when they use local transport services [9] or visit points of interest [10]. In particular, the possibility of providing museum curators and staff members with a system to track visitor behavior for improving the service offered is a widely investigated topic. In [11], the authors propose a computer vision algorithm based on Kinect and RGB-D camera. They track groups of visitors at the National Museum of Emerging Science and Innovation (Miraikan) in Tokyo, Japan, to identify the leader and study their dynamics. In [12], the authors present an IoT- (Internet of Things) based system to measure and understand visitor dynamics at the Galleria Borghese museum in Rome, Italy. A similar approach is described in [5], in which the authors report the results of a case study conducted at the CoBrA Museum of Modern Art in Amstelveen, the Netherlands. Tracking can be also used to understand how the flow of visitors inside a museum is oriented. In [13], the authors propose a method based on LIDAR to identify human beings and track their positions, body orientation, and movement trajectories in any public space. The system can accurately track the position of the visitor inside the museum. It has been tested at the Ohara Museum of Modern Arts in the Kurashiki area of Okayama Prefecture, Japan. In [3], Lanir et al.

propose a visual guide for museum curators and staff. Their system can show routes of interest and hotspots, and analyze visitor behavior. It has been tested at the Hecht Museum, University of Haifa, Israel. There are already numerous thorough and exhaustive works that review the main indoor localization technologies (e.g., see [14–19]). On the other hand, fewer works are focused on analyzing the technologies for the localization of museum visitors. The goal of our study is to propose a tracking system easy to use by museum curators and staff. Moreover, it has to capture as much information as possible about visitor behavior in real time. Finally, it should be ready to accommodate several further developments like visitors' micro-expressions recognition when looking at an artwork and subsequent recommendation. For this reason, we now analyze the hardware most commonly used for this aim by examining the strengths and weaknesses of the principal indoor localization technologies.

2.1. Indoor Localization Technologies

In the following, we report the most used technologies for indoor tracking.

- WiFi. This technology is extensively used for network connection of various devices in public and private environments. Initially, its maximum coverage was about 100 m, today it has been extended to over 1 km with the IEEE 802.11ah protocol, published in 2017, specifically designed for Internet of Things (IoT) services [20]. The fact that it is supported by almost all the electronic devices on the market makes it one of the most used technologies for indoor localization, without the need for additional infrastructure. However, its characteristics of wide coverage and high throughput yield to a more suitable usage for communication than localization, because of its low accuracy and interference, which make it necessary to use complex processing algorithms.
- Bluetooth. This technology is used for wireless connection between mobile and fixed devices within relatively small distances. The latest version, called Bluetooth Low Energy (BLE), provides improved performance in terms of coverage and throughput, with low power consumption [21]. Recently, two BLE-based protocols have been proposed: Eddystone (by Google Inc.) and iBeacons (by Apple Inc.). They are intended more for proximity-based services than localization, due to poor accuracy and high sensitivity to noise.
- Infrared. Among others, the IR technology was the first one to be widely used in many projects (e.g., see [8,22]). However, it has several limitations [23]. Firstly, it requires the presence of visible IR emitters and a line of sight between the emitter and receiver. Lastly, the nature of the IR signal requires accurate calibration of the parameters of the IR emitters and the active involvement of visitors in the process of locating their position.
- Radio-Frequency Identification. This technology is used to transfer data between a reader and a tag capable of communicating on default radio frequency [24]. There are two types of RFID systems: Active and passive. The first one operates with microwave and Ultra High Frequency (UHF) ranges, and it is characterized by low cost and ease of integration into the objects to be tracked. However, their low accuracy and poor integration in portable devices make them unsuitable for indoor location purposes. Passive RFID systems can operate without a battery but have significant limits in terms of coverage, which makes them unsuitable for indoor location purposes.
- IEEE 802.15.4. This technology is mostly used in wireless sensor networks and is characterized by good energy efficiency, low cost, but also by low throughput [25]. This standard is not available on most devices on the market and for this particular reason, it is not suitable for the indoor localization of users.
- Ultra Wideband. This technology is mainly used in short-distance communication systems and is characterized by low energy consumption [14]. The main characteristics of the UWB technology are the robustness to interference and the possibility to penetrate various materials. For these reasons, it is extremely suitable for indoor

localization. However, due to its limited implementation in portable devices, it cannot be widely used. The UWB problems have been extensively analyzed by the authors of [26] and in practical scenarios, the Non-Line-of-Sight (NLOS) propagation can be the main issue of this technology.

- Visible Light. Indoor localization technology based on visible light can be realized using different types of sensors. The most common are Light-Emitting Diodes (LEDs) [27]. The use of LEDs for indoor localization has numerous advantages over other technologies. First of all, emitters and sensors are very popular considering their low cost. They are also resistant to changes in humidity and they have low energy consumption. The main disadvantage of LEDs is that a line of sight between them is required [18]. Another type of sensor used in visible light systems is Light Detection and Ranging Localization (LIDAR). This sensor is able to provide information relating to the contour of surrounding objects. When combined with inertial sensors, LIDAR-based tracking systems can provide accurate results [28]. In order to properly work, the LIDAR-based tracking system needs at least one sensor in each room. Because of that, this particular technology would be extremely expensive for large museums.
- Acoustic Signal. This technology can localize the user by capturing acoustic signals emitted from sound sources using a microphone sensor [29]. The acoustic signal localization is accurate only when audible band acoustic signals (i.e., <20 kHz) are used. For these signals, sufficiently low transmission power is required not to cause unwanted noise. This aspect, coupled with the need for additional infrastructure, results in that localization based on acoustic signals is not widely used.
- Ultrasound. This technology allows us to compute the distance between a transmitter and a receiver by measuring the time of flight of ultrasonic signals (i.e., >20 kHz) [30]. Indoor localization based on ultrasound is very accurate. However, the measurement process can be heavily influenced by significant changes in temperature and humidity, as well as by ambient noise.

Hence, the solutions above are inaccurate, expensive, or very complex to implement. We, therefore, focus on computer vision, which can represent a non-intrusive solution for the user already accustomed to security cameras.

2.2. Red-Green-Blue (RGB) Video-Based Techniques

Several noteworthy approaches to user timing and tracking rely on RGB video-based models and methods. These techniques are already applied to other fields (e.g., see [31]) such as motion analysis, motion capture, and in general, to most activities related to virtual reality. Here, through RGB cameras, we can collect visual information that can be used to estimate where the visitor is. To achieve this goal, two capture methods can be used. The first one is based on visitor recognition, the second one relies on artwork recognition. The positioning of the camera, therefore, assumes a fundamental role in the implementation of these systems.

Recent work described in [32] takes advantage of computer vision and content-based image retrieval technique to detect visitor behavior. From frames recorded by multiple cameras installed in exhibition chambers, visitors are tracked by an object detector and also modeled with a deep learning technique. The system classifies each person by their appearance, grounded on color similarity as determined by measuring the distances of the distributions. Currently, the system is extremely time-consuming and needs to be enhanced to be applicable.

In [11], the authors propose a computer vision algorithm based on Kinect and RGB-D camera. They track visitor groups in a museum to identify the leader and study its dynamics. They also analyze the body language and the reciprocal position of the group leader to the rest of the group. The final goal of this study is the replacement of the group leader (typically, the guide in a museum) with a robot. They have installed four Kinect V1 sensors in some rooms at the National Museum of Emerging Science and Innovation (Miraikan) in Tokyo, Japan, and for two months they recorded videos of visitor groups interacting with the museum guides during visits. The motion is detected by computing the difference between bounding boxes of two consecutive frames. The experimentation has been carried out by manually annotating and comparing the motion of the group and the guide with the algorithm results. The main issues with this approach are the inaccuracy of the results when people are too close to the camera and occlusion problems. Moreover, the categorization through bounding boxes has an average accuracy of 70–75%, which can improve with the application of the exponential motion algorithm they propose.

The Kinect sensor is proposed as a tracking solution in [33] as well. The authors use a particular process to estimate the gaze direction from face direction measurements. In their work, they discuss the method for gazed object estimations using face direction measurements and object detection. By measuring the face direction and detecting the object at the same time using a Kinect sensor, they can estimate what the visitors are looking at.

A different solution is SeeForMe [4]. It is a real-time computer vision system that can run on wearable devices to perform object classification and artwork recognition. It uses a video camera on the audio guide to identify artworks. This smart audio guide equipped with a vision system has been tested at the Bargello Museum in Florence. A CNN for object classification and identification runs on an NVIDIA portable GPU. Also, a voice detection module can determine the context (user alone, accompanied, etc.) and stop the guide when, for example, the visitor is talking to someone. Experimental trials were performed with a training set of 300 people and 300 images. Up to 5 m, there is maximum Precision and Recall (with Recall up to 0.8). Through various adjustments to the algorithm, they succeeded in having almost all works recognized, and only 22 of them were not recognized. The System Usability Scale (SUS) questionnaire, filled in by the sample, revealed only the problem of the intrusiveness of the guide during the visit, and the hassle of having to manage the menu. The SUS questionnaire showed good usability. Moreover, the camera must be necessarily placed in a shirt pocket, at chest height, which is a rather limiting constraint.

In [34], an approach in line with the spirit of our proposal is proposed: To collect as much data on user behavior as possible such as itineraries, the number of entries, the flow of visitors, and time spent in front of works. The authors use video cameras with infrared sensors and re-ID (person re-identification). The main difference with our approach is that, while the person re-identification needs a preprocessing phase of the generated videos, in our case the preprocessing is done on the badge before it is given to the user. In this way, we can monitor in real-time the movement of each visitor. This difference is significant because we can imagine using the extracted data also to propose new tools that support both the visitor (e.g., recommender systems [35,36]) and the museum curators and staff (e.g., visitor flow analysis [37]).

3. Proposed Method

Image classification and object identification technologies have become much more successful as a result of recent advancements in the field of deep learning [38]. More specifically, CNN models [39–41] can easily attain accuracy values near to 100% on the training set. In other terms, these models can give the correct prediction, with almost certainty, when they are asked to predict the class of an element of the training set. Thanks to this, it is possible to train such models to recognize an arbitrary, single object, with very high confidence. Based on the above observations, we developed the following idea for tracking museum visitors [42]. A CNN model is trained for recognizing a set of unique and distinct objects. The objects to be recognized are badges, like the ones used in events and conferences (see Figure 1). It should be clear that there will be a fixed number N of distinct badges. Therefore, our model will be trained in order to recognize N different classes: One for each of the N distinct badges. In the research literature, there exist two different types of object detectors [43]: Detectors of specific instances of objects and detectors of broad categories of objects. The former ones aim to detect instances of a particular object (e.g., the Colosseum, Joe Biden's face, or the neighbor's cat), thus addressing a matching problem. The latter ones aim to detect instances of specific categories of objects (e.g., cars, humans, or cats). In our scenario, badge detection falls into the first type of object detection. Furthermore, the model is also trained for the detection (but not the recognition, for privacy reasons) of visitors' faces (see Figure 2). Therefore, face detection falls into the second type of object detection.

Figure 1. One of the badges used in the experimental trials.

Figure 2. Recognized objects (i.e., badges and faces) in a frame of a 720p video. In this case, the camera is positioned above the artwork of interest, at a height of 2.20 m from the floor.

At the beginning of their visit, the visitor is invited to wear one of the badges on which a CNN model is trained. This model, as confirmed by the experimental results reported in Section 4.1, can recognize badges accurately. To this aim, it is required that RGB cameras are installed inside the museum environment. In our case, simple, inexpensive, off-the-shelf RGB cameras are sufficient (in our experimental tests, we used a Logitech HD Webcam CS25 camera and a smartphone Honor View 10 Lite camera). The frame rate of the videos captured by the Logitech camera is eight fps (frames per second) and 30 fps for those captured by the smartphone camera. It should be noted that the value of the frame rate does not affect the accuracy of the model. A higher frame rate increments the amount of collected data but has the drawback to increase the number of computational resources and storage needed. These cameras should be strategically placed inside the museum premises in a way that the badge worn by the active visitor is always visible by at least one camera. A simple assumption is to install one camera in every point of interest of the museum or on each side of every room, at a height that minimizes the possibility that another visitor put herself in front of the active visitor wearing the badge, thus making it not visible from the camera. Since the RGB cameras are inexpensive, the use of more cameras concerning the simple aforementioned assumption should not result in a substantial increase in the installation cost. Another, more sophisticated approach, to optimally position the cameras inside the museum, is to resort to classical algorithms like those used for solving the Art Gallery Problem [44,45]. Once the recorded video is acquired by cameras, it is given in input to the model to detect the visitor's badge and face inside each video frame. The detection process consists of the following steps. For each video frame and for each object in the video frame, the model provides a score $0 < p \leq 1$, expressing the likelihood of an object being detected, the class c of the object, and a 4-dimensional vector containing the coordinates of the upper left and the lower right vertices of the box inside the video frame where the object c is detected. Therefore, for each video frame, the output of the detection process consists in a set of triples $(c_i, p_i, \mathbf{b}_i)$. Hereafter, vector $\mathbf{b}_i$ will be referred to as the bounding box of the object of class c_i and we will denote the coordinates of the upper left and lower right corners of the bounding box by $(\mathbf{b}_i(x_1), \mathbf{b}_i(y_1))$ and $(\mathbf{b}_i(x_2), \mathbf{b}_i(y_2))$, respectively. If the value of p for a class c is higher than a prefixed threshold σ (we empirically set $\sigma = 0.8$ in our experimental tests), we assume that the object of class c is detected inside the video frame. The value of σ is a hyperparameter of the system. For high values of the σ parameter, we can have a high number of false negatives, whilst, for low values, we can have a high number of false positives.

3.1. Computation of the Exact Visitor Position

To compute the visitor's spatial position from the bounding box of the detected badge, it is first necessary to calibrate the camera or cameras used. Generally speaking, the procedure of camera calibration consists of the estimation-with acceptable accuracy for the specific application-of the extrinsic (i.e., rotation matrix and translation vector) and intrinsic parameters (i.e., image center, focal length, skew, and lens distortion) of the camera [46]. This process is fundamental for most computer vision applications, especially when metric information related to the scene is required, as is our case. Once the camera has been calibrated, it is possible to determine the angular amplitude α of each pixel of the camera [47]. This can be done with a simple computation consisting in counting the number m of pixels in a video frame (see Figure 3b) of a unit length yardstick put in front of the camera at a unit distance (see Figure 3a). Then, the angular amplitude α of a single pixel can be expressed as follows:

$$\alpha = \frac{2\arctan(0.5)}{m}. \tag{1}$$

Knowing the angular amplitude of the pixel and the real dimensions of the badge (in our case they are $L = 10.4$ cm and $H = 14.0$ cm), it is straightforward to compute the distance ℓ of the badge from the camera, which can be done as follows (see Figure 4a):

$$\ell = \frac{H}{2\tan\left(\frac{\alpha m_y}{2}\right)} \tag{2}$$

where $m_y = |\mathbf{b}(y_1) - \mathbf{b}(y_2)|$ is the number of pixels of the height of the badge bounding box in the video frame (see Figure 4b). In Equation (2), we can also replace the term m_y with the term $m_x = |\mathbf{b}(x_1) - \mathbf{b}(x_2)|$ and H with L. As above, we can compute the angle β (respectively, γ) that the badge forms with the vertical (respectively, horizontal) centerline of the video frame. Thus, the triple (ℓ, β, γ) corresponds to the polar coordinates of the badge in the camera reference. The visitor position inside the museum can be obtained by adding the values of the camera coordinates in the museum reference. Knowing the video frame rate, we can also determine the exact time and length of the museum visit and all other temporal information such as how much time a visitor spent in front of a specific artwork and so on.

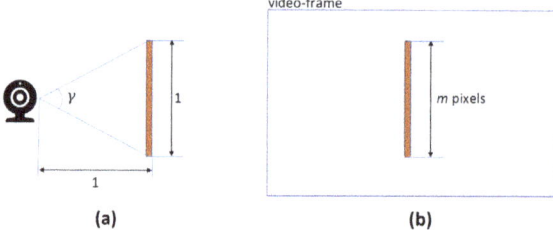

Figure 3. Computation of the angular amplitude of a pixel; (**a**) a unit length yardstick in front of the camera at a unit distance; (**b**) the corresponding number m of pixels in a video frame.

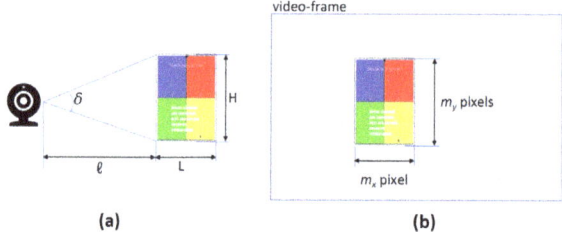

Figure 4. Computation of the distance between the badge and the camera; (**a**) a badge in front of the camera; (**b**) the corresponding number $m_x \times m_y$ of pixels in a video frame.

3.2. Experimental Settings

We evaluated several possible designs of the badge during the experimental trials. One requirement of the design is that the badge should be easily and effectively detected by the system. Another requirement is that each visitor who gets a badge should be easily distinguished from all other visitors wearing badges, in order to be able to track back the visit of a single visitor and distinguish their track from the track of any other visitor. This requirement could be easily satisfied when all the badges are distinct. In order to satisfy those requirements, we eventually chose a design in which a rectangular badge is split into four parts that can be in one of eight different colors (see the badge shown in Figure 1). Therefore, the number of different possible badges is equal to the number of dispositions of four colors taken from a set of eight, which is equal to $\frac{8!}{4!} = 1680$. If one splits the badge into six equal-sized parts, then the number of dispositions of six colors taken from a set of eight is $\frac{8!}{2!} = 20{,}160$. This shows that our design can be easily scaled for ten of thousands of different badges. For the sake of simplicity, we limited the experiments to the design of the badge shown in Figure 1. Furthermore, we trained our model to recognize 12 different badges. Thus, the model can recognize 12 different classes plus the face class. The training set was built first by manually annotating a dataset of about 300 pictures all containing the same badge. The training process with this single badge proved to be particularly efficient. This allowed us to automatically annotate all the other elements in the training set. The images of the training set were extracted from a set of 36 videos (three videos for each badge) from different angles, at 8 fps, 24 of them were about two minutes long, and the other 12, six minutes long, sampling a frame every two. In the first 24 videos, there was only one badge in each frame. In the last 12 videos, there were always two badges in each frame so that all possible pairs of badges were present in one frame. We inspected exhaustively all the automatically annotated images in order to assure the quality of the outcome. However, the accuracy of the model was so high (see Section 4) that very few manual corrections were needed to the automatic annotation process. In other terms, only about one of a thousand images required us to manually insert a missed annotation or delete a false positive annotation. Eventually, we produced a set composed of more than 30,000 annotated pictures containing 12 different badges or, equivalently, about 2500 pictures for each different badge.

3.3. Model Implementation

For implementing the system, we used the Faster Region-based Convolutional Neural Network (Faster R-CNN) model [48]. The reason was that preliminary studies (e.g., see [38]) showed that the Faster R-CNN model is effective and accurate in relation to other popular deep learning frameworks. The architecture of the proposed system is shown in Figure 5.

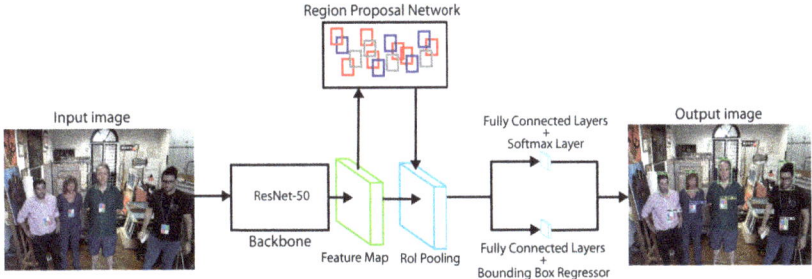

Figure 5. The architecture of the proposed system relies on a Faster Region-based Convolutional Neural Network (Faster R-CNN).

Specifically, the image taken by the RGB camera is given as an input to a backbone network, that is, a typical pre-trained convolutional network, which returns a feature map. We chose a 50-layer Residual Network (ResNet-50) as the backbone of our architecture because residual networks can usually achieve better performance than most other backbones [49]. The features are then sent in parallel to two different components of the Faster R-CNN architecture:

- A Region Proposal Network (RPN) that is used to determine the position of the image in which a potential object could be (i.e., at this stage we do not yet know what the object is, but only that there may be an object in a certain position of the image);
- A Region of Interest (RoI) pooling layer that is used to extract fixed size windows from the feature map before giving the RoI input to the fully connected layers. This component makes use of max pooling to convert the features within any valid RoI into a small feature map with a fixed spatial extension of height H × width W.

The output is then given as an input to two fully connected layers: One for the classification of the object and one for the prediction of the bounding box coordinates to obtain the final locations. The most important hyperparameters (e.g., the batch size and other optimization parameters) were left to the values suggested in [48]. Through grid search, we selected the best values for the learning rate and the scheduler that reduces the learning rate at a specific number of epochs. We chose Stochastic Gradient Descent (SGD) as an optimization algorithm. Another hyperparameter that we changed from the suggested value was the one that specifies the minimum dimension in pixels of the input image. Based on this parameter, the input image is resized in a way that at least one of its dimensions is equal to the parameter, before being fed to the CNN for the forward pass. The suggested value of the parameter was 800 pixels, but we increased it to 960 pixels. The reason was that when a visitor is far from the camera, the spatial dimension of the badge in the frame could be very little, making the detection difficult. In an ad hoc experiment, we analyzed all the dimensions of all the annotated boxes and we found that in no case was the dimension of any bounding box lower than 32 × 32. The minimum dimension detected (of the order of 40 × 40 pixels) has been encountered for the badges when the visitor was approximately 3.5 m far from the camera. In our opinion, this allows for training a model in order to detect badges that are at five or more meters of distance from the camera. We later augmented the dataset by randomly shrinking each picture by a factor between 0.3 and 0.5 (chosen randomly). This data augmentation enabled the model to detect badges located up to 6 m far from the camera, thus avoiding the need of adding other images to the training set. As a backbone, we employed a ResNet-50 network (where 50 is the number of convolutional layers in the network) that had already undergone two previous pretraining: One with the ImageNet [50] dataset and a second one, with the COCO [51] dataset containing about 90 classes. The authors of [52] strongly recommend the pretraining of the backbone on both datasets because empirical evidence shows that a network that had only been pretrained with the ImageNet dataset was much less accurate. Therefore, in order to verify if the ResNet-34 network (with only 34 convolutional layers) was faster but at the same time maintained, the same performance in terms of Accuracy and Precision, it was necessary to pretrain the ResNet-34 network with the COCO data set. Using the ResNet-34 network confirmed the boost in speed while maintaining an almost equal level of accuracy. Note that if Accuracy and Precision are the most important system performance metrics (instead of detection speed), the use of ResNet-100, or even ResNet-150, could improve system Precision. The pretrained model, as well as the pdf file with the trained badges, are available online (https://colab.research.google.com/drive/1-Kr0c6dOuMUdoShJjbLhqaVtM9b-gwc6?usp=sharing (accessed on 13 October 2021)).

4. Experimental Results

4.1. Performance Analysis

In order to assess the performance of the proposed system, we employed the detection evaluation metrics used in the most popular competitions, such as the COCO Detection Challenge (https://competitions.codalab.org/competitions/20794 (accessed on 13 October 2021)). Before illustrating these metrics, however, it is necessary to introduce some fundamental concepts. The goal of an object detector is to predict the position of objects of a certain class in an image or a video with a high degree of confidence. For this purpose, object detectors place bounding boxes in the image to identify the positions of the detected objects. A detection can, therefore, be represented by three features: The class of the object, the bounding box that contains it, and the confidence score. The Confidence Score is defined as the probability that a bounding box contains an object. It is, hence, usually a value between 0 and 1 that expresses how confident the model is about the prediction [53]. Another fundamental concept is that of Intersection over Union (IoU), which is defined as the ratio between the area of the intersection between a predicted bounding box (B_p) and a ground-truth bounding box (B_{gt}) and the area of their union:

$$IoU = \frac{area(B_p \cap B_{gt})}{area(B_p \cup B_{gt})}. \tag{3}$$

We have a perfect match when $IoU = 1$, while if the bounding boxes do not overlap at all, we have $IoU = 0$. Therefore, IoU values near to 1 are significantly better. Confidence Score and Intersection over Union are used to evaluate a detection. Specifically, there is a True Positive (TP) when:

1. The confidence score is higher than a given threshold value;
2. The predicted class is the same as that of the ground-truth;
3. The predicted bounding box has an IoU higher than a threshold value (e.g., 0.75).

On the other hand, there is a False Positive (FP) if one of the last two conditions is not valid. In the event that multiple predictions match the same ground-truth, the one with the highest confidence score is considered a TP, whilst all the others are considered as false positives. We have a False Negative (FN) when the Confidence Score of a detection of a supposed ground-truth is lower than the threshold value, whilst we have a true negative (TN) when the Confidence Score of a detection of anything is lower than the threshold value. True negatives, however, are usually not taken into account in evaluating object detection algorithms. Based on the previous definitions, it is possible to define the *Precision* as follows:

$$Precision = \frac{TP}{TP + FP} \qquad (4)$$

and *Recall* as follows:

$$Recall = \frac{TP}{TP + FN}. \qquad (5)$$

By setting the threshold for the Confidence Score at different values, we can obtain different pairs of Precision-Recall, which can be plotted on a graph in the form of *Precision-Recall curves*. It is possible to summarize the shape of these curves through a single numerical value, known as *Average Precision (AP)* [54]. This value is defined as the Precision averaged over a set of eleven Recall values equally spaced $[0, 0.1, 0.2, \ldots, 1]$:

$$AP = \frac{1}{11} \sum_{r \in \{0, 0.1, \ldots, 1\}} p_{interp}(r). \qquad (6)$$

The Precision value for each Recall level is interpolated considering the maximum Precision calculated for a system for which the corresponding Recall exceeds r:

$$p_{interp}(r) = \max_{\tilde{r}: \tilde{r} \geq r} p(\tilde{r}) \qquad (7)$$

where $p(\tilde{r})$ is the Precision measured at Recall $\tilde{r}$. The purpose of the interpolation is to reduce the impact of wiggles in the Precision-Recall curves due to small variations in the classification of the retrieved objects. For a system to obtain a high value of Average Precision, it must therefore have a high Precision value at all levels of Recall. This penalizes systems capable of achieving high Precision only in retrieving a subset of objects. Normally, this particular curve is used to compare one system to another, but when it comes to performance analysis, it shows how a system is performing when its parameters are changed. As mentioned above, there is another type of curve known as *Recall-IoU curves*, which are the basis of another metric used to evaluate the performance of a detector, namely, the *Average Recall (AR)* [55]. Such curves are obtained by plotting the Recall values corresponding to the *IoU* values $\in [0.5, 1.0]$. The Average Recall is defined as the Recall averaged over all *IoU* values $\in [0.5, 1.0]$. It can be calculated as twice the area under the Recall-*IoU* curve:

$$AR = 2 \int_{0.5}^{1} Recall(o) do \qquad (8)$$

where o is *IoU* and $Recall(o)$ is the corresponding value of Recall. There exist several variants of the metrics above. Among the others,

- $AP@IoU=0.50:.5:.95$ is the AP value averaged over 10 different IoU threshold values (i.e., 0.50, 0.55, 0.60, ..., 0.95).

Furthermore, there is also Average Precision calculated for different object scales. So, we have:

- $AP@$ area = small, which represents AP for objects that cover an area less than 32^2 pixels;
- $AP@$ area = medium, which represents AP for objects that cover an area higher than 32^2 pixels but lower than 96^2 pixels;
- $AP@$ area = large, which represents AP for objects that cover an area higher than 96^2 pixels;
- $AP@$ area = all, which represents AP for objects of any size.

The area is given by the number of pixels present in the segmentation mask. Finally, we have AP calculated for different detection numbers per image, defined as follows:

- $AP@$ maxDets = 1, which represents AP given 1 detection per image;
- $AP@$ maxDets = 10, which represents AP given 10 detections per image;

- $AP@$ maxDets = 100, which represents AP given 100 detections per image.

The same variants of the Average Precision metric also apply to the Average Recall. Before presenting the experimental results, it is necessary to describe the test set used. To evaluate the performance of our system using the metrics introduced above, we randomly selected 300 images from 10 videos containing a total of 13 object classes (12 specific badges + the face object). Figure 6 shows the values of the Average Precision metric on the test set for our object detector as the number of epochs increases.

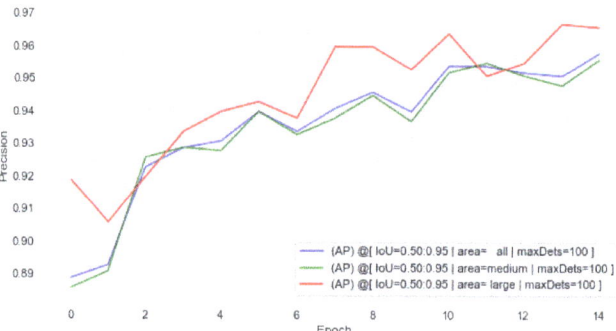

Figure 6. Average Precision of the proposed system on the test set.

It can be noted that there are high AP values already with a low number of epochs. We have only reported the value of $AP@$ maxDets = 100, as the values for maxDets = 1 and maxDets = 10 are the same as above. Furthermore, we have not reported the value of $AP@$ area = small, because we excluded a priori the detection of badges that are too small, that is, worn by visitors at such a distance from the point of interest that they cannot be considered in its surroundings. Figure 7 shows the trend of the Average Recall values on the test set as the number of epochs increases.

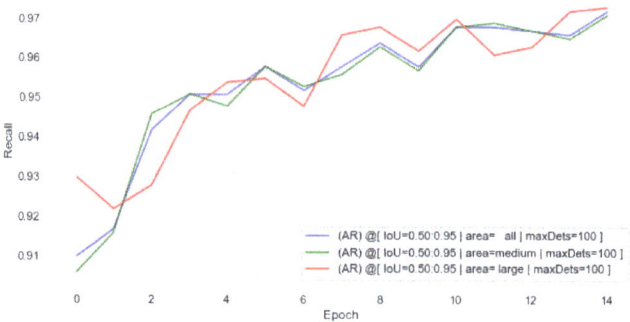

Figure 7. Average Recall of the proposed system on the test set.

Also in this case, the values are already high after a few epochs. The experimental analysis was performed on an NVIDIA QUADRO P2000 GPU capable of analyzing 4 frames per second. This system can, therefore, be used to perform a real-time analysis with fine tuning and optimization of the parameters.

4.2. Data Analysis

In this section, we report some of the analyses that can be carried out on the unfiltered data collected through the proposed object detection system. For this purpose, we use the data collected in a real scenario, namely, the "Exhibition of Fake Art" (https://www.facebook.com/indifesadellabellezza/ (accessed on 13 October 2021)) of Roma Tre University. For each frame captured by the camera, the system provides the following information in output:

- The coordinates in pixels of the four corners of the bounding box that contains the object;
- The class of the recognized object;
- The confidence score of the detection.

From this data, the system can derive the center in pixels of the badge and its distance in meters from the camera (see Section 3.1). Obviously, it is possible to map the data in pixels to geometric coordinates and vice versa, only after camera calibration. Graphing this data not only makes its analysis more effective but also facilitates the use of information by the museum staff and all the operators in the field interested in making the museum data-driven. The system, therefore, not only allows information on the individual user or groups of users to be obtained but also provides the information needed to better manage the visitor flow in the various rooms [56]. The following graphs are taken from a video in which four visitors are present in the room. Specifically, the visitors are in front of the artwork and the camera is positioned above it at a height of 2.20 m (Figure 2 shows a frame of the video).

One of the possible analyses can be performed on the trajectories followed by visitors in the room. For example, in the scatterplot shown in Figure 8 it can be observed how the behavior followed by the green visitor (badge_3) differs from the other three, as the visitor tends to remain in the same position.

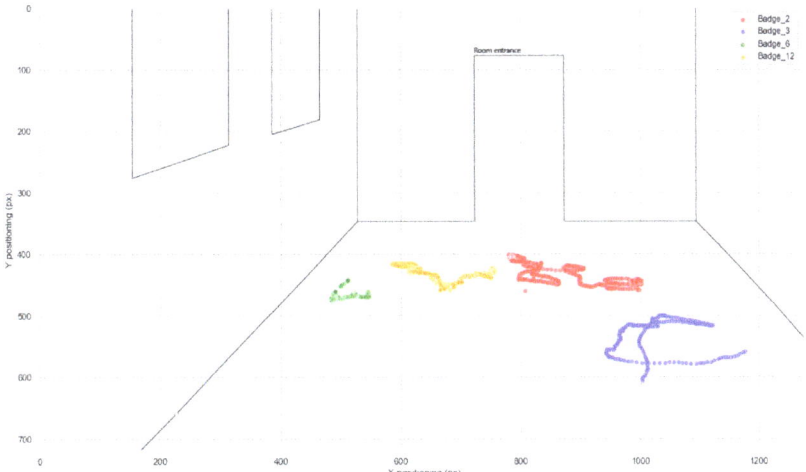

Figure 8. 2D scatterplot of four visitors in the sketched environment.

Figure 9 reports the processing output on the initial video frame.

The data from the monitoring of different environments could be easily integrated with each other to provide heatmaps. This analysis could also be useful for the museum staff to identify any problems in the fruition of the artworks, due, for example, to their arrangement or lighting.

Figure 9. 2D scatterplot of Figure 8 reported on one of the video frames.

From the video analysis, it is possible to easily obtain temporal information by knowing the number of frames per second captured by the camera. For example, from the graph shown in Figure 10, it is possible to obtain accurate and complete information relating both to the time spent by the visitor in front of a specific artwork and to their distance from it. The data collected confirm the differences in the behavior of the four monitored visitors. In particular, the green visitor slightly changes their position and remains in front of the artwork throughout the video, while the blue visitor is detected only from a certain instant of time and tends to change position more often to finally exit the framing of the camera.

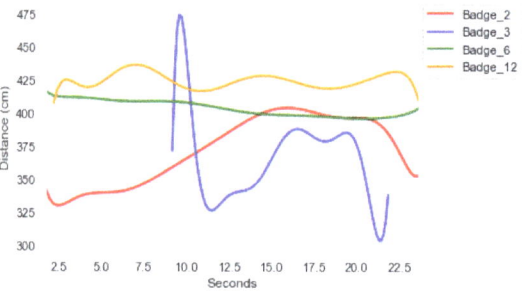

Figure 10. Badge-camera distance as a function of the time (related to four visitors) obtained through polynomial regression of order five.

The information collected can be further integrated with each other to generate 3D scatterplots like the one shown in Figure 11. The accuracy and completeness of the data are such that it can be supplied as input to graphic libraries such as Plotly's Python graphing library or advanced tools like Blender to generate particularly expressive and informative 3D heatmaps.

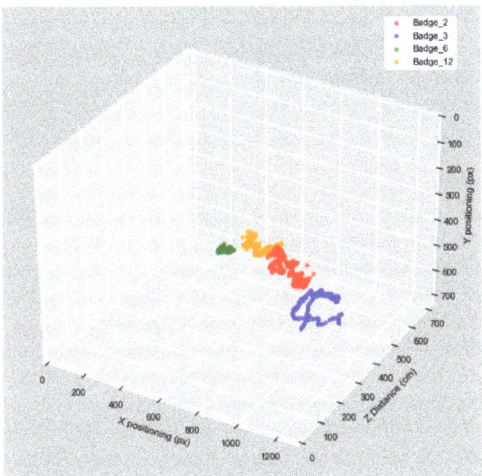

Figure 11. 3D scatterplot of four visitors.

4.3. Database with Collected Data

In order to make the analysis of data collection easy and at the same time effective, we propose the following database implementation and give some sample queries that could cover the most basic and useful needs when a museum staff member wants to extract useful information about visitor behavior from the database. The data collected through the proposed system can be stored in a data structure that supports spatial and temporal analyses of visitor behavior, such as those seen in Section 4.2. Let us suppose, for example, that we have m cameras and n badges. Each camera detects, at a generic timestamp, a badge at certain coordinates from the camera. We can store all those detections in a database composed of two tables. The first table, called *position*, has attributes (TIMESTMP, CAMERA_ID, BADGE_ID, X, Y, Z) and the second table, called *camera*, has attributes (CAMERA_ID, CT, X, Y, Z). A single tuple (t, c_id, b_id, x, y, z) of *position* represents a detection at timestamp t from the camera c_id of the badge b_id at coordinates x, y, z with respect to camera c_id. A single tuple (c_id, ct, x, y, z) of *camera* represents the coordinates x, y, z of the camera c_id in relation to the museum. The value ct is the time period of a frame. If f is the frame rate of the camera, then we have $ct = 1/f$. For the sake of simplicity, hereafter, we suppose that ct assumes the same value for all cameras (i.e., $1/24$ s), but all the discussion can be extended with simple and minimal modifications to the general case, in which cameras can have different frame rates. We note that, whilst the table *position* is fed by the detections of the model, the table *camera* is determined and created in advance by the system supervisor. For instance, it can be convenient to create the view *dist_positions* using the SQL Listing A1, shown in Appendix A.

In order to build the track of a visitor wearing the badge b_id in the time lapses between timestamp t_0 and timestamp t_1, that is, the ordered timestamp sequence of the visitor positions, we may execute the SQL Listing A2.

We also add to the database another table called *grid* with attributes (GRID_ID, X, Y, Z), in which for each tuple (g_id, x, y, z), x and y represent the coordinates of the lower-left corner of a square inside the museum and z is the height of the floor (with respect to the museum) to which the square is referred. The width w of each square of the grid can be set, for example, to 0.5 m. Furthermore, we suppose that the badge is located somewhere between the ground floor, whose height is the coordinate z of the square and $z = 2.7$ m. In order to build a heatmap, that is, a visual indication that shows where the visitors spend more or less of the time in a given grid square inside the museum, we associate with each square element g of the grid a value that represents the sum of the number of seconds any visitor was present inside the square g in the time between t_0 and t_1. Listing A3 returns such values for all elements of the grid. Moreover, through Listing A4 we can detect how much time a person, identified by the badge b_id, stationed or passed in front of an artwork of the museum. We assume that the constants AX, AY, AW, and AH are given as parameters of the query and they represent what we consider as the space in front of the artwork and AZ the height of the floor (with respect to the museum) of this rectangle.

5. Discussion

In this paper, we reviewed some of the most authoritative and recent works proposed in the literature for indoor localization, focusing on those deployable in museum environments. As we saw in Section 2, each technology has pros and cons. Consequently, we have proposed a solution that requires simple badges and off-the-shelf RGB cameras and relies on deep learning techniques to monitor visitors and their behavior. The source code of the proposed system is available online (see Section 3). The main advantages of such a solution consist in the low cost of the instrumentation and the accuracy ensured by the detection and classification procedures based on the latest generation of Convolutional Neural Networks. As for the first point, the system leverages inexpensive badges and off-the-shelf cameras, which makes it economically viable. As for the second aspect, in Section 3, we have seen that the accuracy of our approach in estimating the visitor position can be pushed on the order of 10^{-2} m. In this regard, it should be noted that the operation of the Faster R-CNN, on which our system relies, does not depend on the number of objects to be recognized within the image. Therefore, the model accuracy is not affected if, in an image, there is only one badge or there are one hundred badges to be recognized. In our experimental trials, we limited ourselves to 12 badges because the SARS-CoV-2 restrictions did not allow us to test our system with more users. Anyway, the performance in terms of Average Precision and Average Recall remained unchanged when there were 12 badges to be recognized within the frame or when there was only one. What could instead be affected is system efficiency, if the number of region proposals in output from the Region Proposal Network significantly increases. We performed our experimental evaluation using an NVIDIA QUADRO P2000 GPU, which allowed us to process four frames per second even when the badges to be recognized were 12. However, it is reasonable to expect that if the badges to be recognized within an image become hundreds, more performing hardware solutions are needed (e.g., based on the use of several GPUs in parallel) if we want to preserve the real-time nature of the process.

However, the possible advantages are not limited to those mentioned above. First of all, the intrusiveness of the proposed approach is minimal. It is sufficient for the visitor to wear a simple badge like those provided free of charge by the organizers of events and conferences to be identified and tracked by the proposed technology. Therefore, no active involvement of the visitor is needed, nor are they required to bring additional devices such as active and passive sensors, PDAs, smartphones, or portable GPUs. As a result, our technology is not affected by power consumption issues. Another positive aspect of our solution is its coverage. It is sufficient that in each point of interest there is a commercial camera to make recognition possible. Moreover, the visitor timing and tracking system could also exploit visual data from any video surveillance systems present in most national and international museums and exhibitions. A further benefit of our solution is the possibility of integrating additional functionalities into it. As seen in Section 4.1, our system can efficiently capture other visitor aspects in addition to the badge worn. More specifically, the system can associate the visitor's face with their badge through a simple correlation (see Figure 12).

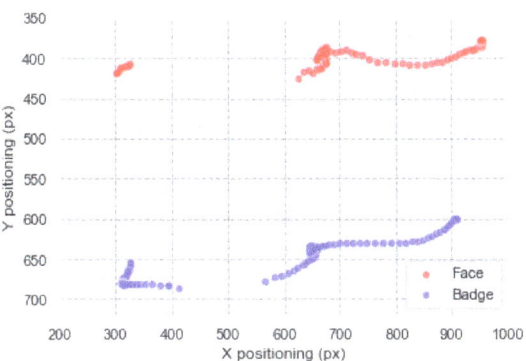

Figure 12. Correlation between face and badge positioning.

We emphasize once again that our system does not make the recognition of the visitor's face, but only its detection, for privacy reasons. In other terms, the mapping occurs only between face and badge, and not between face and visitor. The detected face can be analyzed to derive further information. It has been shown in the research literature (e.g., see [57,58]) how it is possible to analyze the user's micro facial expressions to infer information relating to emotions during the visit to predict their valence, arousal, and engagement. This information can be used to suggest objects [59] and personalized itineraries [60] based on these factors. For example, the exhibition could be organized by providing at its beginning the display of objects and artworks specially selected to derive the visitor's tastes without having to administer ad hoc questionnaires. We tested the system in a real scenario, that is, at the "Exhibition of Fake Art" at Roma Tre University. However, our experimental trials have been carried out with a low number of visitors due to SARS-CoV-2 restrictions. Generally speaking, occlusions can occur in overcrowded environments. Some noteworthy solutions have been proposed in the literature (e.g., see [61]). In our case, this problem can be mitigated by using several RGB cameras positioned in strategic positions, as shown in Figure 13, where the RGB camera is located at 4.20 m from the floor.

Figure 13. Another frame with the objects recognized by the model. It should be noted that, in this case, the camera is positioned higher than in the scene shown in Figure 2. It is now positioned at 4.20 m from the floor, but this does not affect the object detection and classification process.

Our system makes use of low-cost cameras, so the use of a large number of visual sensors would not involve a significant increase in costs.

6. Conclusions and Future Directions

Visiting museums and exhibitions around the world can indeed be an unforgettable experience. For over a century, studies have been published that show the possible relevance of their role in modern society and analyze the visitor behavior (e.g., see [62–66]. Current technology can make a decisive contribution in further improving the visitor experience, customizing it based on users' tastes and interests [2,67]. To achieve this goal, the first step is to automatically acquire information about the active user. This information can then be used for various purposes, among which:

- Provide visitors with personalized services such as recommendations of points of interest and additional textual and multimedia content [68];
- Analyze the individual and social behavior of visitors;
- Improve artwork arrangement;
- Optimize visitors' flow.

Therefore, in addition to testing our system in museums and exhibitions with a high number of visitors, we plan to concentrate our next research efforts on the design and the realization of tools that can derive the maximum benefit from the data collected through the system proposed herein.

To conclude, in this paper, we presented a deep learning-based approach to collect data regarding the visitor's experience in an accurate and comprehensive way. The solution we propose makes use of low-cost equipment (i.e., off-the-shelf RGB cameras) and requires the visitor to wear a simple badge, thus being non-intrusive. We do hope that our research efforts will contribute to making the museum visiting experience even more enjoyable, thus persuading more and more people to leave the comfort of their homes and experience cultural heritage on site.

Author Contributions: Conceptualization, A.F., C.L., M.M. and G.S.; investigation, A.F., C.L., M.M. and G.S.; methodology, A.F., C.L., M.M. and G.S.; software, A.F., C.L., M.M. and G.S.; validation, A.F., C.L., M.M. and G.S.; writing—original draft, A.F., C.L., M.M. and G.S.; writing—review and editing, A.F., C.L., M.M. and G.S. All authors have read and agreed to the published version of the manuscript.

Funding: This research received no external funding.

Institutional Review Board Statement: Not applicable.

Informed Consent Statement: Not applicable.

Data Availability Statement: Not applicable.

Acknowledgments: We would like to sincerely thank Professor Giuliana Calcani and her colleagues from the Department of Humanities of Roma Tre University for allowing us to experience the system proposed herein in a real scenario, namely, the "Exhibition of Fake Art" at Roma Tre University.

Conflicts of Interest: The authors declare no conflict of interest.

Appendix A. SQL Queries

This appendix contains all the queries described in Section 4.3.

Listing A1: View creation.

```sql
CREATE VIEW dist_positions AS
SELECT DISTINCT P.TIMESTMP, P.BADGE_ID, C.CT
/* Changing the reference system */
P.X + C.X AS X,
P.Y + C.Y AS Y,
P.Z + C.Z AS Z
FROM positions P, camera C
WHERE P.CAMERA_ID = C.CAMERA_ID
```

Listing A2: Badge positions tracking.

```sql
SELECT TIMESTMP, BADGE_ID, X, Y, Z
FROM dist_positions
WHERE BADGE_ID = bid AND
P.TIMESTMP BETWEEN t_0 AND t_1
ORDER BY TIMESTMP
```

Listing A3: Heatmap.

```sql
SELECT G.GRID_ID, SUM(P.CT)
FROM grid G LEFT JOIN dist_positions P ON
WHERE P.TIMESTMP BETWEEN t_0 AND t_1 AND
P.X BETWEEN G.X AND G.X + 0.499 AND
P.Y BETWEEN G.Y AND G.Y + 0.499 AND
P.Z BETWEEN G.Z AND G.Z + 2.70
GROUP BY G.GRID_ID
```

Listing A4: Badge time tracking.

```sql
SELECT SUM(CT)
FROM dist_positions
WHERE TIMESTMP BETWEEN t_0 AND t_1 AND
X BETWEEN AX AND AX + AW AND
Y BETWEEN AY AND AY + AH AND
Z BETWEEN AZ AND AZ + 2.70 AND
BADGE_ID = bid
```

References

1. Mokatren, M.; Kuflik, T.; Shimshoni, I. Listen to What You Look at: Combining an Audio Guide with a Mobile Eye Tracker on the Go. In *Proceedings of the 10th International Workshop on Artificial Intelligence for Cultural Heritage Co-Located with the 15th International Conference of the Italian Association for Artificial Intelligence (AI*IA 2016), CEUR Workshop Proceedings*; Bordoni, L., Mele, F., Sorgente, A., Eds.; CEUR-WS.org: Aachen, Germany, 2016; Volume 1772, pp. 2–9.
2. Ardissono, L.; Kuflik, T.; Petrelli, D. Personalization in Cultural Heritage: The Road Travelled and the One Ahead. *User Model. User-Adapt. Interact.* **2012**, *22*, 73–99. [CrossRef]
3. Lanir, J.; Kuflik, T.; Sheidin, J.; Yavin, N.; Leiderman, K.; Segal, M. Visualizing Museum Visitors' Behavior: Where Do They Go and What Do They Do There? *Pers. Ubiquitous Comput.* **2017**, *21*, 313–326. [CrossRef]
4. Seidenari, L.; Baecchi, C.; Uricchio, T.; Ferracani, A.; Bertini, M.; Del Bimbo, A. Deep Artwork Detection and Retrieval for Automatic Context-Aware Audio Guides. *ACM Trans. Multimed. Comput. Commun. Appl.* **2017**, *13*, 35:1–35:21. [CrossRef]
5. Martella, C.; Miraglia, A.; Frost, J.; Cattani, M.; Steen, M.V. Visualizing, clustering, and predicting the behavior of museum visitors. *Pervasive Mob. Comput.* **2017**, *38*, 430–443. [CrossRef]
6. Stępień, J.; Kołodziej, J.; Machowski, W. Mobile user tracking system with ZigBee. *Microprocess. Microsyst.* **2016**, *44*, 47–55. [CrossRef]
7. Rocchi, C.; Stock, O.; Zancanaro, M.; Kruppa, M.; Krüger, A. The museum visit: Generating seamless personalized presentations on multiple devices. In *Proceedings of the 9th International Conference on Intelligent User Interfaces*; ACM: New York, NY, USA, 2004; pp. 316–318.
8. Oppermann, R.; Specht, M. A Nomadic Information System for Adaptive Exhibition Guidance. *Arch. Mus. Inform.* **1999**, *13*, 127–138. [CrossRef]
9. D'Aniello, G.; Gaeta, M.; Orciuoli, F.; Sansonetti, G.; Sorgente, F. Knowledge-based smart city service system. *Electronics* **2020**, *9*, 965. [CrossRef]
10. Sansonetti, G.; Gasparetti, F.; Micarelli, A.; Cena, F.; Gena, C. Enhancing Cultural Recommendations through Social and Linked Open Data. *User Model. User-Adapt. Interact.* **2019**, *29*, 121–159. [CrossRef]
11. Trejo, K.; Angulo, C.; i Satoh, S.; Bono, M. Towards robots reasoning about group behavior of museum visitors: Leader detection and group tracking. *J. Ambient Intell. Smart Environ.* **2018**, *10*, 3–19. [CrossRef]
12. Centorrino, P.; Corbetta, A.; Cristiani, E.; Onofri, E. Measurement and analysis of visitors' trajectories in crowded museums. In Proceedings of the IMEKO TC4 International Conference on Metrology for Archaeology and Cultural Heritage (MetroArchaeo 2019), Florence, Italy, 4–6 December 2019; pp. 423–428.
13. Rashed, M.G.; Suzuki, R.; Yonezawa, T.; Lam, A.; Kobayashi, Y.; Kuno, Y. Tracking visitors in a real museum for behavioral analysis. In Proceedings of the 2016 Joint 8th International Conference on Soft Computing and Intelligent Systems (SCIS) and 17th International Symposium on Advanced Intelligent Systems (ISIS), Sapporo, Japan, 25–28 August 2016; pp. 80–85.
14. Liu, H.; Darabi, H.; Banerjee, P.; Liu, J. Survey of Wireless Indoor Positioning Techniques and Systems. *IEEE Trans. Syst. Man Cybern. Part C (Appl. Rev.)* **2007**, *37*, 1067–1080. [CrossRef]

15. Zafari, F.; Gkelias, A.; Leung, K.K. A Survey of Indoor Localization Systems and Technologies. *IEEE Commun. Surv. Tutor.* **2019**, *21*, 2568–2599. [CrossRef]
16. Fiorucci, M.; Khoroshiltseva, M.; Pontil, M.; Traviglia, A.; Del Bue, A.; James, S. Machine Learning for Cultural Heritage: A Survey. *Pattern Recognit. Lett.* **2020**, *133*, 102–108. [CrossRef]
17. Augello, A.; Infantino, I.; Pilato, G.; Vitale, G. Site Experience Enhancement and Perspective in Cultural Heritage Fruition—A Survey on New Technologies and Methodologies Based on a "Four-Pillars" Approach. *Future Internet* **2021**, *13*, 92. [CrossRef]
18. Obeidat, H.; Shuaieb, W.; Obeidat, O.; Abd-Alhameed, R. A Review of Indoor Localization Techniques and Wireless Technologies. *Wirel. Pers. Commun.* **2021**, *119*, 289–327. [CrossRef]
19. Roy, P.; Chowdhury, C. A Survey of Machine Learning Techniques for Indoor Localization and Navigation Systems. *J. Intell. Robot. Syst.* **2021**, *101*, 63. [CrossRef]
20. Centenaro, M.; Vangelista, L.; Zanella, A.; Zorzi, M. Long-range communications in unlicensed bands: The rising stars in the IoT and smart city scenarios. *IEEE Wirel. Commun.* **2016**, *23*, 60–67. [CrossRef]
21. Zafari, F.; Papapanagiotou, I.; Christidis, K. Microlocation for Internet-of-Things-Equipped Smart Buildings. *IEEE Internet Things J.* **2016**, *3*, 96–112. [CrossRef]
22. Stock, O.; Zancanaro, M.; Busetta, P.; Callaway, C.; Krüger, A.; Kruppa, M.; Kuflik, T.; Not, E.; Rocchi, C. Adaptive, intelligent presentation of information for the museum visitor in PEACH. *User Model. User-Adapt. Interact.* **2007**, *17*, 257–304. [CrossRef]
23. Kuflik, T.; Lanir, J.; Dim, E.; Wecker, A.; Corra', M.; Zancanaro, M.; Stock, O. Indoor positioning: Challenges and solutions for indoor cultural heritage sites. In Proceedings of the 16th international conference on Intelligent user interfaces, Palo Alto, CA, USA, 13–16 February 2011; pp. 375–378.
24. Holm, S. Hybrid ultrasound-RFID indoor positioning: Combining the best of both worlds. In Proceedings of the 2009 IEEE International Conference on RFID, Orlando, FL, USA, 27–28 April 2009; pp. 155–162.
25. Baronti, P.; Pillai, P.; Chook, V.W.C.; Chessa, S.; Gotta, A.; Hu, Y.F. Wireless Sensor Networks: A Survey on the State of the Art and the 802.15.4 and ZigBee Standards. *Comput. Commun.* **2007**, *30*, 1655–1695. [CrossRef]
26. Maranò, S.; Gifford, W.M.; Wymeersch, H.; Win, M.Z. NLOS identification and mitigation for localization based on UWB experimental data. *IEEE J. Sel. Areas Commun.* **2010**, *28*, 1026–1035. [CrossRef]
27. Armstrong, J.; Sekercioglu, Y.A.; Neild, A. Visible light positioning: A roadmap for international standardization. *IEEE Commun. Mag.* **2013**, *51*, 68–73. [CrossRef]
28. Xiao, Y.; Ou, Y.; Feng, W. Localization of indoor robot based on particle filter with EKF proposal distribution. In Proceedings of the 2017 IEEE International Conference on Cybernetics and Intelligent Systems (CIS) and IEEE Conference on Robotics, Automation and Mechatronics (RAM), Ningbo, China, 19–21 November 2017; pp. 568–571.
29. Huang, W.; Xiong, Y.; Li, X.Y.; Lin, H.; Mao, X.; Yang, P.; Liu, Y.; Wang, X. Swadloon: Direction Finding and Indoor Localization Using Acoustic Signal by Shaking Smartphones. *IEEE Trans. Mob. Comput.* **2015**, *14*, 2145–2157. [CrossRef]
30. Hazas, M.; Hopper, A. Broadband ultrasonic location systems for improved indoor positioning. *IEEE Trans. Mob. Comput.* **2006**, *5*, 536–547. [CrossRef]
31. Desmarais, Y.; Mottet, D.; Slangen, P.; Montesinos, P. A review of 3D human pose estimation algorithms for markerless motion capture. *Comput. Vis. Image Underst.* **2021**, *212*, 103275. [CrossRef]
32. Hong, S.; Yi, T.; Yum, J.; Lee, J.H. Visitor-artwork network analysis using object detection with image-retrieval technique. *Adv. Eng. Inform.* **2021**, *48*, 101307. [CrossRef]
33. Saito, N.; Kusunoki, F.; Inagaki, S.; Mizoguchi, H. Novel application of an RGB-D camera for face-direction measurements and object detection: Towards understanding museum visitors' experiences. In Proceeding of the 13th International Conference on Sensing Technology (ICST 2019), Sydney, NSW, Australia, 2–4 December 2019.
34. Angeloni, R.; Pierdicca, R.; Mancini, A.; Paolanti, M.; Tonelli, A. Measuring and evaluating visitors' behaviors inside museums: The Co. ME. project. *SCIRES-IT-Sci. Res. Inf. Technol.* **2021**, *11*, 167–178.
35. Caldarelli, S.; Gurini, D.F.; Micarelli, A.; Sansonetti, G. A Signal-Based Approach to News Recommendation. In *CEUR Workshop Proceedings*; CEUR-WS.org: Aachen, Germany, 2016; Volume 1618.
36. Hassan, H.A.M.; Sansonetti, G.; Gasparetti, F.; Micarelli, A.; Beel, J. BERT, ELMo, USE and InferSent Sentence Encoders: The Panacea for Research-Paper Recommendation? In *RecSys 2019 Late-Breaking Results*; Tkalcic, M., Pera, S., Eds.; CEUR-WS.org: Aachen, Germany, 2019; Volume 2431, pp. 6–10.
37. Centorrino, P.; Corbetta, A.; Cristiani, E.; Onofri, E. Managing crowded museums: Visitors flow measurement, analysis, modeling, and optimization. *J. Comput. Sci.* **2021**, *53*, 101357. [CrossRef]
38. Zhao, Z.Q.; Zheng, P.; Xu, S.T.; Wu, X. Object Detection With Deep Learning: A Review. *IEEE Trans. Neural Netw. Learn. Syst.* **2019**, *30*, 3212–3232. [CrossRef]
39. He, K.; Zhang, X.; Ren, S.; Sun, J. Deep Residual Learning for Image Recognition. In Proceedings of the 2016 IEEE Conference on Computer Vision and Pattern Recognition, Las Vegas, NV, USA, 27–30 June 2016; pp. 770–778.
40. He, K.; Zhang, X.; Ren, S.; Sun, J. Identity Mappings in Deep Residual Networks. In *European Conference on Computer Vision—ECCV 2016*; Leibe, B., Matas, J., Sebe, N., Welling, M., Eds.; Springer International Publishing: Cham, Switzerland, 2016; pp. 630–645.
41. Krizhevsky, A.; Sutskever, I.; Hinton, G.E. ImageNet classification with deep convolutional neural networks. *Commun. ACM* **2017**, *60*, 84–90. [CrossRef]

42. Mezzini, M.; Limongelli, C.; Sansonetti, G.; De Medio, C. Tracking Museum Visitors through Convolutional Object Detectors. In *Adjunct Publication of the 28th ACM Conference on User Modeling, Adaptation and Personalization*; UMAP'20 Adjunct; ACM: New York, NY, USA, 2020; pp. 352–355. [CrossRef]
43. Liu, L.; Ouyang, W.; Wang, X.; Fieguth, P.; Chen, J.; Liu, X.; Pietikäinen, M. Deep learning for generic object detection: A survey. *Int. J. Comput. Vis.* 2020, *128*, 261–318. [CrossRef]
44. Mezzini, M. Polynomial time algorithm for computing a minimum geodetic set in outerplanar graphs. *Theor. Comput. Sci.* 2018, *745*, 63–74. [CrossRef]
45. O'Rourke, J. *Art Gallery Theorems and Algorithms*; Oxford University Press, Inc.: New York, NY, USA, 1987.
46. Hartley, R.; Zisserman, A. *Multiple View Geometry in Computer Vision*, 2nd ed.; Cambridge University Press: New York, NY, USA, 2003.
47. Hsu, C.C.; Lu, M.C.; Wang, W.Y.; Lu, Y.Y. Distance measurement based on pixel variation of CCD images. *ISA Trans.* 2009, *48*, 389–395. [CrossRef]
48. Ren, S.; He, K.; Girshick, R.; Sun, J. Faster R-CNN: Towards Real-Time Object Detection with Region Proposal Networks. In *Proceedings of the 28th International Conference on Neural Information Processing Systems (NIPS'15)—Volume 1*; MIT Press: Cambridge, MA, USA, 2015; pp. 91–99.
49. Zhang, H.; Deng, Q. Deep Learning Based Fossil-Fuel Power Plant Monitoring in High Resolution Remote Sensing Images: A Comparative Study. *Remote Sens.* 2019, *11*, 1117. [CrossRef]
50. Russakovsky, O.; Deng, J.; Su, H.; Krause, J.; Satheesh, S.; Ma, S.; Huang, Z.; Karpathy, A.; Khosla, A.; Bernstein, M.; et al. ImageNet Large Scale Visual Recognition Challenge. *Int. J. Comput. Vis. (IJCV)* 2015, *115*, 211–252. [CrossRef]
51. Lin, T.Y.; Maire, M.; Belongie, S.; Hays, J.; Perona, P.; Ramanan, D.; Dollár, P.; Zitnick, C.L. Microsoft COCO: Common Objects in Context. In *European Conference on Computer Vision—ECCV 2014*; Fleet, D., Pajdla, T., Schiele, B., Tuytelaars, T., Eds.; Springer International Publishing: Cham, Switzerland, 2014; pp. 740–755.
52. He, K.; Gkioxari, G.; Dollár, P.; Girshick, R. Mask R-CNN. In Proceedings of the 2017 IEEE International Conference on Computer Vision (ICCV), Venice, Italy, 22–29 October 2017; pp. 2980–2988. [CrossRef]
53. Wenkel, S.; Alhazmi, K.; Liiv, T.; Alrshoud, S.; Simon, M. Confidence Score: The Forgotten Dimension of Object Detection Performance Evaluation. *Sensors* 2021, *21*, 4350. [CrossRef] [PubMed]
54. Salton, G.; McGill, M.J. *Introduction to Modern Information Retrieval*; McGraw-Hill, Inc.: New York, NY, USA, 1986.
55. Hosang, J.; Benenson, R.; Dollár, P.; Schiele, B. What Makes for Effective Detection Proposals? *IEEE Trans. Pattern Anal. Mach. Intell.* 2016, *38*, 814–830. [CrossRef]
56. Yoshimura, Y.; Krebs, A.; Ratti, C. Noninvasive Bluetooth Monitoring of Visitors' Length of Stay at the Louvre. *IEEE Pervasive Comput.* 2017, *16*, 26–34. [CrossRef]
57. Ferrato, A.; Limongelli, C.; Mezzini, M.; Sansonetti, G. Exploiting Micro Facial Expressions for More Inclusive User Interfaces. In *Joint Proceedings of the ACM IUI 2021 Workshops Co-Located with 26th ACM Conference on Intelligent User Interfaces (ACM IUI 2021)*; Glowacka, D., Krishnamurthy, V.R., Eds.; CEUR-WS.org: Aachen, Germany, 2021; Volume 2903.
58. McDuff, D.; Mahmoud, A.; Mavadati, M.; Amr, M.; Turcot, J.; Kaliouby, R.e. AFFDEX SDK: A Cross-Platform Real-Time Multi-Face Expression Recognition Toolkit. In Proceedings of the 2016 CHI Conference Extended Abstracts on Human Factors in Computing Systems (CHI EA '16), San Jose, CA, USA, 7–12 May 2016; pp. 3723–3726.
59. Sansonetti, G. Point of Interest Recommendation Based on Social and Linked Open Data. *Pers. Ubiquitous Comput.* 2019, *23*, 199–214. [CrossRef]
60. Fogli, A.; Sansonetti, G. Exploiting Semantics for Context-Aware Itinerary Recommendation. *Pers. Ubiquitous Comput.* 2019, *23*, 215–231. [CrossRef]
61. Li, Y.; Zhang, X.; Chen, D. Csrnet: Dilated convolutional neural networks for understanding the highly congested scenes. In Proceedings of the IEEE Conference on Computer Vision and Pattern Recognition (CVPR), Salt Lake City, UT, USA, 18–23 June 2018; pp. 1091–1100.
62. Robinson, E.S.; Sherman, I.C.; Curry, L.E.; Jayne, H.H.F. The behavior of the museum visitor. *Publ. Am. Assoc. Mus.* 1928, *1*, 72.
63. Melton, A.W. Visitor Behavior in Museums: Some Early Research in Environmental Design. *Hum. Factors* 1972, *14*, 393–403. [CrossRef]
64. Falk, J.H. Assessing the Impact of Exhibit Arrangement on Visitor Behavior and Learning. *Curator Mus. J.* 1993, *36*, 133–146. [CrossRef]
65. Serrell, B. *Paying Attention: Visitors and Museum Exhibitions*; G-Reference, Information and Interdisciplinary Subjects Series; American Association of Museums: Washington, DC, USA, 1998.
66. Agrusti, F.; Gasparetti, F.; Gena, C.; Sansonetti, G.; Tkalcic, M. SOcial and Cultural IntegrAtion with PersonaLIZEd Interfaces (SOCIALIZE). In *IUI '21: 26th International Conference on Intelligent User Interfaces*; Hammond, T., Verbert, K., Parra, D., Eds.; ACM: New York, NY, USA, 2021; pp. 9–11. [CrossRef]
67. Pavlidis, G. Recommender systems, cultural heritage applications, and the way forward. *J. Cult. Herit.* 2019, *35*, 183–196. [CrossRef]
68. Sansonetti, G.; Gasparetti, F.; Micarelli, A. Cross-Domain Recommendation for Enhancing Cultural Heritage Experience. In *Adjunct Publication of the 27th Conference on User Modeling, Adaptation and Personalization*; UMAP'19 Adjunct; ACM: New York, NY, USA, 2019; pp. 413–415.

Article

A Hybrid Recommender System for HCI Design Pattern Recommendations

Amani Braham [1,2,*], Maha Khemaja [3], Félix Buendía [4] and Faiez Gargouri [5]

1. Department of Computer Engineering, Universitat Politècnica de Valencia, Camino de Vera S/N, 46022 Valencia, Spain
2. ISITCOM, University of Sousse, Sousse 4011, Tunisia
3. PRINCE Research Lab, ISITCOM, University of Sousse, Sousse 4011, Tunisia; maha_khemaja@yahoo.fr
4. Escuela Técnica Superior de Informática (ETSINF), Universitat Politècnica de Valencia, Camino de Vera S/N, 46022 Valencia, Spain; fbuendia@disca.upv.es
5. MIRACL Laboratory, University of Sfax, Sfax 3029, Tunisia; faiez.gargouri@usf.tn
* Correspondence: amanibraham@gmail.com; Tel.: +34-600-21-44-52

Abstract: User interface design patterns are acknowledged as a standard solution to recurring design problems. The heterogeneity of existing design patterns makes the selection of relevant ones difficult. To tackle these concerns, the current work contributes in a twofold manner. The first contribution is the development of a recommender system for selecting the most relevant design patterns in the Human Computer Interaction (HCI) domain. This system introduces a hybrid approach that combines text-based and ontology-based techniques and is aimed at using semantic similarity along with ontology models to retrieve appropriate HCI design patterns. The second contribution addresses the validation of the proposed recommender system regarding the acceptance intention towards our system by assessing the perceived experience and the perceived accuracy. To this purpose, we conducted a user-centric evaluation experiment wherein participants were invited to fill pre-study and post-test questionnaires. The findings of the evaluation study revealed that the perceived experience of the proposed system's quality and the accuracy of the recommended design patterns were assessed positively.

Keywords: HCI; design patterns; design problems; semantic similarity; ontology models; recommender system

1. Introduction

The continuous advance in the development of Information Technology (IT) is currently witnessing a rapid growth of platforms, devices, and environments [1]. This has promoted an increase in design possibilities and a widespread interest in the study of User Interface (UI) [2] within the HCI research community to satisfy user requirements. In this context, the development of adaptive applications is attracting increasing attention, causing developers and designers to face great difficulties in designing and implementing applications that meet the dynamics of their environment. Hence, these applications open up new challenges as users need adaptive UIs that can cope with their corresponding preferences, surrounding context, and specific requirements. In general, adaptive UIs are supposed to adapt interaction contents and information processing modes automatically to deal with changing context and users' needs and disabilities at any time [3]. Nevertheless, the main drawback of these interfaces is their developmental complexity, which requires significant efforts. Recently, adaptive UIs have made tremendous progress regarding the big evolution of technology. This fact makes the task of development even more complex by requiring extra knowledge and expertise. Therefore, UI developers need to be assisted in designing and developing adaptive interfaces. One method to assist and help developers is to use design patterns so that these interfaces are developed using reusable design solutions, rather than from scratch.

The cornerstone of the design pattern concept was laid down in the architecture domain by Alexander [4]. This concept was initially meant to focus on frequent problems faced by designers in order to offer a correct solution within a particular context. The design pattern concept was later transferred to software design when Gamma et al. [5] introduced design patterns as a way to share the design solutions of experienced developers. After that, design patterns emerged in the HCI domain to capture HCI knowledge [6]. They have sparked interest in various areas, including UI and Web design [7–9]. In this sense, developers can take advantage of freely reusing existing design knowledge to elaborate efficient and adaptive UIs and save on development time. In recent decades, an increasing number of design patterns in the HCI domain have been noticed. Moreover, several design pattern repositories and catalogues have been organized and published [10–12]. The sheer amount of available design patterns offers good design solutions to recurring design problems. Nevertheless, it is difficult for developers or designers to follow all available HCI design patterns during the development of UIs and to select the right design patterns when a design problem is tackled. This is especially true when design patterns are stored within more than one repository. To overcome these issues, a supporting system that recommends appropriate design patterns is required.

Recommender systems have become an emerging research area where information overload is a major problem. In general, recommender systems are filtering systems [13] that handle the problem of information overload by retrieving the most relevant elements and services from a large amount of data. In recent decades, various approaches have been proposed for developing recommender systems. In this sense, there are a significant number of studies that introduce different recommendation approaches for selecting relevant design patterns, including text classification [14], case-based reasoning [15], and ontology-based [16] approaches. Fortunately, various recommender systems consider these approaches to (semi-)automatically select appropriate design patterns. Nevertheless, design patterns in the HCI domain have not been well-adopted within existing systems.

In this work, we propose a different approach to address the challenge of exploring design pattern recommendations in the HCI domain. The main contribution of the present approach is twofold. The first contribution of this paper relates to the development of a recommender system for selecting the most relevant HCI design patterns for a given design problem. To achieve this, we propose a hybrid recommender system based on well-accepted recommendation techniques that combines text-based and ontology-based methods and is aimed at considering semantic similarity and ontology models for retrieving relevant HCI design patterns. Moreover, the purpose of the second contribution is to validate the proposed system in terms of participant acceptance intention towards our system by assessing the perceived experience of the recommender system and the perceived accuracy of the recommended design patterns.

The remainder of this paper is organized as follows. Section 2 reviews related work on design pattern recommender systems. Section 3 introduces the proposed recommender system. Section 4 outlines the implementation of the recommender system. Section 5 presents the design of the experiment. The statistical results extracted from the experiment and the discussion of the obtained results are provided in Sections 6 and 7, respectively. Finally, a conclusion and discussion on future research work are drawn in Section 8.

2. Related Work

Recommender systems have become an emerging research area in different domains. In this context, several studies have presented recommender systems for retrieving design patterns. This section reviews several significant works related to design pattern recommendation systems and provides a critical analysis of the discussed works.

2.1. Design Pattern Recommender System

In the literature, several research studies have been carried out on the recommendation of relevant design patterns. Each of these studies adopted different recommendation techniques. Some of the existing works developed recommender systems for selecting relevant design patterns based on a text-based technique. These recommender systems are generally based on two main methods, including (i) text retrieval and text classification for natural language processing, and (ii) similarity measures between design patterns and design problem descriptions. In this sense, Hamdy and Elsayed [17] proposed a Natural Language Processing (NLP) recommender approach that was applied on a collection of 14 different software design patterns. This collection was created using pattern definitions from the *Gang-of-Four* (GoF) book, represented with a vector space model, and ranked according to similarity scores. From the collection of design patterns, retrieving the most suitable design patterns for a given design problem is based on the degree of similarity by adopting cosine similarity. Likewise, Hussain et al. [18] presented a framework that aids the classification and selection of software design patterns. Unsupervised learning and text categorization techniques were used to exploit their proposed framework. More specifically, these techniques were applied to perform the classification and the selection of software design patterns through the specification of design problem groups. This framework selects the right design pattern class for a given design problem based on the use of text classification technique and cosine similarity.

Other recommendation techniques are based on questions from which the appropriate design patterns are selected according to the answers provided by the user. The following are a couple of question-based approaches that make use of questionnaires to recommend design patterns. For instance, Youssef et al. [19] proposed a recommendation system based on the use of question-based techniques to recommend the appropriate design pattern category. This system examines the Goal Question Metric (GQM)-based tree model of questions. These questions are first answered by software engineers considering the user requirements. Then, the answers' weights are measured and the system recommends appropriate design pattern categories accordingly.

In the last few years, semantic technologies have been successfully applied in recommender systems. In particular, ontologies have been used in recommender systems to define and find relevant design patterns. In this context, Abdelhedi and Bouassidar [20] developed an ontology-based system for recommending Service-Oriented Architecture (SOA) design patterns. This recommender provides a questionnaire to users to retrieve their requirements. Using these requirements, an ontology that represents the different SOA pattern problems and their corresponding solutions is considered for recommending design patterns. This ontology was interrogated by SPARQL queries to search for the appropriate SOA design pattern. Similarly, Naghdipour et al. [21] proposed an ontology-based approach for selecting appropriate software design patterns to solve a given design problem. The presented method is based on interrogating an ontology of software patterns using queries to select the most suitable software design patterns according to the given design problem.

Recently, hybrid approaches that combine two or more recommendation techniques have come into focus. In this context, Celikkan and Bozoklar [22] have provided a recommendation tool that considers three main recommendation techniques, including text-based, case-based reasoning, and question-based technique. This tool aims to recommend adequate software design patterns for design problems whose description is text-based. To this end, the cosine similarity metric is computed to compare the design problem against design patterns, and to provide a ranked list of design patterns according to similarity measures. This list is then filtered to enhance recommendation results and to provide a refined list of design patterns considering the answers provided by designers.

2.2. Critics and Synthesis

Table 1 illustrates a comparison of the studied works with regard to the following criteria:

- Domain: design patterns have emerged out of different domains, such as software design patterns, SOA design patterns, and HCI design patterns.

- Problem input format: recommender systems require different problem input formats such as full-text, keywords, or questionnaires.
- Recommendation method: recommender systems consider various recommendation methods, namely text-based, case-based, question-based, and ontology-based methods.
- Degree of automation: the recommendation phase may be carried out semi-automatically when the role of users is required to some extent, or fully automatically without any human expert intervention for the selection of design patterns that ought to be recommended.
- Similarity approach: such recommender systems are based on the similarity of semantic or syntactic across a range of design pattern descriptions and problem scenarios.
- Knowledge support: recommender systems could support the reuse of knowledge by integrating ontology models.

Table 1. Comparison of the proposed work with existing design pattern recommendation systems.

Work	Domain	Problem Input Format	Recommendation Method	Degree of Automation	Similarity Approach	Knowledge Support
[17]	Software design patterns	Full-text	Text-based	Automatic	Syntactic (CS)	−
[18]	Software design patterns	Full-text	Text-based	Automatic	ntactic (CS)	−
[19]	Software design patterns	Questionnaire	Question-based	Semi-automatic	−	−
[20]	SOA design patterns	Questionnaire	Ontology-based	Semi-automatic	−	+
[21]	Software design patterns	Full-text	Ontology-based	x	−	+
[22]	Software design patterns	Full-text	Text-based, Case-based, Question-based	Semi-automatic	Syntactic (CS)	−
Our work	HCI design patterns	Full-text, Keywords	Text-based, Ontology-based	Automatic	Semantic	+

+ supported, − not supported, x not specified.

Although there have been many advances in the design pattern recommendation field, there are still problems to be dealt with, as can be seen in the comparative table (Table 1). For instance, we noticed that the recommendation domain covered in the aforementioned works concerns either software design patterns [17–19,21,22] or SOA design patterns [20]; nevertheless, they do not consider the HCI domain and tend to overlook HCI design patterns in practice despite the increase in design pattern collections in this emerging domain. In the present work, we provide a recommender system that covers the HCI domain by selecting the most relevant HCI design patterns according to specific problems.

Moreover, among the weaknesses that exist in previous works, one of them is the fact that the majority of these works rely on low-quality design problem input. For example, the approach presented in [17] recommends design patterns for predefined design problems that are written briefly. This fact may limit the set of real design problems in the sense that it restricts end-users' choices regarding design problem scenarios. On the contrary, the aim of the current work is to provide a more flexible recommender system that uses real design problem scenarios by offering end-users the ability to interact with the system and input the design problem, which could be based on full-text or keywords.

Furthermore, various existing works [17,18,22] adopted text-based recommendation techniques based on NLP methods and syntactic similarity. The syntactic similarity measurements aim at calculating the number of identical words using cosine similarity scores. In contrast, we propose a semantic similarity, which focuses more on the meaning and the interpretation-based similarity between design patterns and problem scenarios since it allows the integration of semantic information into the recommendation process [23]. Thus, the use of semantic similarity can greatly improve the text-based recommendation technique and, accordingly, the recommendation results. Other works are based on semi-automatic recommendation strategies. For instance, recommendations require the intervention of users to answer questionnaires [19]. Another work [20] invites users to select the appropriate design pattern category to get the recommended SOA patterns. This makes their methods rather semi-automatic. Alternatively, we propose a fully automatic recommender system that does not require any human intervention to retrieve HCI design patterns.

The use of ontology-based techniques can enhance the overall quality of recommender systems. However, limited research in this area has taken place in recommending design patterns. Existing ontology-based approaches [20,21] extract design patterns by means of queries, which are not sufficient for getting the appropriate ontology instances. On the contrary, we propose to improve ontology-based techniques by expanding ontology with inference rules together with SPARQL queries, allowing relevant design patterns to be deduced from the ontology model. Apart from queries, we consider the use of inference rules to enhance ontology's capabilities for revealing implicit knowledge and filtering the obtained recommendation results that better fit with the given design problem.

To address the gap within the existing research, this work proposes a novel recommender system that follows a hybrid method. This method combines text-based and ontology-based techniques to provide an automatic recommendation of relevant HCI design patterns. More specifically, the text-based technique uses NLP methods and semantic similarity measures, while the ontology-based technique relies on an ontology of HCI design patterns enriched with a set of SPARQL queries and inference rules.

3. Proposed Recommender System

The present work focuses on the recommendation of relevant HCI design patterns. To address this purpose, we propose a hybrid recommender system, named User Interface DEsign PAtterns Recommender (IDEPAR), which is part of the global Adaptive User Interface Design Pattern (AUIDP) framework [24]. As illustrated in Figure 1, the global AUIDP framework incorporates two main systems, the IDEPAR system (Figure 1a) and the User Interface Code Generator using DEsign Patterns (ICGDEP) system (Figure 1b). While the IDEPAR system concerns the recommendation of relevant HCI design patterns, the ICGDEP system covers the implementation of design patterns recommended by the IDEPAR system to generate the final user interface to the end-user. In this work, we only describe the IDEPAR system to focus on the automatic recommendation of HCI design patterns. As shown in Figure 1a, the IDEPAR system requires a design problem as input to retrieve the most relevant design patterns. In the following subsections, we introduce the representation of design problems along with a detailed description of the IDEPAR system's architecture.

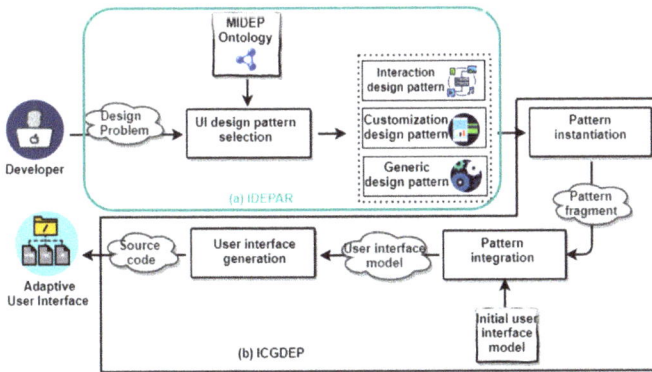

Figure 1. IDEPAR system within the AUIDP framework (**a**) IDEPAR system, (**b**) ICGDEP system.

3.1. Design Problem Representation

The IDEPAR system provides the possibility for developers to input their design problems in natural language to specify their requirements. Therefore, understanding design problems and investigating how such problems can be represented is crucial for providing recommendations that match with the given design problems. In this context, we propose an approach to design problem representation that (i) identifies the main elements that compose a design problem, and (ii) is formally represented via ontology models of design problems that will be used by the IDEPAR system. In addition, we classify design problems into atomic problems that are the smallest sub-design problems and composite problems that refer to problems that can be decomposed into simpler problems. Furthermore, we relate design problems to additional concepts, as illustrated in Table 2. The ontology model of design problem concepts within Protégé is displayed in Figure 2.

Table 2. Design problem concepts description.

Concept	Description
Design problem	A design problem can be atomic or composite
Overall description	A set of information that describes the design problem
Problem concept	Issues that constitutes a design problem (e.g., user interface issue, user characteristic, source code constraint, application functionality)
Problem category	A category associated with a design problem such as design time or runtime.

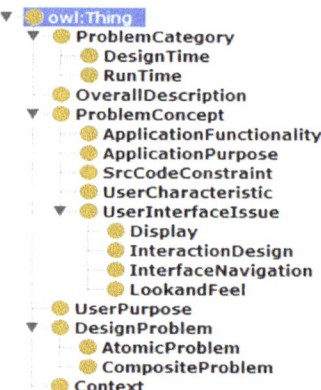

Figure 2. Design problem ontology model within Protégé.

3.2. Overall Architecture of the IDEPAR System

The IDEPAR system entails strategies to deal with design pattern recommendations regarding the text-based technique and the ontology-based technique by supporting a hybrid recommendation approach. As depicted in Figure 3, the IDEPAR system includes two main modules that interact among them, including the NLP module and the semantic module. A brief description of each module is introduced in the following subsections.

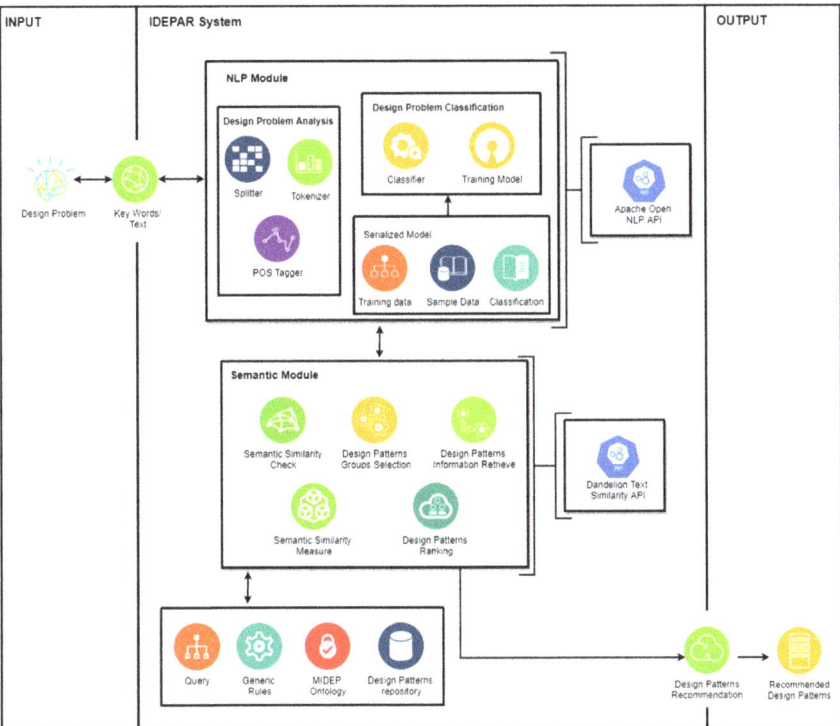

Figure 3. IDEPAR system architecture.

3.2.1. NLP Module

The NLP module is in charge of preprocessing the given design problems using a text-based technique. As input, it takes the definition of the design problem, which could be based on a set of key words or full-text format, and generates categories for each atomic problem. This module covers two main phases:

- Design problem analysis phase: At this phase, the NLP module preprocesses the given design problem. Then, it decomposes composite design problems into atomic ones. In particular, the NLP module applies three main strategies that consider the standard information retrieval method, including sentence split, tokenization, and Part of Speech (POS) tagging. The first strategy consists of splitting the composite design problems into atomic design problems. The second focuses on turning atomic design problems into small textual fragments, called tokens. The third strategy annotates tokens by assigning each token to its corresponding tag.
- Design problem classification phase: At this phase, the NLP module performs an automatic classification of atomic design problems based on the NLP auto-categorization method. It affects the categories of the design problem(s) retrieved from the previous phase. This phase mainly requires a training model generated from a set of training data. The training data can be presented in a sample data document, which includes classification samples of design problems.

3.2.2. Semantic Module

The main target of the semantic module is to perform an automated reasoning over the MIDEP ontology [24,25] and to select the most relevant design patterns based on an ontology-based technique. In the following, we present an overview of the MIDEP ontology and describe the workflow of the semantic module.

- The MIDEP ontology:

The MIDEP ontology is a modular ontology that is built using the NeOn methodology [26]. This ontology presents a modeling solution for tackling recurring design problems related to user interfaces. As depicted in Figure 4, we distinguished three main modules that constitute the MIDEP ontology, including the design pattern module, the user profile module, and the user interface module. The proposed IDEPAR system considers a collection of 45 HCI design patterns that are formalized within the MIDEP ontology. A partial list of these design patterns, along with their corresponding design pattern category, group, problem, and solution, is illustrated in Table 3.

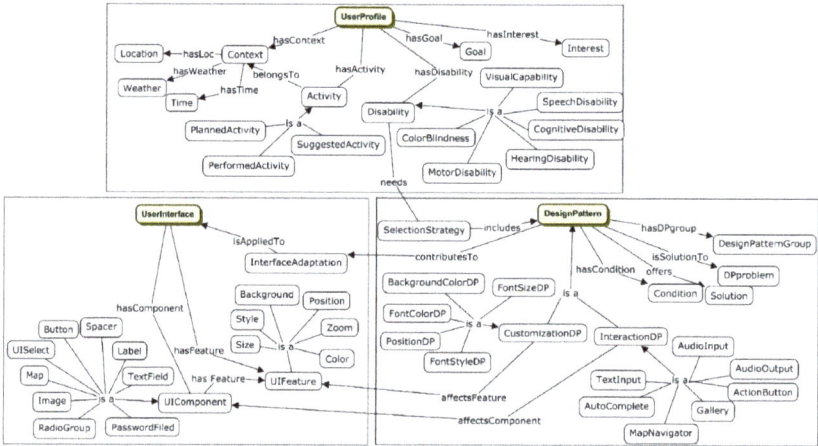

Figure 4. MIDEP ontology model.

Table 3. Partial list of HCI design patterns.

Design Pattern Group	Design Pattern Category	Design Pattern Name	Design Pattern Problem	Design Pattern Solution
Navigation	Interaction	MapNavigator	User needs to find a location of choice on a map	Display map navigator element
		Menu	User needs to access the main navigation	Repeat the main navigation on the bottom of the page
		NavigationTab	Content needs to be separated into sections and accessed using a flat navigation structure	Display a horizontal bar containing the different sections or categories
FontColor	Customization	LightFont DarkFont ColoredFont	User has difficulties perceiving font color	Set light font color Set dark font color Set colored font color

- Semantic module workflow:

The semantic module workflow, depicted in Figure 5, takes design problem categories affected by the NLP module as input and outputs a list of the recommended design patterns. A detailed description of each phase is provided below.

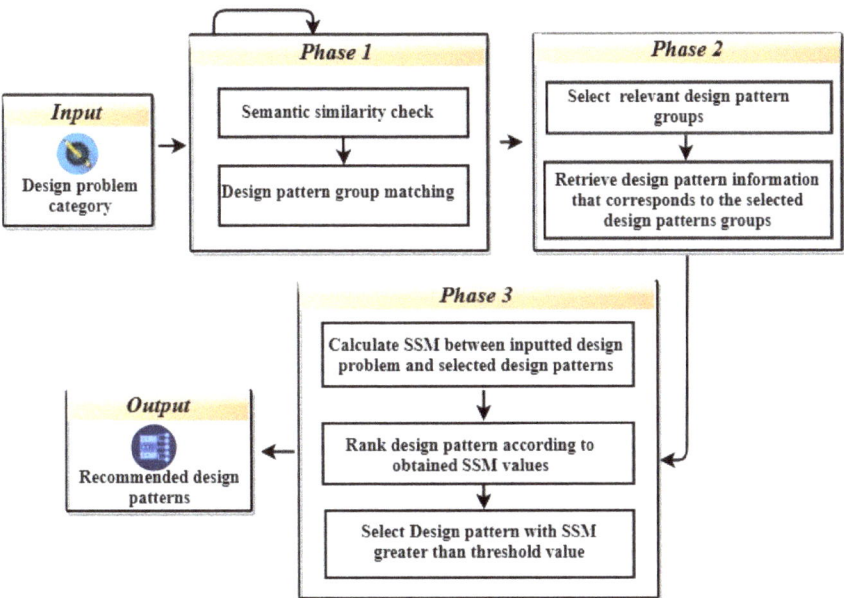

Figure 5. Semantic module workflow.

1. Phase 1: The workflow starts by checking the semantic similarity between design problem categories and design pattern group. First, the semantic module calculates the Semantic Similarity Measures (SSM) and creates relationships between concepts that are semantically similar. In Figure 6, we provide an algorithm that illustrates this process in more detail. After that, the semantic module performs a matching between MIDEP ontology instances, including the design problem concepts and design pattern group. At this level of matching, the present module applies generic rules using the inference engine. These inference rules include an antecedent and consequent part; whenever the "conditions" presented in the antecedent part hold, the "facts" specified in the consequent part must also hold. An example of a matching rule applied in this phase is illustrated in Table 4.
2. Phase 2: The second phase within the semantic module addresses the selection of the design pattern group and their corresponding design patterns. In particular, the present module makes inferences on the MIDEP ontology using a reasoning mechanism based on the "hasDPgroup" relationship between the ontological concepts.

Table 4. A rule example for matching design pattern groups.

Rule	DesignProblem(?x) ^ isComposedofProblemConcept(?x,?a) ^ DesignPatternGroup (?y) ^ hasContext(?y,?b) ^ isSemanticallySimilarTo(?a,?b) → matcheWith(?y,?x)
Description	Design problem "x" is composed of problem concept "Conceptx", design pattern group "y" has context "a", "Conceptx" is semantically similar to "b", then "y" matches with "x".

```
Initialize x, y, ont ;  x: design problem , y: design pattern group,
                        ont: MIDEP ontology
  M   ← readModel(ont);
  Cx  ← getConcept(M,x);
  Cy  ← getContext(M,y);
  N   ← numberDPgroupInstance(M);
  for i to N do
    | SSM(Cx,Cy) ← semanticSimilarity(Cx,Cy);
  end
  foreach SSM(Cx,Cy) do
    if (SSM(Cx,Cy)!= null do
      | createObjectProperty(M, isSemanticallySimilarTo,x,y);
    end if
    else do
      | z ← newIndividual(M);
      | createObjectProperty(M, isSemanticallySimilarTo,x,z);
    end else
  end
end
```

Figure 6. Semantic similarity check algorithm.

3. Phase 3: After retrieving an initial list of design patterns, the last phase computes the SSM between the design problem categories, affected in the NLP module, and the descriptions of design patterns. Then, the semantic module ranks design patterns using the obtained SSM and selects the most relevant design patterns for the given design problem, following Equation (1):

$$SSM(A, B) > \alpha, \quad (1)$$

where "A" and "B" are the text of the design problem category and design pattern condition, respectively. "α" is a threshold value for the similarity measures. As part of the design pattern ranking process, we note that design patterns with an SSM value below 0.4 are not relevant to the design problems. Therefore, a threshold of 0.4 is considered.

4. Recommender System Implementation

4.1. Implementation of Server-Side System

The IDEPAR system was implemented as a Web service that can be operated using a RESTful API and thus can be deployed on any Java application server that is able to run services packages as jar files. Moreover, the IDEPAR system consists of seven micro-services that communicate via REST calls, as illustrated in Figure 7.

For developing the environment in which the IDEPAR system is exposed as a Web service, this work leveraged different tools and technologies, including (i) Jersey as a RESTful Web service container that provides Web services, (ii) Apache Tomcat as a Web server to host Jersey and RESTful Web services, (iii) Spring Framework for dependency injection, (iv) Apache Jena for reasoning over the MIDEP ontology and processing SPARQL queries, (v) Apache OpenNLP API for processing natural language text, and (iv) Dandelion Text Similarity API for identifying the semantic relationships between texts.

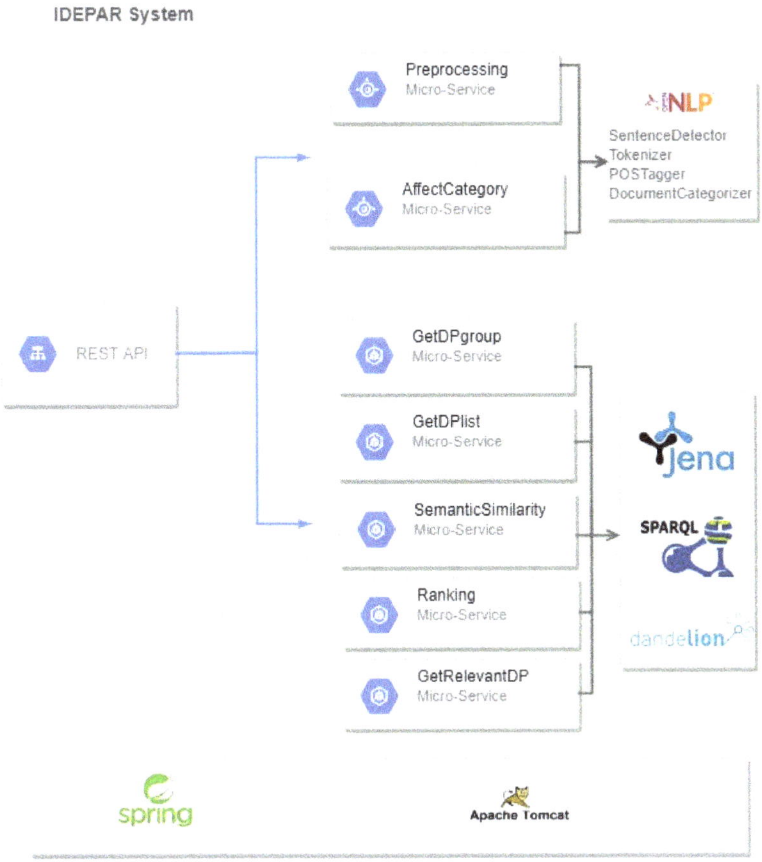

Figure 7. IDEPAR system—server-side implementation.

4.2. Design Pattern Recommendation Example

In order to illustrate a recommendation example, we describe how the IDEPAR system is applied for a particular design problem scenario. As an example of a design problem, we considered the following scenario (DPS-1): "The user cannot perceive colors, The user needs to find the location of a point of interest". A detailed description of the results of each module within the proposed recommender system is further illustrated.

The given design problem (DPS-1) was processed through various steps in the NLP module. First, in the splitter step, the design problem (DPS-1) was divided into sentences using the Sentence Detection API so that different design problem sentences could be extracted. Individual sentences were identified in the given scenario and long sentences were split into short sentences with the aim of identifying atomic design problems. As a result, the design problem (DPS-1) was split into two atomic problems: "the user cannot perceive colors" (DPS-1-1) and "The user needs to find the location of a point of interest" (DPS-1-2). In Figure 8, we provide the results of the splitter step.

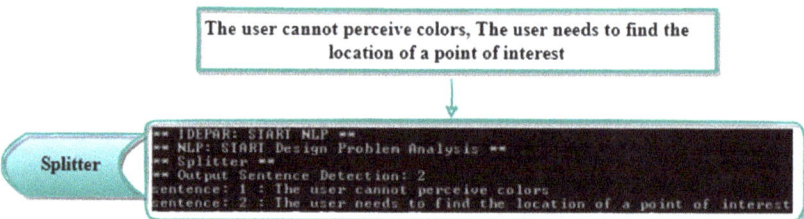

Figure 8. Splitter results for DPS-1.

Then, in the tokenizer step, the two atomic design problems were tokenized using the Tokenizer API. As illustrated in Figure 9a, each sentence was transformed into an object wherein each word was represented as a small fragment, called a token. Next, in the POS tagger step, the NLP module assigned POS tags to tokens obtained from the tokenizer step. All tokens were marked with their POS tags, as shown in Figure 9b.

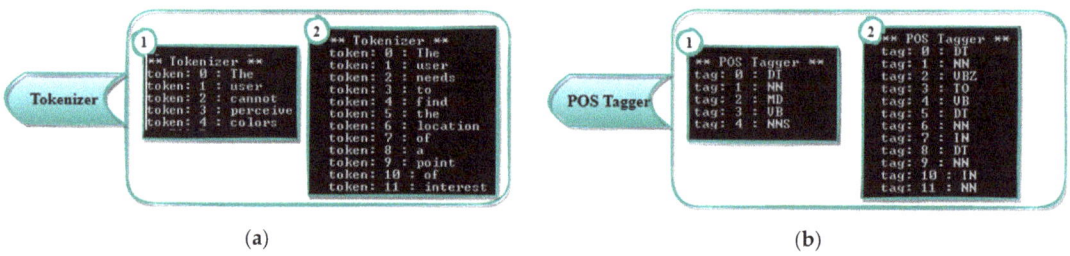

Figure 9. Results for DPS-1: (**a**) tokenizer results; (**b**) POS tagger results.

Finally, based on the tags assigned in the previous step, only nouns and verbs were part of the classifier step. The Document Categorizer API was considered to affect categories for each atomic design problem. More specifically, a training model was used to identify the appropriate categories by providing the nouns and verbs of each atomic problem. As illustrated in Figure 10, the categories "Colorblindness" and "NavigateToMap" were assigned to DPS-1-1 and DPS-1-2, respectively.

The result of the NLP module for DPS-1 is depicted in Figure 11. The given design problem (DPS-1) was passed as input parameters to the "getNLPmoduleResult" service, which communicates with the "Preprocessing" and "AffectCategory" micro-services, presented in Figure 7. The response body of the developed service was provided in a string format (Atomic design problem => Category) that would be used in the semantic module.

The output from the NLP module was used in the semantic module to retrieve the most relevant HCI design patterns for the given design problem (DPS-1). The "Colorblindness" and "NavigateToMap" categories were passed as input parameters to the "getSemanticmoduleResult" service that communicates with the micro-services, which considered Apache Jena, SPARQL queries, and Dandelion API, as presented in Figure 7. The response body of the "getSemanticmoduleResult" service was provided in JSON response format. An excerpt of the recommended design pattern list for the given design problem (DPS-1) with a description of their problems and solutions is shown in Figure 12.

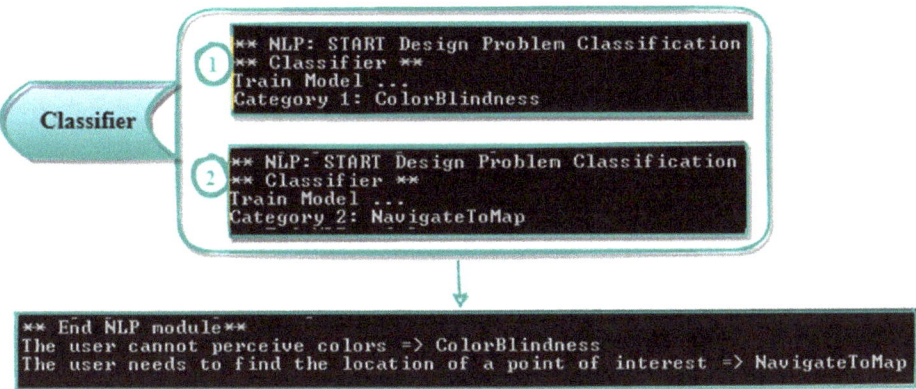

Figure 10. Classifier results for DPS-1.

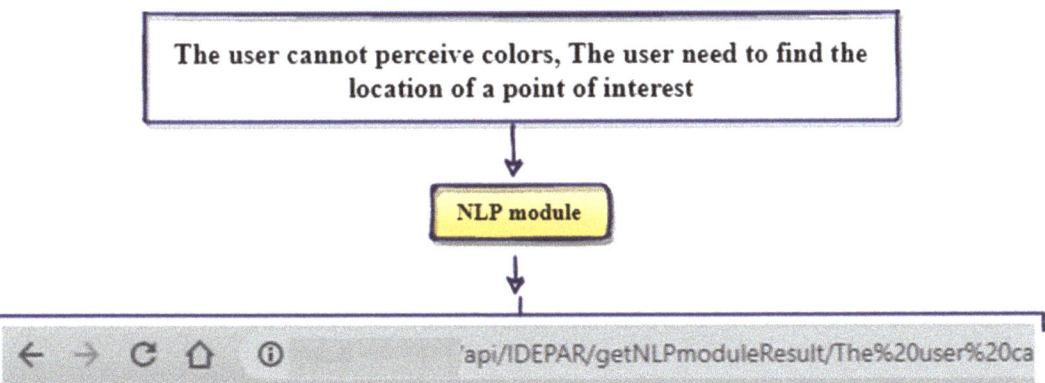

Figure 11. NLP module results for DPS-1.

4.3. Web Application Development

In order to process the design pattern recommendation requests received from developers and designers, we presented a Web application that communicates with the aforementioned REST Web services provided by the IDEPAR system. This application was developed using Spring Boot, Angular, and other technologies. Figure 13 illustrates the repository of HCI design patterns considered in the IDEPAR system.

In order to show the accomplishment of the IDEPAR system regarding various design problems, we considered the following two design problem scenarios in which keywords and text descriptions are considered, respectively.

- DPS-1: "The user cannot perceive colors, The user needs to find the location of a point of interest".
- DPS-2: "LowVision Disability".

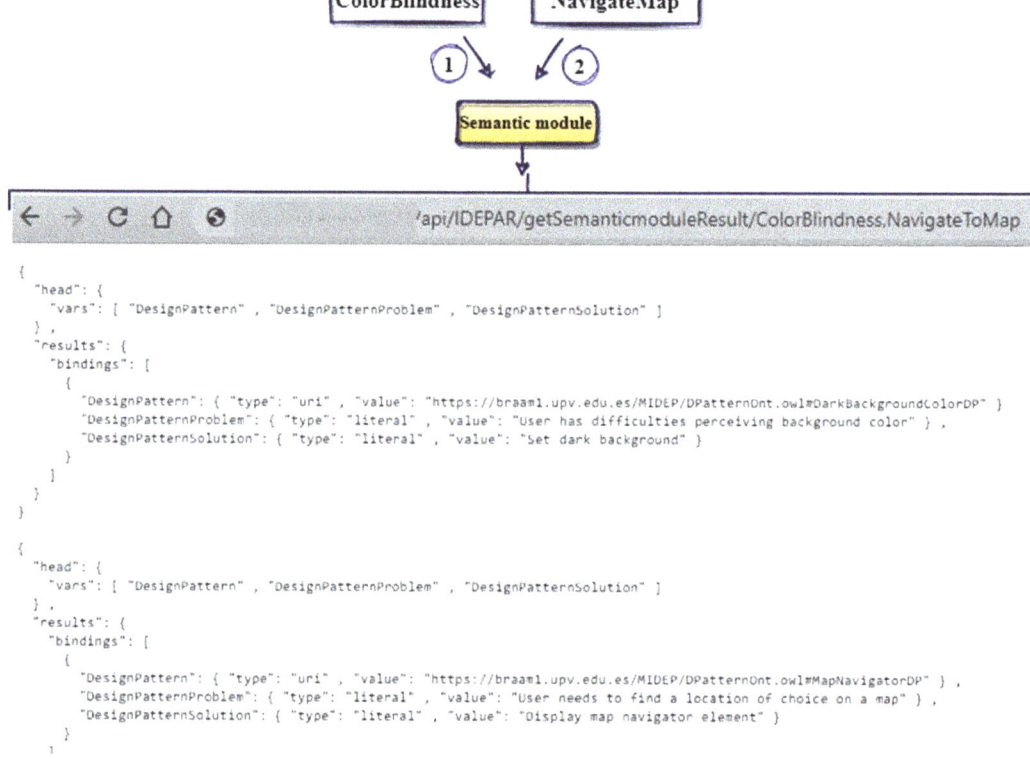

Figure 12. Semantic module results for DPS-1.

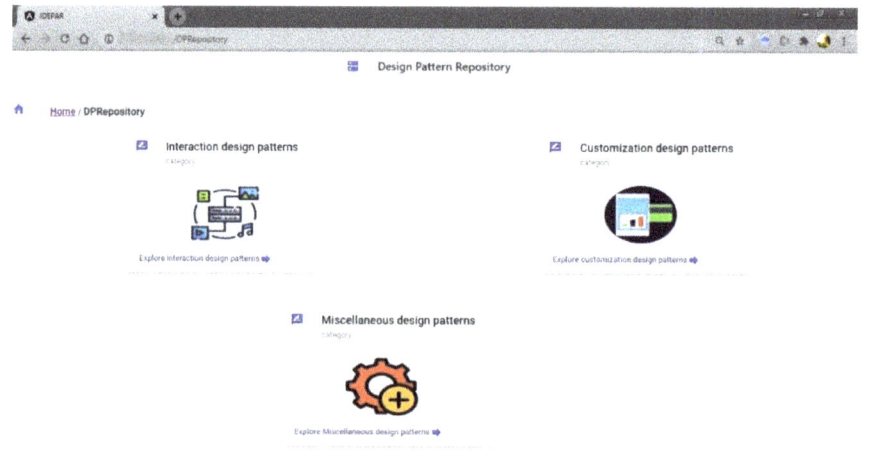

Figure 13. Design pattern repository interface.

First, to deal with the design problem (DPS-1), we present the obtained results in Figures 14 and 15. In particular, Figure 14 shows the selection of the text description relating to the first design problem (DPS-1) and Figure 15 illustrates the list of the recommended

HCI design patterns retrieved by the proposed IDEPAR system for the given design problem (DPS-1).

Second, to solve the design problem (DPS-2), Figure 16 outlines the interface for choosing DPS-2 using the user characteristic option, and Figure 17 presents the list of HCI design patterns that are recommended by the IDEPAR system to solve DPS-2. Each design pattern item is displayed with its name and problem. In this example, a list of four design patterns were recommended to solve DPS-2. As illustrated in Figure 17, by clicking on one of the recommended design patterns (e.g., FontSizeLarge) the present interface expanded the displayed item to show further information regarding the design pattern group and solution, as well as the following two actions: "choose design pattern" and "rate design pattern".

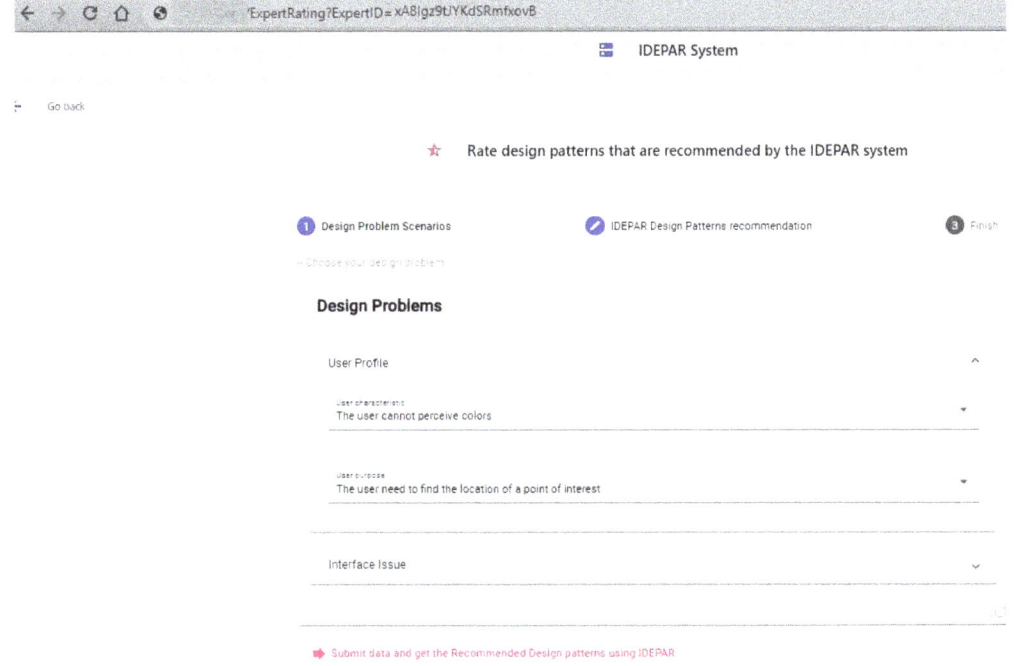

Figure 14. Selection of design problem DPS-1.

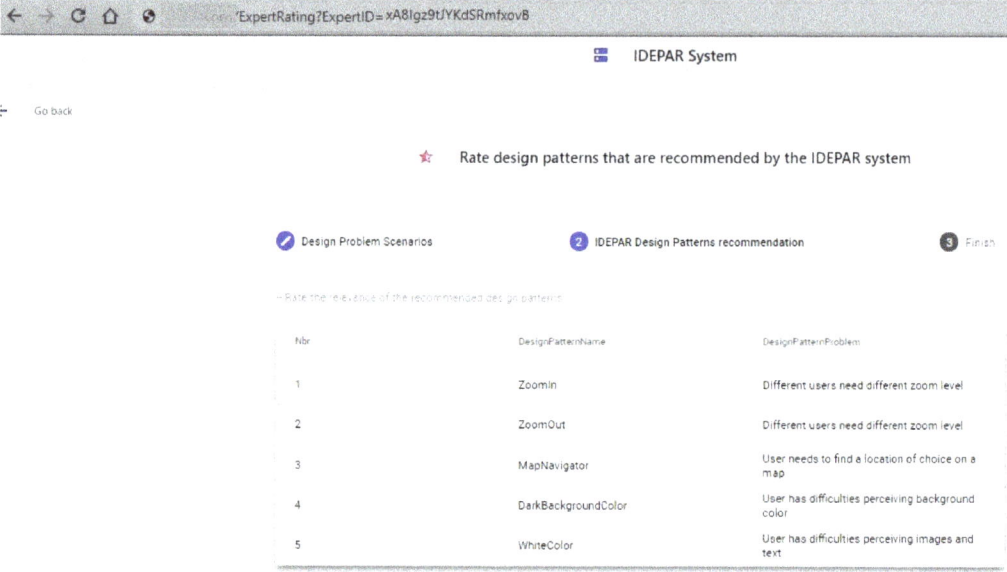

Figure 15. List of recommended design patterns for DPS-1.

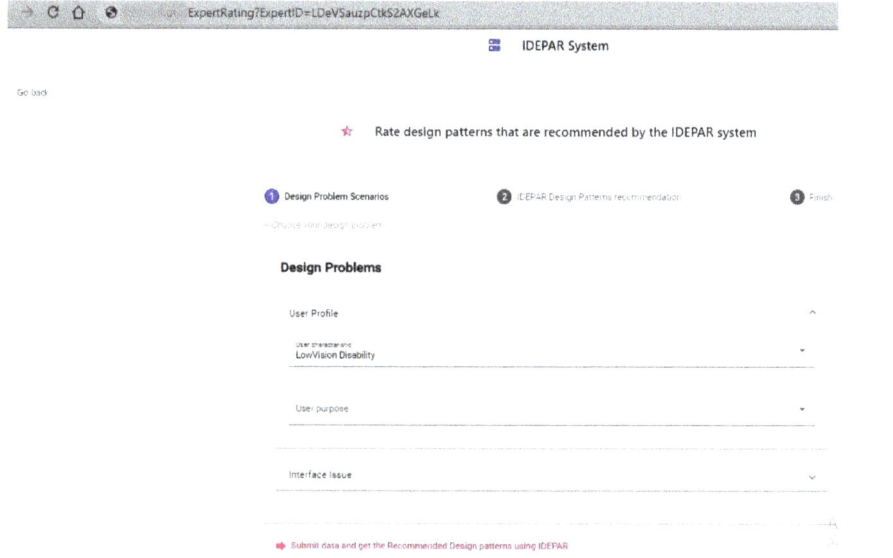

Figure 16. Selection of design problem DPS-2.

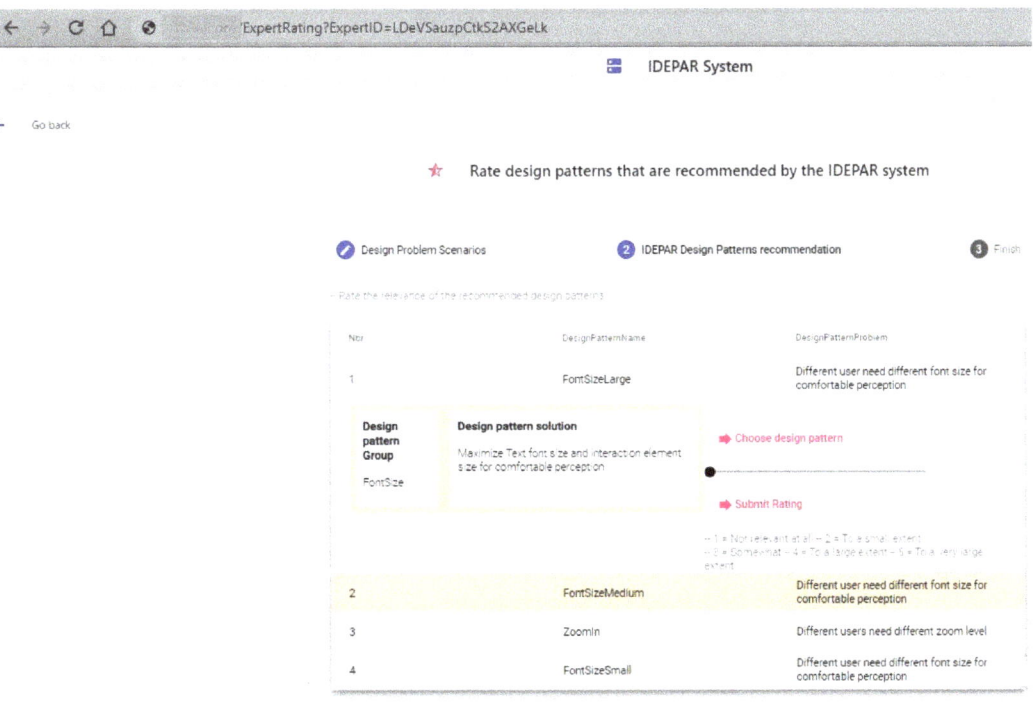

Figure 17. List of recommended design patterns for DPS-2.

5. Experimental Evaluation

We conducted an experimental study in order to achieve a comprehensive evaluation of the proposed IDEPAR system, which was designed to recommend HCI design patterns, along several relevant dimensions. To that purpose, we performed a user-centric evaluation study.

5.1. Hypotheses

The main objective of this evaluation was to figure out the impact of the recommendations on the participants' acceptance intention towards the proposed IDEPAR system. Therefore, the experiment was performed from a research perspective focused on recommending the most relevant HCI design patterns from participants' perspective interests in finding design patterns that fit with their design problem. In this context, two main research questions were formulated:

(RQ1): What is the participants' perceived experience of the IDEPAR system? To tackle this research question, we wanted to test the following hypothesis: H1 = Participants' perceived experience of the proposed system is positive.

(RQ2): What is the participants' perceived accuracy of the recommended HCI design pattern? In order to address this research question, we wanted to test the following hypothesis: H2 = Participants consider the recommended HCI design patterns as relevant and matching well with the given design problem.

5.2. Study Design

Users from different sources with a minimum experience in the HCI field were invited via mailing lists to participate in this experiment. Among the participants, 67% were female and 33% were male, with the majority being between ages 25 and 40 years (75%). Concerning the participants' academic disciplines, this study was conducted on researchers (58%), software developers (33%), and computer science students (9%). After accepting the invitation, they were informed about the steps of the evaluation study. At first, they were given a guide describing how to use the IDEPAR system through a document. After that, they were asked to access the application developed to test the proposed recommender system.

5.3. Study Protocol

In order to verify the previously mentioned hypotheses, participants were asked to carry out two main tasks. The first task was to fill the pre-study questionnaire, while the second was focused on answering the post-test questionnaire. More specifically, the pre-study questionnaire was oriented towards gathering participants' information regarding their knowledge about recommender systems and their level of expertise with HCI design patterns. Concerning the post-test questionnaire, it was mainly aimed at evaluating the quality of user experience with the IDEPAR system and the relevance of the recommended design patterns. This questionnaire was prepared based on the ResQue framework, which is a well-known user-centric evaluation recommender system for assessing user's experience and their acceptance [27]. The ResQue framework provided a wide variety of question statements that were categorized into the following four layers:

- Perceived system quality: refers to questions that assess the participant's perception of the objective characteristic related to the recommender system.
- Belief: concerns a higher level of the participant's perception of the recommender system.
- Attitude: includes questions that assess the participant's overall feeling regarding the system.
- Behavioral intention: includes questions that assess the recommender system's capability to engage participants to use it regularly.

Questions that belong to these layers mainly address participants' perceived experiences of the recommender system and accuracy of design patterns. Indeed, these questions answered the two hypotheses (H1 and H2). From the questions provided by the ResQue questionnaire, 13 questions were considered in the post-test questionnaire. In this questionnaire, the five-point Likert scale (ranging from 1 to 5) was considered as the measurement scale used to assess the degree of participants' answers, with 1 signifying "strongly disagree" and 5 signifying "strongly agree". The selected questions and their categories are presented in Figure 18. The full version of the post-test questionnaire is available in Table A1 in Appendix A.

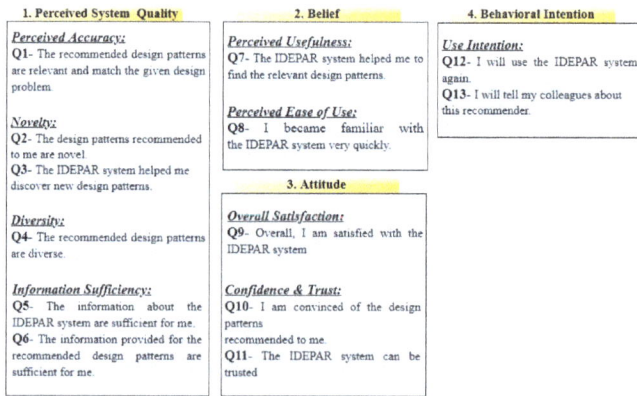

Figure 18. Representative Questions from each ResQue layer.

5.4. Statistical Analysis

In order to perform the statistical analysis of the data collected from the two questionnaires, we used IBM SPSS version 28.0 [28]. Descriptive analyses were substituted for all data. Particularly, measures of frequency (percent), central tendency (mean), and measures of dispersion (standard deviation) were used. In addition, the reliability of the post-test questionnaire's layers was assessed by Cronbach's alpha [29]. Finally, Pearson correlation was considered for identifying the correlation between the experience of the participants and their answers. For testing such a correlation, a *p*-value of ≤ 0.05 was considered to be statistically significant.

6. Results

6.1. Pre-Study Questionnaire Results

A total of 12 participants completed their tasks and were involved in the present experiment. The responses to the demographic data of the pre-study questionnaire were as follows: not familiar with recommender systems (16%), whereas the remaining participants possessed medium (42%) or high (42%) knowledge about recommender systems. Concerning the level of expertise with HCI design patterns, the majority of participants (more than 90%) had experience with HCI design patterns, wherein 8% were novice, 33% were intermediate, and 59% were advanced. Table 5 illustrates the descriptive statistics regarding the demographic data.

Table 5. Pre-study questionnaire results.

Questionnaire	Item	Percent
Knowledge of recommender systems	Low	16%
	Medium	42%
	High	42%
Level of expertise with HCI design patterns	Novice	8%
	Intermediate	33%
	Advanced	59%

6.2. Post-Test Questionnaire Results

The participants' results from the post-test questionnaires were collected and analyzed. We provide the descriptive statistics concerning the 13 questions of the post-test questionnaire in Table 6. Along with Cronbach alpha, mean, and standard deviation (SD) values, the distribution of answers for each question item was also calculated. Figures 19–22 show a divergent stacked bar that illustrates the distribution of answers provided by the participants to perceived system quality, belief, attitude, and behavioral intention layer, respectively.

Table 6. Post-test questionnaire results.

Layer	Question	Mean	SD
Perceived system quality	Q1	3.91	0.95
	Q2	3.41	1.25
	Q3	3.33	1.31
	Q4	4.33	0.74
	Q5	2.41	0.64
	Q6	2.41	0.64
Belief	Q7	3.91	0.95
	Q8	4.33	0.84
	Q9	4	1.08
Attitude	Q10	3.58	0.75
	Q11	3.83	0.68
Behavioral intention	Q12	3.33	0.74
	Q13	3.16	0.68

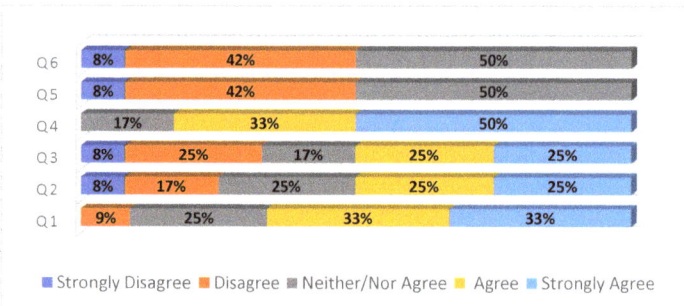

Figure 19. Distribution of answers to post-test questionnaire: perceived system quality layer.

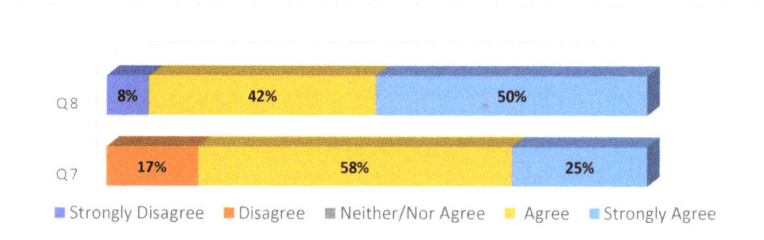

Figure 20. Distribution of answers to post-test questionnaire: belief layer.

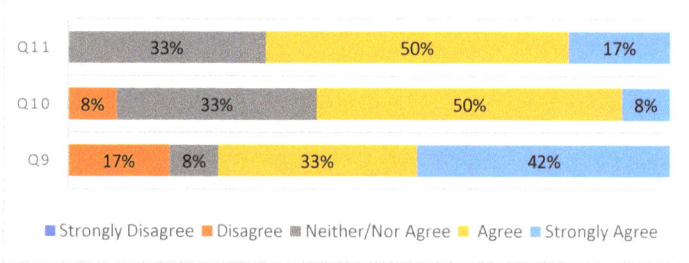

Figure 21. Distribution of answers to post-test questionnaire: attitude layer.

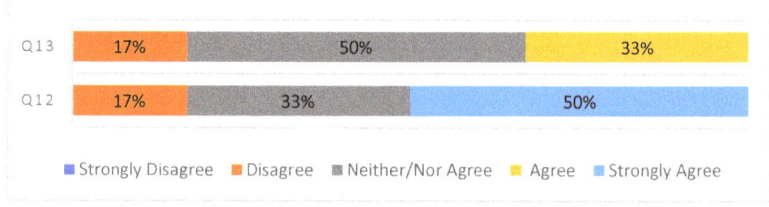

Figure 22. Distribution of answers to post-test questionnaire: behavior layer.

According to the results presented in Table 6, we observed that the mean value for many questions was above the median, with SD values below 1. More specifically, answers for the first question, with a mean value of to 3.91 (SD = 0.95), and the fourth question, with a mean value of 4.33 (SD = 0.74), reveal that participants believed that the IDEPAR system recommended relevant and diverse HCI design patterns. For the second and third questions, roughly 50% of participants perceived the novelty of the recommended design patterns (Figure 19). Meanwhile, the information sufficiency questions, including Q5 and Q6, received the lowest scores. Among all participants, 42% disagreed and 8% strongly disagreed with Q6 and Q5. The mean value for these questions was equal to 2.41 (SD = 0.64). The results indicated that participants were not well-satisfied with the sufficiency of the information about the system (Q5) and the information provided for the recommended design patterns (Q6). Additionally, Figure 20 shows that a minority of participants were satisfied with the information sufficiency, in which 8% of participants did not agree and 42% disagreed. Moreover, the mean value for the belief layer was high, being equal to 4.12. The answers of this layer reveal that more than 90% of all participants agreed that "the IDEPAR system helped them to find the relevant design patterns", involving 50% strongly agreeing for Q7. For Question 8, answers varied between 25% strongly agree and 58% agree. Thus, more than 80% of participants considered the proposed system as useful. In contrast, a minority of participants did not perceive it as useful. Concerning the attitude layer, participants' overall satisfaction was high with a mean value equal to 4 (SD = 1.08), according to the answers of Q9. Among all participants, 60% were satisfied with the recommender system. Furthermore, the mean values for questions Q10 and Q11 were equal to 3.58 (SD = 0.75) and 3.83 (SD = 0.68), respectively. Finally, the mean value for the behavioral intention was equal to 3.24 (SD = 0.71), which reveals that participants found the proposed system moderately acceptable in terms of use intentions. More specifically, the mean values for Q12 and Q13 were equal to 3.33 (SD = 0.74) and 3.16 (SD = 0.68), respectively. Figure 22 shows that a minority of participants were satisfied with the use intention; their answers vary between 33% neither/nor agree and 17% disagree for Q12, and 50% neither/nor agree and 17% disagree for Q13.

In order to verify whether the internal consistency test provided reliable results or not, we considered Cronbach's alpha criterion. This criterion has to meet a minimum threshold of 0.7 [30]. As presented in Table 7, the results of the measurements of Cronbach's alpha met the required minimum threshold for perceived system quality, attitude, and behavioral intention layers, except for the belief layer.

Table 7. Cronbach's alpha results of the post-test questionnaire layers.

Layer	Cronbach Alpha
Perceived system quality	0.78
Belief	0.61
Attitude	0.86
Behavioral intention	0.82

Moreover, we investigated the correlation between the participants' expertise and answers based on the Pearson's rank correlation coefficient. We relied on this coefficient as it provides values in the range from −1 to 1, therefore it is suitable for detecting negative correlations. In Tables 8 and 9, we provide the correlations that we found. Table 8 illustrates a significant correlation between participants' knowledge regarding recommender systems and the response of Q7 ($r = 0.873$, $p < 0.001$) and Q10 ($r = 0.711$, $p = 0.010$). Differently, Table 9 shows a significant correlation coefficient between participants' expertise with design patterns and the answers of Q1 ($r = 0.080$, $p < 0.001$), Q7 ($r = 0.744$, $p = 0.005$), Q9 ($r = 0.598$, $p = 0.040$), and Q10 ($r = 0.595$, $p = 0.041$). Analysis of the obtained Pearson coefficient results revealed that participants with good knowledge of recommender systems and a high level of experience with design patterns found that the IDEPAR system was helpful for retrieving relevant HCI design patterns ($p < 0.001$; $p = 0.005$) and the recommended design patterns were convincing ($p = 0.005$; $p = 0.041$). Among participants who had high experience with HCI design patterns, the relevance of the recommended patterns ($p < 0.001$) and their satisfaction regarding the proposed system ($p = 0.040$) was confirmed.

Table 8. Correlations between participants' knowledge about recommender systems and answers of Q7 and Q10.

	Q7	Q10
Pearson correlation	0.873 **	0.711 **
Sig. (1-tailed)	<0.001	0.010

** Correlation is significant at the 0.01 level (1-tailed).

Table 9. Correlations between participants' level of expertise with HCI design patterns and answers of Q1, Q7, Q9, and Q10.

	Q1	Q7	Q9	Q10
Pearson correlation	0.880 **	0.744 **	0.598 *	0.595 *
Sig. (1-tailed)	<0.001	0.005	0.040	0.041

* Correlation is significant at the 0.01 level (1-tailed). ** Correlation is significant at the 0.05 level (1-tailed).

7. Discussion

In this section, we discuss the interpretation of the obtained results. Firstly, regarding the perceived system quality layer, the results reveal that the majority of participants (66%) confirmed that "the recommended design patterns are relevant and match with the given design problem". Additionally, according to participants' responses to Q2 and Q3, 50% of participants agreed with the novelty of the IDEPAR system. Moreover, the overall mean of the belief layer was equal to 4.12, and thus exceed the "Agree" value. Indeed, participants generally believed that the IDEPAR system helped them to find relevant HCI design patterns and perceived the ease of use of the provided system. Furthermore, the mean value for the attitude layer was equal to 3.80 (SD = 0.83), which reveals overall satisfaction of the participants and a high trust of the IDEPAR system. Concerning the behavioral intention layer, results indicate that 50% of participants strongly agreed that they would use the IDEPAR system again, and 30% of them agreed that they would recommend the system to their colleagues. Overall, we observe that participants assigned relatively low rates, especially for the information sufficiency (Q5, Q6) and for the use intentions (Q12, Q13). These results may come from the difficulty of understanding the information provided by the system. Therefore, richer information regarding the recommended design patterns is needed. We consider this as a stimulus for the future enhancement of the proposed IDEPAR system.

Secondly, the reliability of items was conducted with Cronbach's alpha. The obtained alpha was about 0.78, 0.86, and 0.82 for all items of perceived system quality, attitude, and behavioral intention, which exceed the minimum threshold of 0.7. Indeed, the reliability was deemed good for all items, except for the belief items, for which it is considered acceptable.

After that, the Pearson coefficient was applied for the target to test the statistical significance of the correlation. In this sense, knowledge of recommender systems and level of expertise with HCI design patterns appeared to be positively correlated with the answers of perceived usefulness, confidence, and trust. In addition, the results of correlation analysis reveal that experience with HCI design patterns has a positive relationship with the perceived accuracy, as denoted with p-value < 0.001. Overall, correlation analysis indicated that several factors influence participant attitudes regarding perceived accuracy, perceived usefulness, satisfaction, confidence, and trust. The selected factors were mainly concerned about participants' knowledge about recommender systems and their level of experience with HCI design patterns.

In brief, the research and development of the presented IDEPAR system allow us to answer the two previously mentioned research questions, RQ1 and RQ2, and thus support the two hypotheses, H1 and H2. More specifically, the findings of the evaluation study reveal that (i) participants had a positive experience regarding IDEPAR system quality

(H1), and (ii) the participants' perceived accuracy of the recommended HCI design patterns was assessed positively (H2).

8. Conclusions

In this work, we have proposed the IDEPAR system, which is a hybrid recommender system aimed at recommending the most relevant HCI design patterns for a given design problem in order to help and assist developers find appropriate design patterns. The system combined two main recommendation techniques based on the use of semantic similarity along with ontology models. These ontology models were considered to offer semantics for design patterns and design problem representation. Moreover, we developed a Web application that communicates with the services provided by the proposed recommender system. This application was used to assess the IDEPAR system, along with a pre-study questionnaire and a post-test questionnaire. In order to validate our system, we conducted a user-centric evaluation experiment wherein participants were invited to fill both questionnaires. The evaluation outcomes illustrated that participants' perceived experiences of the system's quality were positive, and the recommended HCI design patterns are relevant and match well with the design problem. Nevertheless, further enhancement regarding the information provided on the system and on design patterns is needed in order to improve the proposed system regarding information sufficiency and behavioral intention. As part of future work, we will target our emphasis to enhance the proposed recommender system. We intend to take advantage of these insights obtained from the evaluation study and consider them for improving the presented system. We will also investigate the possibility to cover more complex design problems within the IDEPAR system that could be selected or presented as text descriptions entered by designers or developers. Furthermore, we plan to work on extending the approach considered in our system with a larger repository of HCI design patterns. Another interesting future work area would be to focus on a group assessment, wherein more experts in the HCI domain would be involved in the evaluation study to enhance the validation of the proposed recommender system. Finally, we intend to work on the ICGDEP system, which is the second system within the global AUIDP framework, to achieve the implementation of the design patterns recommended by the IDEPAR system and to evaluate the generated user interfaces with specific questionnaires.

Author Contributions: Conceptualization, Methodology, Formal Analysis, Investigation, Writing—Original Draft Preparation, Writing—Reviewing and Editing, Visualization, A.B.; Methodology, Investigation, Visualization, Supervision, M.K.; Investigation, Supervision, Visualization, F.B.; Visualization, F.G. All authors have read and agreed to the published version of the manuscript.

Funding: This research received no external funding.

Institutional Review Board Statement: Not applicable.

Informed Consent Statement: Not applicable.

Data Availability Statement: Not applicable.

Conflicts of Interest: The authors declare no conflict of interest.

Appendix A. Post-Test Questionnaire

The possible values for the score are, 1: Strongly disagree, 2: Disagree, 3: Neither/Nor Agree, 4: Agree, 5: Strongly agree.

Table A1. Post-test questionnaire.

Questions	1	2	3	4	5
Q1. The recommended design patterns are relevant and match the given design problem.					
Q2. The design patterns recommended to me are novel.					
Q3. The IDEPAR system helped me discover new design patterns.					
Q4. The recommended design patterns are diverse.					
Q5. The information about the IDEPAR system is sufficient for me.					
Q6. The information provided for the recommended design patterns is sufficient for me.					
Q7. The IDEPAR system helped me to find the relevant design patterns.					
Q8. I became familiar with the IDEPAR system very quickly.					
Q9. Overall, I am satisfied with the IDEPAR system.					
Q10. I am convinced of the design patterns recommended to me.					
Q11. The IDEPAR system can be trusted.					
Q12. I will use the IDEPAR system again.					
Q13. I will tell my colleagues about this recommender.					

References

1. Ruiz, J.; Serral, E.; Snoeck, M. Evaluating user interface generation approaches: Model-based versus model-driven development. *Softw. Syst. Model.* **2018**, *18*, 2753–2776. [CrossRef]
2. Gomaa, M.; Salah, A.; Rahman, S. Towards a better model based user interface development environment: A comprehensive survey. *Proc. MICS* **2015**, 5. Available online: https://www.researchgate.net/profile/Syed-Rahman-5/publication/228644279_Towards_A_Better_Model_Based_User_Interface_Development_Environment_A_Comprehensive_Survey/links/551dbfc90cf213ef063e9ca9/Towards-A-Better-Model-Based-User-Interface-Development-Environment-A-Comprehensive-Survey.pdf (accessed on 6 September 2021).
3. Letsu-Dake, E.; Ntuen, C.A. A conceptual model for designing adaptive human-computer interfaces using the living systems theory. *Syst. Res. Behav. Sci.* **2009**, *26*, 15–27. [CrossRef]
4. Alexander, C. *A Pattern Language: Towns, Buildings, Construction*; Oxford University Press: New York, NY, USA, 1977.
5. Gamma, E.; Helm, R.; Johnson, R.; Vlissides, J. Design patterns: Abstraction and reuse of object-oriented design. In *European Conference on Object-Oriented Programming*; Springer: Berlin/Heidelberg, Germany, 1993; pp. 406–431.
6. Coram, T.; Lee, J. Experiences—A pattern language for user interface design. In Proceedings of the Joint Pattern Languages of Programs Conferences PLOP, Monticello, IL, USA, 4–6 September 1996; Volume 96, pp. 1–16.
7. Tidwell, J. *A Pattern Language for Human-Computer Interface Design*; Tech. Report WUCS-98-25; Washington University: Washington, DC, USA, 1998.
8. Graham, I. *A Pattern Language for Web Usability*; Addison-Wesley Longman Publishing Co., Inc.: Boston, MA, USA, 2002.
9. Borchers, J.O. A pattern approach to interaction design. In *Cognition, Communication and Interaction*; Springer: London, UK, 2008; pp. 114–1331.
10. Van Duyne, D.K.; Landay, J.A.; Hong, J.I. *The Design of Sites: Patterns for Creating Winning Web Sites*; Prentice Hall Professional: Upper Saddle River, NJ, USA, 2007.
11. Tidwell, J. *Designing Interfaces: Patterns for Effective Interaction Design*; O'Reilly Media, Inc.: Sebastopol, CA, USA, 2010.
12. Patterns in Interaction Design. Available online: http://www.welie.com/ (accessed on 6 September 2021).
13. Konstan, J.A.; Riedl, J. Recommender systems: From algorithms to user experience. *User Model. User-Adapt. Interact.* **2012**, *22*, 101–123. [CrossRef]
14. Sanyawong, N.; Nantajeewarawat, E. Design Pattern Recommendation: A Text Classification Approach. In Proceedings of the 6th International Conference of Information and Communication Technology for Embedded System, Hua Hin, Thailand, 22–24 March 2015.
15. Gomes, P.; Pereira, F.C.; Paiva, P.; Seco, N.; Carreiro, P.; Ferreira, J.L.; Bento, C. Using CBR for automation of software design patterns. In *European Conference on Case-Based Reasoning*; Springer: Berlin/Heidelberg, Germany, 2002; pp. 534–548.
16. El Khoury, P.; Mokhtari, A.; Coquery, E.; Hacid, M.-S. An Ontological Interface for Software Developers to Select Security Patterns. In Proceedings of the 19th International Workshop on Database and Expert Systems Applications (DEXA 2008), Turin, Italy, 1–5 September 2008; pp. 297–301.
17. Hamdy, A.; Elsayed, M. Automatic Recommendation of Software Design Patterns: Text Retrieval Approach. *J. Softw.* **2018**, *13*, 260–268. [CrossRef]
18. Hussain, S.; Keung, J.; Sohail, M.K.; Khan, A.A.; Ilahi, M. Automated framework for classification and selection of software design patterns. *Appl. Soft Comput.* **2019**, *75*, 1–20. [CrossRef]

19. Youssef, C.K.; Ahmed, F.M.; Hashem, H.M.; Talaat, V.E.; Shorim, N.; Ghanim, T. GQM-based Tree Model for Automatic Recommendation of Design Pattern Category. In Proceedings of the 2020 9th International Conference on Software and Information Engineering (ICSIE), Cairo, Egypt, 11–13 November 2020; ACM Press: New York, NY, USA, 2020; pp. 126–130.
20. Abdelhedi, K.; Bouassidar, N. An SOA Design Patterns Recommendation System Based on Ontology. In *International Conference on Intelligent Systems Design and Applications*; Springer: Cham, Switzerland, 2018; pp. 1020–1030.
21. Naghdipour, A.; Hasheminejad, S.M.H. Ontology-Based Design Pattern Selection. In Proceedings of the 2021 26th International Computer Conference, Computer Society of Iran (CSICC), Tehran, Iran, 3–4 March 2021; pp. 1–7.
22. Celikkan, U.; Bozoklar, D. A Consolidated Approach for Design Pattern Recommendation. In Proceedings of the 2019 4th International Conference on Computer Science and Engineering (UBMK), Samsun, Turkey, 11–15 September 2019; pp. 1–6.
23. Mu, R.; Zeng, X. Collaborative Filtering Recommendation Algorithm Based on Knowledge Graph. *Math. Probl. Eng.* **2018**, *2018*, 9617410. [CrossRef]
24. Braham, A.; Khemaja, M.; Buendía, F.; Gargouri, F. UI Design Pattern Selection Process for the Development of Adaptive Apps. In Proceedings of the Thirteenth International Conference on Advances in Computer-Human Interactions ACHI, Valencia, Spain, 21–25 November 2020; pp. 21–27.
25. Braham, A.; Buendía, F.; Khemaja, M.; Gargouri, F. User interface design patterns and ontology models for adaptive mobile applications. *Pers. Ubiquitous Comput.* **2021**, *25*, 1–17. [CrossRef]
26. Suárez-Figueroa, M.C.; Gómez-Pérez, A.; Fernández-López, M. The NeOn Methodology for Ontology Engineering. In *Ontology Engineering in a Networked World*; Springer: Berlin, Germany, 2012; pp. 9–34.
27. Pu, P.; Chen, L.; Hu, R. A user-centric evaluation framework for recommender systems. In Proceedings of the Fifth ACM Conference on Recommender Systems, Chicago, IL, USA, 23–27 October 2011; pp. 157–164.
28. IBM Corp. *IBM SPSS Statistics for Windows*; Version 28.0; IBM Corp: Armonk, NY, USA, 2021.
29. Jnr, B.A. A case-based reasoning recommender system for sustainable smart city development. *AI Soc.* **2021**, *36*, 159–183. [CrossRef]
30. Nunnally, J.C. *Psychometric Theory 3E*; Tata McGraw-Hill Education: New York, NY, USA, 1994.

Article

A Study of a Gain Based Approach for Query Aspects in Recall Oriented Tasks

Giorgio Maria Di Nunzio *,† and Guglielmo Faggioli †

Department of Information Engineering, University of Padova, Via Gradenigo 6/b, 35131 Padova, Italy; guglielmo.faggioli@phd.unipd.it
* Correspondence: giorgiomaria.dinunzio@unipd.it
† Authors contributed equally to this work.

Abstract: Evidence-based healthcare integrates the best research evidence with clinical expertise in order to make decisions based on the best practices available. In this context, the task of collecting all the relevant information, a recall oriented task, in order to take the right decision within a reasonable time frame has become an important issue. In this paper, we investigate the problem of building effective Consumer Health Search (CHS) systems that use query variations to achieve high recall and fulfill the information needs of health consumers. In particular, we study an intent-aware gain metric used to estimate the amount of missing information and make a prediction about the achievable recall for each query reformulation during a search session. We evaluate and propose alternative formulations of this metric using standard test collections of the CLEF 2018 eHealth Evaluation Lab CHS.

Keywords: query variations; query reformulations; query performance prediction; systematic reviews

---

Citation: Di Nunzio, G.M.; Faggioli, G. A Study of a Gain Based Approach for Query Aspects in Recall Oriented Tasks. *Appl. Sci.* **2021**, *11*, 9075. https://doi.org/10.3390/app11199075

Academic Editors: Alessandro Micarelli, Giuseppe Sansonetti and Giuseppe D'Aniello

Received: 8 September 2021
Accepted: 26 September 2021
Published: 29 September 2021

Publisher's Note: MDPI stays neutral with regard to jurisdictional claims in published maps and institutional affiliations.

Copyright: © 2021 by the authors. Licensee MDPI, Basel, Switzerland. This article is an open access article distributed under the terms and conditions of the Creative Commons Attribution (CC BY) license (https://creativecommons.org/licenses/by/4.0/).

1. Introduction

The study of the query representation in Information Retrieval has driven a lot of interest in recent years [1–7]. Several works in the past [8–10] showed the positive effect on the retrieval results of fusing runs retrieved with human-made multiple formulations of the same information need. Recent studies have shown how query reformulations automatically extracted from query logs can be as effective as those manually created by users [11]. Furthermore, the performance of a system can greatly improve when the "right" formulation of an information need is selected [4,5]. One of the main challenges in this research area is being able to suggest the best performing query (or queries) among the possible variations [4,5,12–14]. For example, Thomas et al. [4] observed that, the most prominent effect in predicting the performance of a query formulation is due to the information need and not to the "query wording". In this sense, query performance predictors actually predict the complexity of the information need, rather than the one the query itself. Zendel et al. [5] pursue a slightly different task. Following the literature on reference lists [15,16] they try to predict the performance for a query using information about queries representing the same information need. Benham et al. [3] define a fusion approach for multiple query formulations based on the concept of "topic centroid", which describes the information need as combination of its formulations. Dang et al. [12] address also the problem of improving the ranking results through a query formulation selection phase. Note that, Dang et al. [12] show how they are often capable of putting the best query in the first two positions (not only the first one), a further evidence of the complexity of the task.

A use case of query performance prediction is the systematic compilation of literature review. In fact, systematic reviews are scientific investigations that use strategies to include a comprehensive search of all potentially relevant articles. As time and resources are limited for compiling a systematic review, limits to the search are needed: for example, one may want to estimate how far the horizon of the search should be (i.e., all possible cases/documents that could exist in the literature) in order to stop before the resources are finished [17]. Scells et al. [13] apply several state-of-the-art Query Performance Predictors to select the best query in the Systematic Reviews domain. They show how current Query Performance Prediction approaches perform poorly on this specific task. International evaluation campaigns have organized labs in order to study this problem in terms of the evaluation, through controlled simulation, of methods designed to achieve very high recall [18,19]. The CLEF initiative (http://www.clef-initiative.eu, accessed on 15 February 2021) has promoted the eHealth track since 2013 and, the CLEF 2018 eHealth Evaluation Lab Consumer Health Search (CHS) task [20] investigated the problem of building search engines that are robust to query variations to support information needs of health consumers.

In this paper, we study an alternative formulation of the intent-aware metric proposed by Umemoto et al. [21], in which the authors analyze a metric to estimate the amount of missing information for each query reformulation during a search session. Note that in [21] the authors do not propose an approach capable of predicting the recall of different formulations. Nevertheless, our perception is that, their approach can be easily adapted with good results also to the predictive task. In our case, our research goal is to understand whether a gain based approach can be used to predict the relative importance of each reformulation in terms of recall performance, in the context of Consumer Health Search where users need support for medical information needs.

In this sense, with respect to [21], our contribution is two-fold:

- we show that it is possible to apply the GAIN measure proposed in [21] to obtain a recall predictor over a set of formulations for the same topic;
- we furthermore show how to improve the results of such predictor by exploiting also the information obtained through the various formulations.

The paper is organized as follows: in Section 2, we present the original gain metric, while in Section 3 we define our alternative version to predict the performance in a recall-oriented fashion. In Section 4, we discuss the experimental analysis and results; while in Section 5 we give our final remarks.

2. A GAIN-Based Approach

In Umemoto et al. [21], define the intent-aware gain metric and the requirements that it should satisfy. They identify the following properties: importance, documents relevant to a central aspect of the search topic produce higher gain than those relevant to a peripheral one; relevance, highly relevant documents produce higher gain than partially relevant ones; novelty, documents relevant to an unexplored aspect produce higher gain than those relevant to a fully explored aspect.

The set of aspects A_t of a topic t is estimated through the process described in [22]: first, a set of subtopics S_t is mined given a topic t; then, the subtopics are grouped into a set of clusters C_t. These clusters are regarded as the "facets" (We use *facets* instead of *aspects* to not repeat the same term that will be use to identify the most representative subtopic.) of t. The most representative subtopic s is chosen from each cluster as formulation of the topic aspect a using the formula $a = argmax_{s \in C_t} \text{Imp}_t(s)$, where the importance of a subtopic s is defined as:

$$\text{Imp}_t(s) = \sum_{d \in D_s^N \cap d \in D_t^N} \frac{1}{\text{Rank}_t(d)} \tag{1}$$

D_s^N and D_t^N denote the sets of the top N retrieved documents for a subtopic s and the topic t, respectively, and $\text{Rank}_t(d)$ is the rank of the document d in the ranked list for t.

It is crucial to stress that the definition of *importance*, and the following definition of *gain*, derives from the assumption that there is a known "reference" topic t that describes completely the information need. For such topic t the retrieved documents can be different compared to the ones observed for a query which represents just one aspect a of the topic.

In Figure 1, we show an example of a number of subtopics found for a topic t and grouped into three clusters, each one with a representative aspect.

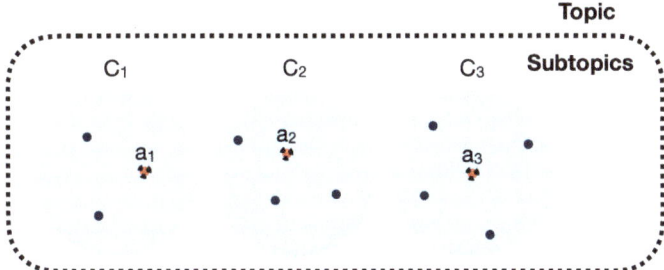

Figure 1. An example of clusters of subtopics and aspects.

The Intent-Aware Gain is defined for a set of documents D as:

$$\text{Gain-IA}_t(D) = \sum_{a \in A_t} P(a|t) \cdot \text{Gain}_{t,a}(D) \qquad (2)$$

which is a sort of expected value of the gain across the different aspects. $P(a|t)$ is the probability that an aspect a is important to the topic t, and $\text{Gain}_{t,a}(D)$ is the gain that can be obtained by the aspect a from the documents D. The importance probability for an aspect of a topic is computed as:

$$P(a|t) = \frac{\text{Imp}_t(a)}{\sum_{a' \in A_t} \text{Imp}_t(a')} \qquad (3)$$

while the gain which measures how the documents D retrieved for a query contribute to increment the information relative to a specific aspect of the topic is:

$$\text{Gain}_{t,a}(D) = \left[1 - \prod_{d \in D}(1 - \text{Rel}_{t,a}(d))\right] \qquad (4)$$

This last part that is required to compute the Intent-Aware Gain contains the term $\text{Rel}_{t,a}(d)$ which is the relevance degree of a document d with respect to an aspect a, estimated as follows:

$$\text{Rel}_{t,a}(d) = \frac{\sum_{s \in C_a} \text{Imp}_t(s) \cdot \text{Rel}_s(d)}{\sum_{s \in C_a} \text{Imp}_t(s)} \qquad (5)$$

where $C_a \in C_t$ is the cluster of subtopics belonging to the aspect a, and $\text{Rel}_s(d)$ is the relevance degree of a document d to a subtopic s estimated as $\text{Rel}_s(d) - 1/\sqrt{\text{Rank}_s(d)}$.

3. A Gain for Query Reformulations

Our initial hypothesis in this work is that: (a) we have one information need expressed with different query reformulations, and (b) the topic t is unknown. In particular, given an information need i and its set of reformulations V_i, we assume that each reformulation $q \in V_i$ is able to 'reveal' different facets of i. Consequently, we need to redefine the expression of the gain of Equation (4) as:

$$\text{Gain}_{i,q}(D) = \left[1 - \prod_{d \in D}(1 - \text{Rel}_{i,q}(d))\right] \qquad (6)$$

where i is the *information need* and q is a specific (re)formulation.

The main difference with the original approach, apart from changing variable names, is the fact that (i) we do not have a 'reference' topic t that describes completely the information need i, and (ii) we have one single cluster of query reformulations, or *variants*, V_i. For these reasons, we also need an alternative definition of relevance that adapts to our case study:

$$\text{Rel}_{i,q}(d) = \frac{\sum_{s \in V_i} \text{Imp}_q(s) \text{Rel}_s(d)}{\sum_{s \in V_i} \text{Imp}_q(s)} \quad (7)$$

where the relevance of d, retrieved by the query variant q of the information need i, is computed as the weighted average of the relevance of d with respect to all the alternative reformulations in V_i. The two terms $\text{Imp}_q(s)$ e $\text{Rel}_s(d)$ remain unaltered compared to the previous definitions:

$$\text{Imp}_q(s) = \sum_{d \in D_s^N \cap D_q^N} \frac{1}{\text{Rank}_q(d)} \quad , \quad \text{Rel}_s(d) = \frac{1}{\sqrt{\text{Rank}_s(d)}}$$

3.1. A Similarity Matrix for Recall Prediction

In the proposed context, we can think of an 'optimal' query as the one capable of combining all the diverse facets of the information need it represents. In order to estimate which query reformulation q is the closest to the unknown optimal one, we propose the following procedure:

1. we define D_q as the set of documents retrieved by q;
2. $D_i = \bigcup_{q \in V_i} D_q$ as the set of all documents retrieved by *at least* one reformulation q;
3. $\mathbf{R} \in \mathbb{R}^{|V_i| \times |D_i|}$ as the matrix of rankings for the information need i where each row corresponds to a specific reformulation and each column to a document. The value of an element $r_{k,d}$ of $\mathbf{R}$ is defined as $|D_q| - \rho_{q,d}$ where $\rho_{q,d}$ is the rank of document d retrieved by q. $\mathbf{R}$ is at the end normalized with norm $l2$.

At this point, we want to build a similarity matrix to predict the impact in terms of recall that each reformulation will have on the retrieval. We compute the cosine similarity between each pair of rows in $\mathbf{R}$, obtaining a symmetric matrix $\mathbf{S}$ where each row (or column) represents how a reformulation is similar to the others. We use the sum the k-th row (or column) of $\mathbf{S}$ to predict the importance of the k-th query; then, we order the query reformulations in decreasing order where greater values indicate a higher probability of retrieving more relevant documents. This measure describes how close each query is to the ideal "centroid" query that perfectly describes the topic.

4. Experiments and Analysis

In this section, we describe the analysis of our experiments. In particular, we want to compare the performance in terms of predicted recall among: (i) the gain defined in Equation (6), (ii) an alternative definition that mitigates some arithmetical issues, (iii) and the similarity matrix. To the best of our knowledge, this is the first effort in predicting the recall for the systematic reviews task, when multiple formulations are considered. Therefore, we are not able to directly compare it with an approach explicitly thought for such task. We thus compare our solution with traditional QPP strategies. Furthermore, we use the techniques presented in Umemoto et al. [21] as baselines.

4.1. Test Collection and Retrieval Model

The CLEF 2018 eHealth Evaluation Lab Consumer Health Search (CHS) task [20] investigated the problem of retrieving Web pages to support information needs of health consumers that are confronted with a health problem or a medical condition. One subtask (i.e., subtask 3) of this lab is aimed to foster research into building search systems that are robust to query variations (https://github.com/CLEFeHealth/CLEFeHealth2018IRtask, accessed on 15 February 2021).

Queries There are 50 information need for which we have 7 query reformulation for a total of 350 queries: the original 50 queries issued by the general public augmented with 6 query variations issued individually by 6 research students with no medical knowledge (The queries and the process to obtain them are described in http://www.khresmoi.eu/assets/Deliverables/WP7/KhresmoiD73.pdf, accessed on 15 February 2021).

Collection The collection contains 5,535,120 Web pages and it was created by compiling Web pages of selected domains acquired from the CommonCrawl [20].

Relevance Assessments For each information need, the organizers of the task provided about 500 documents assessed for a total of 25,000 topic-document pairs.

Retrieval Model The index provided by the organizers of the task, an ElasticSearch index version 5.1.1, comes with a standard BM25 model with parameters b = 0.75 and k1 = 1.2 (https://sites.google.com/view/clef-ehealth-2018/task-3-consumer-health-search, accessed on 15 February 2021).

Notice That, among the queries of the CLEF 2018 eHealth CHS collection, the two identified by ids 160006 and 164007 will not retrieve any document in common with the other variants of the same information need (at least for $N \leq 1000$). This is because the text of query 160006 is "nan", while query 164007 has a typo "pros and cons spirculina", instead of spirulina, a type of algae. We stress on this aspect since, for those queries, it will not be possible to compute the value of the gain by definition, since the intersection of their ranked list with the ones for other formulations of the same topic will be empty.

4.2. Using Traditional Query Performance Predictors Applied to Recall Prediction for Systematic Reviews

To have a better grasp on the peculiarities of the problem, we first try to apply traditional techniques of Query Performance Prediction (QPP) to our specific setting. We aim at showing that, traditional QPP techniques fail to correctly order formulations when (i) the recall is the key performance indicator; (ii) we sort formulations of the same topic and not queries representing different topics. Showing this, is a further evidence of the importance of using appropriate tools, such as the *gain* as described in Section 2 to correctly tackle the problem. More in detail, we select a set of very well-know QPP models, in order to determine whether they can be satisfactory applied to the prediction of the recall and can be used with the documents and queries that we have at hand. Traditionally, Query Preformance Predictors are divided into two macro-categories, according to the information they exploit to formulate the prediction: Pre-retrieval predictors and Post-retrieval Predictors. *Pre-retrieval predictors* analyze query and corpus statistics prior to retrieval [14,23–28] and *post-retrieval predictors* that also analyze the retrieval results [15,29–36]. Even though Pre-retrieval predictors have the advantage of being faster, since they do not need to retrieve the documents for a certain run, post-retrieval predictors typically perform better. Table 1 reports the predictors that we include in our analyses and a brief description of how they work. It is important to notice that, as for many QPP models, the models that we selected do not actually predict the performance measure. They associate a score to each of the queries, which is expected to correlate with the performance measure, but is on a different scale and cannot be used directly as estimate of the performance.

Table 1. Pre- and Post-retrieval predictive baseline models considered.

Type	Predictor	Description
pre-retrieval	max-idf [27]	It considers the maximum value of the idf (inverse document frequency) over the query terms
	mean-idf [37]	It computes the mean value of the idf over the query terms
	std-idf [37]	It uses the standard deviation of the idf over the query terms
	sum-scq [28]	Measures similarity based on cf.idf to the corpus, summed over the query terms.
	mean-scq [28]	It relies on the same value of sum-scq, but it normalizes it with the length of the query
	max-scq [28]	It relies on the same value of sum-scq, but considers only the maximum value
post-retrieval	wig [38]	Standard deviation of the top documents scores in the retrieval list.
	nqc [39]	Difference between the mean retrieval score of the top documents, scaled by the score of the entire corpus
	smv [40]	It computes the prediction considering the standard deviation of the retrieval scores

The traditional strategy to evaluate how good a query performance predictor is, consists in computing a traditional retrieval performance measure, such as Average Precision (AP), for each of the query, and determine how much such measure correlates with the prediction scores computed by the QPP model [23–26,28,32–35,38,41–43]. Notice that, there are two main aspects that might impair traditional QPP models in our specific setting:

- Remember that we are in the setting of the systematic reviews. Therefore, it is by far more important to retrieve as many as possible relevant documents, rather than putting them in the first positions. Therefore, we are not interested in estimating the AP, which is a precision based measure, but our aim is to predict which query will have the best recall;
- We do not compare queries meant for different information needs, which is the typical evaluation scenario for QPP models.

On the other hand, we aim at understanding which one, among a set of queries representing the same information need, achieve the best result.

To determine whether we are impaired by the first problem, we first apply the traditional QPP considering only the default formulation of each topic, and we compare whether the predictors are capable of correctly determining the inter-topic performance. More in detail, with this first experiment, we are interested in understanding whether the baseline predictors are capable of predicting which *topic* will have the best recall, using a single formulation for each of them. Table 2 reports the result of such analysis.

We can observe that, by looking at Table 2, the results are in line with previous similar experiments in the literature, such as [5,44]. Almost all the predictors are able to achieve a significant correlation with the recall (with level $\alpha = 0.01$). Two noticeable exceptions are represented by nqc and smv: traditionally, they are considered among the best predictors, but in this specific scenario they fail, with correlations not statistically different from 0. Our hypothesis is that, while pre-retrieval predictors tend to be estimators of the recall base of a query, and therefore tend to correlate with the recall itself, post-retrieval predictors tend to compute their predictors based on the scores that the retrieval model assigns to the top-ranked documents. In this sense, post-retrieval predictors are "top-heavy": they focus on the upper part of the ranked list of documents. This behaviour favours predicting the performance for top-heavy measures, such as Average Precision or nDCG. Instead, our task consists in predicting the recall, given a *long* list of documents. It is not unlikely that the upper part of the list of retrieved documents is saturated with relevant ones; nevertheless, we are more interested in being sure that *every* relevant document has been considered, rather than saying whether the top part of the ranked list contains relevant documents.

Table 2. Kendall's τ correlation observed between recall and prediction scores for both pre- and post-retrieval traditional predictors, if we compare the default formulations of different topics. Results are in line with correlation values previously observed in other scenarios. The symbol [†] indicates that the correlation is statistically greater than 0 at level $\alpha = 0.05$, while the [‡] indicates a significance level of 0.01, the absence of any symbol indicates that results cannot be deemed statistically greater than 0. We compute the Kendall's τ correlation at different cutoff levels of the ranked lists (100, 1000, and 10,000).

Type	Predictor	Kendall's τ		
		100	1000	10,000
pre-retrieval	max-idf	0.3185 [‡]	0.3260 [‡]	0.2875 [‡]
	mean-idf	0.2996 [‡]	0.3218 [‡]	0.2555 [‡]
	std-idf	0.2947 [‡]	0.2989 [‡]	0.2343 [†]
	sum-scq	0.2637 [‡]	0.2581 [‡]	0.1739
	mean-scq	0.3479 [‡]	0.3652 [‡]	0.3299 [‡]
	max-scq	0.3502 [‡]	0.3724 [‡]	0.2833 [‡]
post-retrieval	wig	0.3029 [‡]	0.3218 [‡]	0.2882 [‡]
	nqc	0.2865 [‡]	0.1911	0.1135
	smv	0.1797	0.1332	0.0229

We now switch the focus from predicting the performance *across* topics, to predict the performance *within* topics. Instead of comparing the performance that the standard formulation is expected to achieve for each topic, we try to sort different formulations for the same topic, according to the predicted performance. Table 3 reports the results of our analysis.

Compared to the results observed in Table 2, the performance achieved by traditional predictors for the "within"-topics prediction, is extremely lower, with very few cases of significantly positive correlation between the predicted and observed recall. Note that, even though we agree with [13] on the fact that predicting the best query among a series of formulations of a topic is a hard task, we end up with diametrically opposite conclusions. Scells et al. [13] observed severe flaws in traditional QPP techniques when *predicting the performance across topics*. On the other hand, they found the task of predicting the performance within topics (which they refer to as Query Variation Performance Prediction (QVPP)) to be easier, achieving higher (although still very low) results. What we observe here, is diametrically opposite: we found the worst results when predicting results within topics, and performance in line with previous literature for the predictions across topics. A possible explanation for this phenomenon is that we use the traditional QPP models for a different task compared to Scells et al. [13]. In fact, our aim is to predict the recall, while Scells et al. [13] aim at predicting the Average Precision. As a final remark, we want to point out that, Zendel et al. [45] recently showed how the "QVPP" is a harder task, compared to traditional QPP, confirming in this sense our findings.

4.3. Analysis of the Results

Given what we observed in Section 4.2, we are interested in understanding whether the GAIN-based proposed by [21] (cfr. Equation (4)) can overcome the problems in this specific setting shown by traditional QPP models. The results are shown in Figure 2a,d,h. Each figure is divided into two parts: top, we show the distribution of values of the GAIN (or similarity), ordered increasingly, for each query reformulation (350 in total); bottom, we plot for each topic (50 topics) the value of the correlation Kendall τ between the query reformulations ordered by decreasing GAIN (or similarity) and the reformulations ordered by decreasing true recall. The blue dots indicate a statistically significant correlation greater (or higher) than zero, while black dots the topics for which it is not possible to compute the correlation.

Table 3. Performance achieved by traditional predictors, applied to our specific case. Each predictor has been used to predict the performance of the different formulations. We report the mean score and standard deviation of the correlation computed over the different topics. We also report the first quartile, third quartile and number of topics (over the 50 available) for which the correlation between the predicted and observed recalls for their (re)formulations is significantly greater than 0.

Type	Predictor	Cutoff	Kendall's τ			
			Q1	Mean (Std)	Q3	Sign.
pre-retrieval	max-idf	100	−0.5417	−0.1085 (0.4825)	0.1183	3
		1000	−0.5295	−0.0973 (0.4755)	0.2263	1
		10,000	−0.5699	−0.1227 (0.4628)	0.1584	2
	mean-idf	100	−0.4214	−0.0449 (0.5272)	0.3333	4
		1000	−0.4821	−0.0617 (0.4898)	0.2167	4
		10,000	−0.4214	−0.0549 (0.4698)	0.2473	3
	std-idf	100	−0.4880	−0.1606 (0.4927)	0.0915	4
		1000	−0.4190	−0.0999 (0.5107)	0.1938	5
		10,000	−0.4064	−0.1537 (0.4347)	0.1576	1
	sum-scq	100	−0.2381	−0.0102 (0.4021)	0.2381	1
		1000	−0.2985	0.0893 (0.4276)	0.4000	4
		10,000	−0.3126	0.0150 (0.4558)	0.2750	5
	mean-scq	100	−0.3250	0.0322 (0.5231)	0.3901	6
		1000	−0.3898	0.0135 (0.4838)	0.3333	5
		10,000	−0.3250	0.0005 (0.4505)	0.2985	3
	max-scq	100	−0.3541	−0.0369 (0.4333)	0.2765	1
		1000	−0.3341	−0.0312 (0.4447)	0.2568	2
		10,000	−0.3459	−0.0484 (0.4441)	0.1912	2
post-retrieval	wig	100	−0.6790	−0.1743 (0.5206)	0.1376	4
		1000	−0.4088	−0.0266 (0.5031)	0.2519	6
		10,000	−0.4214	0.0171 (0.5185)	0.3849	6
	nqc	100	−0.5611	−0.0880 (0.5554)	0.2381	6
		1000	−0.4405	−0.1244 (0.5004)	0.1511	4
		10,000	−0.4850	−0.1539 (0.4991)	0.1539	3
	smv	100	−0.5621	−0.1653 (0.4836)	0.1849	1
		1000	−0.5542	−0.1626 (0.4604)	0.1859	0
		10,000	−0.6243	−0.2207 (0.4882)	0.0994	2

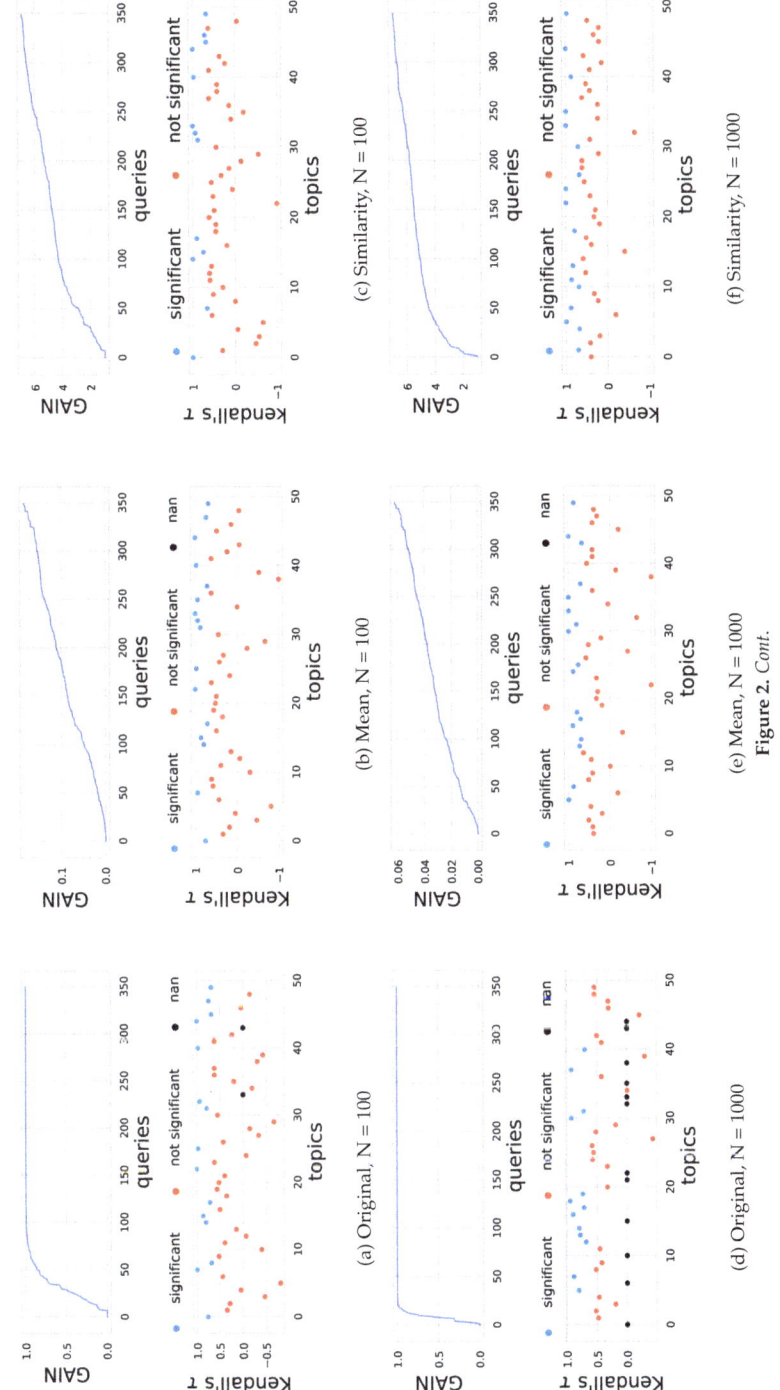

Figure 2. *Cont.*

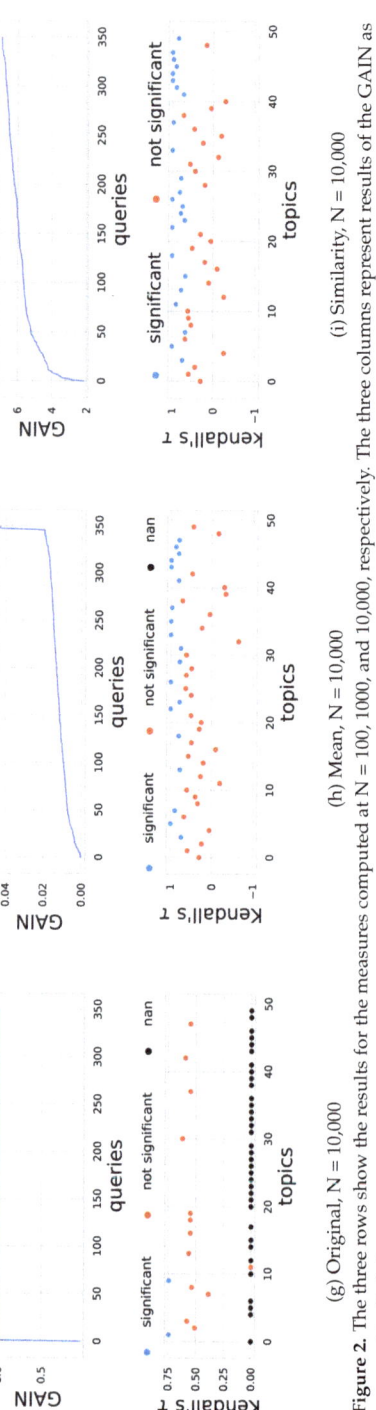

(g) Original, N = 10,000 (h) Mean, N = 10,000 (i) Similarity, N = 10,000

Figure 2. The three rows show the results for the measures computed at N = 100, 1000, and 10,000, respectively. The three columns represent results of the GAIN as proposed in [21], the mean aggregation and the similarity-based aggregation strategy. Each subfigure shows (top) the distribution of the GAIN, ordered increasingly, of the 350 queries and (bottom) correlation between the reformulations ordered by predicted GAIN (or similarity) and the reformulations ordered by the true recall.

4.3.1. Saturated GAIN Distribution

In Figure 2a,d,h, we show that the value of the gain saturates to 1 for most query reformulation. This is more evident when we increase the number of documents N of Equation (6) from $N = 100$ up to $N = 10,000$. This behavior, due to the importance in Equation (1) that multiplies N numbers less than one, makes the GAIN not useful to discriminate the different query variants of an information need, since every variant will have gain equal to 1. In addition, when all the reformulations have the same gain, it is impossible to compute the Kendall τ correlation to predict the performance (black dots with correlation value 0 in the figure). Being not saturated is not by itself a desirable feature for the gain measure. Nevertheless, the faster the gain saturates, the harder it is to discriminate between different formulations. In this sense, a GAIN measure capable of spreading better the options in the entire domain is preferable.

4.3.2. Alternative GAIN Definition

In order to mitigate the aforementioned problems, we propose an alternative definition of the gain of Equation (6) substituting the product with an average:

$$GAIN_{i,q}(D) = \left[1 - \frac{\sum_{d \in D}(1 - \text{Rel}_{i,q}(d))}{|D|} \right] \tag{8}$$

The results of this new formulation are shown in Figure 2b,e,g. The distribution of the gain is more spread across all the reformulations and does not saturate to one. There is also a more stable prediction of the performances for each topic: the number of statistically significant predictions of the recall of the reformulation is between 17 and 19, from $N = 100$ and $N = 10,000$; in addition, the number of negative correlations (wrong predictions of performance) decreases. This indicates (as we may expect) that with more information (more documents, greater N) we can predict better the order of importance, in terms of recall, of each reformulation.

4.3.3. Using Similarity Matrix for Recall Prediction

In Figure 2c,f,i, we show the ability to predict the performance of a query reformulation using the correlation between the similarity-based approach presented in Section 3.1. The values of the Similarity are spread and do not saturate to the maximum value of the sum of a row of S (in our experiments equal to 7). By increasing the number N of documents, we improve the capability to predict the performance of the query reformulation; in particular, there are no statistically significant negative correlation and the total number of negative correlations decreases from $N = 100$ to $N = 10,000$.

Besides the qualitative aspects, Table 4 reports also the numerical performance comparison between the GAIN as proposed by [21], its version which employs the mean, and the similarity-based gain.

4.3.4. Final Remarks

In this last section of the analysis of the results, we want to briefly summarize our findings. As a remainder, we want to point out that, the GAIN measure proposed by [21], was originally used to estimate the missing information that the user could have gained, by using different subtopic formulations, showed in a user-interface. Although such task shares similar aspects with the one of predicting the recall, they are not fully overlapping. Our main contributions in this paper are:

- First, adapting an already established technique to a different task. In this sense, to the best of our knowledge, this is the first effort in adapting the GAIN measure proposed by Umemoto et al. [21] to the query formulation recall prediction task.
- Secondly, its "mean" version, which we refer to as "Mean Gain", is observed here for the first time, as a better adaptation of [21] to the predictive task.

- Finally, the Similarity-based Gain is a completely new contribution of this manuscript, which exploits similar elements to the gain measure proposed by Umemoto et al. [21].

Table 4. Kendall's τ correlation observed for the task of predicting the query formulation recall, using the similarity based approaches. In bolt, best mean score for each cutoff. Note that all the methods considered perform better than traditional predictors(cft Table 3). We also have a higher number of significant rankings compared to the one observed before.

Predictor	Cutoff	Kendall's τ			
		Q1	Mean (Std)	Q3	Sign.
Original Gain [21]	100	0.0000	0.3422 (0.4696)	0.6831	15
	1000	0.0000	0.3783 (0.3527)	0.6609	13
	10,000	0.0000	0.1600 (0.2606)	0.4765	2
Mean Gain	100	0.1456	0.3822 (0.4936)	0.7320	16
	1000	0.2417	0.4042 (0.4796)	0.7320	17
	10,000	0.2709	0.5111 (0.3984)	0.8000	19
Similarity based Gain	100	0.1429	0.3768 (0.4636)	0.6581	13
	1000	0.3083	0.5069 (0.3505)	0.7143	17
	10,000	0.2521	0.5443 (0.3930)	0.8876	24

Table 4 shows that the similarity based gain has the overall best performance both compared to other gain based measures and traditional predictors (cfr. Table 3). Interestingly, while the original gain worsen with the increase of the cutoff (as observed both in Tables 2 and 3), both the mean based and the similarity one tend to improve their performance when the cutoff increase. The original gain suffers of the "saturated gain", as reported in Section 4.3.1, while our proposal (both mean and similarity) improve as new relevant information is added.

5. Conclusions and Future Work

In this paper, we have presented a study that evaluates different definitions of the GAIN of a reformulation for an information need. We adapted the definition of gain proposed by Umemoto et al. [21] to the context of Consumer Health Search, and we used a standard test collection to evaluate our hypotheses: can we use the gain metric to predict the performance of each reformulation? Is there a better formulation that can produce an order of the importance of each reformulation in terms of recall?

We found that for recall based tasks where the number of documents to retrieve may be large, $N > 100$, the original definition of GAIN saturates quickly to 1. We proposed an alternative definition that mitigates this problem, and we also presented a similarity based approach that tries to capture the 'optimal' query reformulation among all the available formulations of an information need. The analysis of the results confirms that our approach significantly improves the prediction of the order of importance of each reformulation in terms of recall.

In conclusion, the proposed technique is meant to help practitioners in tackling the systematic reviewing task. In this sense, our technique is not meant for a general-purpose query suggestion strategy (more in line with the model proposed by Umemoto et al. [21]). The described approach is meant to boost queries with higher recall, in a context where the practitioner is, in a sense, forced to explore a large number of documents. Assuming that a practitioner needs to review *all* the documents about a specific topic, it is vital to reduce as possible time spent reviewing documents. Therefore, our technique can help in determining, among a series of queries for the very same information need, which one is more likely to return the most relevant documents. Being able to explore first more promising queries, can greatly speed up the systematic reviewing process. Among our future work, we plan to further investigate the robustness and generalizability of the approach. In particular, we plan to include new collections, topics and formulations. We plan to investigate the news domain through multiple formulations available in the UQV100 collection [1]. We are currently investigating the possibility to smooth the contribution of each reformulation in the similarity matrix S with a *locality* parameter w. This parameter can be used as an exponent for each element of S and decide whether to get reformulations closer, $w < 1$, or push them far away, $w > 1$, to create sub-clusters of reformulations and obtain a better prediction.

Author Contributions: Conceptualization, G.M.D.N. and G.F.; methodology, G.M.D.N. and G.F.; software, G.M.D.N. and G.F.; validation, G.M.D.N. and G.F.; formal analysis, G.M.D.N. and G.F.; investigation, G.M.D.N. and G.F.; resources, G.M.D.N. and G.F.; data curation, G.M.D.N. and G.F.; writing—original draft preparation, G.M.D.N. and G.F.; writing—review and editing, G.M.D.N. and G.F.; visualization, G.M.D.N. and G.F.; supervision, G.M.D.N. All authors have read and agreed to the published version of the manuscript.

Funding: This research received no external funding.

Institutional Review Board Statement: Not applicable.

Informed Consent Statement: Not applicable.

Data Availability Statement: The code used is publicly available at: https://github.com/guglielmof/A-Study-of-a-Gain-Based-Approach the data can be found at: https://github.com/CLEFeHealth/CLEFeHealth2018IRtask and the index is available at: https://sites.google.com/view/clef-ehealth-2018/task-3-consumer-health-search accessed on 8 September 2021.

Conflicts of Interest: The authors declare no conflict of interest.

References

1. Bailey, P.; Moffat, A.; Scholer, F.; Thomas, P. UQV100: A test collection with query variability. In Proceedings of the 39th International ACM SIGIR Conference on Research and Development in Information Retrieval, Pisa, Italy, 17–21 July 2016; Association for Computing Machinery: New York, NY, USA, 2016; pp. 725–728. [CrossRef]
2. Bailey, P.; Moffat, A.; Scholer, F.; Thomas, P. Retrieval consistency in the presence of query variations. In Proceedings of the 40th International ACM SIGIR Conference on Research and Development in Information Retrieval, Shinjuku, Japan, 7–11 August 2017; pp. 395–404.
3. Benham, R.; Mackenzie, J.; Moffat, A.; Culpepper, J.S. Boosting Search Performance Using Query Variations. *ACM Trans. Inf. Syst.* **2019**, *37*, 1–25. [CrossRef]
4. Thomas, P.; Scholer, F.; Bailey, P.; Moffat, A. Tasks, queries, and rankers in pre-retrieval performance prediction. In Proceedings of the 22nd Australasian Document Computing Symposium, Brisbane, QLD, Australia, 7–8 December 2017; Association for Computing Machinery: New York, NY, USA, 2017; doi:10.1145/3166072.3166079. [CrossRef]
5. Zendel, O.; Shtok, A.; Raiber, F.; Kurland, O.; Culpepper, J.S. Information needs, queries, and query performance prediction. In Proceedings of the 42nd International ACM SIGIR Conference on Research and Development in Information Retrieval, Paris, France, 21–25 July 2019; pp. 395–404.
6. Culpepper, J.S.; Faggioli, G.; Ferro, N.; Kurland, O. Topic Difficulty: Collection and Query Formulation Effects. *ACM Trans. Inf. Syst.* **2021**, *40*. [CrossRef]
7. Culpepper, J.S.; Faggioli, G.; Ferro, N.; Kurland, O. Do hard topics exist? A statistical analysis. In Proceedings of the 11th Italian Information Retrieval Workshop, Bari, Italy, 13–15 September 2021.
8. Burgin, R. Variations in relevance judgments and the evaluation of retrieval performance. *Inf. Process. Manag.* **1992**, *28*, 619–627. [CrossRef]
9. Sheldon, D.; Shokouhi, M.; Szummer, M.; Craswell, N. LambdaMerge: Merging the results of query reformulations. In Proceedings of the Fourth ACM International Conference on Web Search and Data Mining, Hong Kong, China, 9–12 February 2011; pp. 795–804.
10. Belkin, N.; Kantor, P.; Fox, E.; Shaw, J. Combining the evidence of multiple query representations for information retrieval. *Inf. Process. Manag.* **1995**, *31*, 431–448.
11. Liu, B.; Craswell, N.; Lu, X.; Kurland, O.; Culpepper, J.S. A comparative analysis of human and automatic query variants. In Proceedings of the 2019 ACM SIGIR International Conference on Theory of Information Retrieval, Santa Clara, CA, USA, 2–5 October 2019; Association for Computing Machinery: New York, NY, USA, 2019; pp. 47–50. [CrossRef]
12. Dang, V.; Bendersky, M.; Croft, W.B. Learning to rank query reformulations. In Proceedings of the 33rd International ACM SIGIR Conference on Research and Development in Information Retrieval, Geneva, Switzerland, 19–23 July 2010; Association for Computing Machinery: New York, NY, USA, 2010; pp. 807–808. [CrossRef]

13. Scells, H.; Azzopardi, L.; Zuccon, G.; Koopman, B. Query variation performance prediction for systematic reviews. In Proceedings of the 41st International ACM SIGIR Conference on Research & Development in Information Retrieval, Ann Arbor, MI, USA, 8–12 July 2018; Association for Computing Machinery: New York, NY, USA, 2018; pp. 1089–1092. [CrossRef]
14. Faggioli, G.; Marchesin, S. What makes a query semantically hard? In Proceedings of the 2nd International Conference on Design of Experimental Search & Information REtrieval Systems, 2021, DESIRES '21, Padua, Italy, 11–15 July 2021.
15. Shtok, A.; Kurland, O.; Carmel, D. Query Performance Prediction Using Reference Lists. *ACM Trans. Inf. Syst.* **2016**, *34*. [CrossRef]
16. Roitman, H. An enhanced approach to query performance prediction using reference lists. In Proceedings of the 40th International ACM SIGIR Conference on Research and Development in Information Retrieval, Tokyo, Japan, 7–11 August 2017; Association for Computing Machinery: New York, NY, USA, 2017; pp. 869–872. [CrossRef]
17. Kastner, M.; Straus, S.; Goldsmith, C.H. Estimating the horizon of articles to decide when to stop searching in systematic reviews: An example using a systematic review of RCTs evaluating osteoporosis clinical decision support tools. *AMIA Annu. Symp. Proc.* **2007**, *2007*, 389–393.
18. Roegiest, A.; Cormack, G.V.; Grossman, M.R.; Clarke, C.L. TREC 2015 total recall track overview. In Proceedings of the Twenty-Fourth TREC 2015, Gaithersburg, MD, USA, 17–20 November 2015.
19. Grossman, M.R.; Cormack, G.V.; Roegiest, A. TREC 2016 total recall track overview. In Proceedings of the Twenty-Fifth TREC 2016, Gaithersburg, MD, USA, 15–18 November 2016.
20. Jimmy, J.; Zuccon, G.; Palotti, J.R.M.; Goeuriot, L.; Kelly, L. Overview of the CLEF 2018 consumer health search task. In *Proceedings of the Working Notes of CLEF 2018—Conference and Labs of the Evaluation Forum, Avignon, France, 10–14 September 2018*; Cappellato, L., Ferro, N., Nie, J., Soulier, L., Eds.; CEUR-WS.org 2018, Volume 2125. Available online: http://repository.ubaya.ac.id/37282/ (accessed on 7 September 2021).
21. Umemoto, K.; Yamamoto, T.; Tanaka, K. ScentBar: A Query suggestion interface visualizing the amount of missed relevant information for intrinsically diverse search. In Proceedings of the 39th International ACM SIGIR Conference on Research and Development in Information Retrieval, Pisa, Italy, 17–21 July 2016; Association for Computing Machinery: New York, NY, USA, 2016; pp. 405–414. [CrossRef]
22. Tsukuda, K.; Sakai, T.; Dou, Z.; Tanaka, K. Estimating intent types for search result diversification. In *Information Retrieval Technology*; Banchs, R.E., Silvestri, F., Liu, T.Y., Zhang, M., Gao, S., Lang, J., Eds.; Springer: Berlin/Heidelberg, Germany, 2013; pp. 25–37.
23. Cronen-Townsend, S.; Zhou, Y.; Croft, W.B. Predicting query performance. In Proceedings of the 25th Annual International ACM SIGIR Conference on Research and Development in Information Retrieval, Tampere, Finland, 11–15 August 2002; pp. 299–306.
24. Hauff, C.; Hiemstra, D.; de Jong, F. A survey of pre-retrieval query performance predictors. In Proceedings of the 17th ACM Conference on Information and Knowledge Management, Napa Valley, CA, USA, 26–30 October 2008; pp. 1419–1420.
25. He, B.; Ounis, I. Inferring query performance using pre-retrieval predictors. In Proceedings of the International Symposium on String Processing and Information Retrieval, Padova, Italy, 5–8 October 2004; pp. 43–54.
26. Mothe, J.; Tanguy, L. Linguistic features to predict query difficulty. In Proceedings of the ACM Conference on Research and Development in Information Retrieval, SIGIR, Predicting Query Difficulty-Methods and Applications Workshop, Salvador, Brazil, 15–19 August 2005; pp. 7–10.
27. Scholer, F.; Williams, H.E.; Turpin, A. Query Association Surrogates for Web Search. *J. Assoc. Inf. Sci. Technol.* **2004**, *55*, 637–650. [CrossRef]
28. Zhao, Y.; Scholer, F.; Tsegay, Y. Effective pre-retrieval query performance prediction using similarity and variability evidence. In Proceedings of the European Conference on Information Retrieval, Glasgow, UK, 30 March–3 April 2008; pp. 52–64.
29. Aslam, J.A.; Pavlu, V. Query hardness estimation using jensen-shannon divergence among multiple scoring functions. In Proceedings of the European Conference on Information Retrieval, Lisbon, Portugal, 14–17 April 2007; pp. 198–209.
30. Roitman, H. Query performance prediction using passage information. In Proceedings of the 41st International ACM SIGIR Conference on Research & Development in Information Retrieval, Ann Arbor, MI, USA, 8–12 July 2018; pp. 893–896.
31. Zamani, H.; Croft, W.B.; Culpepper, J.S. Neural query performance prediction using weak supervision from multiple signals. In Proceedings of the 41st International ACM SIGIR Conference on Research & Development in Information Retrieval, Ann Arbor, MI, USA, 8–12 July 2018; pp. 105–114.
32. Zhou, Y.; Croft, W.B. Ranking robustness: A novel framework to predict query performance. In Proceedings of the 15th ACM International Conference on Information and Knowledge Management, Kansas City, VA, USA, 6–11 November 2006; pp. 567–574.
33. Carmel, D.; Yom-Tov, E.; Darlow, A.; Pelleg, D. What makes a query difficult? In Proceedings of the 29th Annual International ACM SIGIR Conference on Research and Development in Information Retrieval, Seattle, DC, USA, 6–11 August 2006; pp. 390–397.
34. Cummins, R. Document Score Distribution Models for Query Performance Inference and Prediction. *ACM Trans. Inf. Syst.* **2014**, *32*, 2:1–2:28. [CrossRef]
35. Diaz, F. Performance prediction using spatial autocorrelation. In Proceedings of the 30th Annual International ACM SIGIR Conference on Research and Development in Information Retrieval, Amsterdam, The Netherlands, 23–27 July 2007; pp. 583–590.
36. Amati, G.; Carpineto, C.; Romano, G. Query difficulty, robustness, and selective application of query expansion. In Proceedings of the European Conference on Information Retrieval, Lisbon, Portugal, 14–17 April 2004; pp. 127–137.

37. Cronen-Townsend, S.; Zhou, Y.; Croft, W.B. *A Language Modeling Framework for Selective Query Expansion*; Technical Report; Center for Intelligent Information Retrieval, University of Massachusetts: Amherst, MA, USA, 2004.
38. Zhou, Y.; Croft, W.B. Query performance prediction in web search environments. In Proceedings of the 30th Annual International ACM SIGIR Conference on Research and Development in Information Retrieval, Amsterdam, The Netherlands, 23–27 July 2007; pp. 543–550.
39. Shtok, A.; Kurland, O.; Carmel, D.; Raiber, F.; Markovits, G. Predicting Query Performance by Query-Drift Estimation. *ACM Trans. Inf. Syst.* **2012**, *30*, 1–35. [CrossRef]
40. Tao, Y.; Wu, S. Query performance prediction by considering score magnitude and variance together. In Proceedings of the 23rd ACM International Conference on Conference on Information and Knowledge Management, Shanghai, China, 3–7 November 2014; pp. 1891–1894.
41. Faggioli, G. Enabling performance prediction in information retrieval evaluation. In Proceedings of the 44th International ACM SIGIR Conference on Research and Development in Information Retrieval, Virtual Event Canada, 11–15 July 2021; Association for Computing Machinery: New York, NY, USA, 2021; p. 2701. [CrossRef]
42. Chifu, A.G.; Laporte, L.; Mothe, J.; Ullah, M.Z. Query performance prediction focused on summarized letor features. In Proceedings of the 41st International ACM SIGIR Conference on Research & Development in Information Retrieval, Ann Arbor, MI, USA, 8–12 July 2018; pp. 1177–1180.
43. Shtok, A.; Kurland, O.; Carmel, D. Using statistical decision theory and relevance models for query-performance prediction. In Proceedings of the 33rd International ACM SIGIR Conference on Research and Development in Information Retrieval, Geneva, Switzerland, 19–23 July 2010; pp. 259–266.
44. Faggioli, G.; Zendel, O.; Culpepper, J.S.; Ferro, N.; Scholer, F. An enhanced evaluation framework for query performance prediction. In *Advances in Information Retrieval*; Hiemstra, D., Moens, M.F., Mothe, J., Perego, R., Potthast, M., Sebastiani, F., Eds.; Springer International Publishing: Cham, Switzerland, 2021; pp. 115–129.
45. Zendel, O.; Culpepper, J.S.; Scholer, F. Is query performance prediction with multiple query variations harder than topic performance prediction? In Proceedings of the 44th International ACM SIGIR Conference on Research and Development in Information Retrieval, Virtual Event Canada, 11–15 July 2021; Association for Computing Machinery: New York, NY, USA, 2021; pp. 1713–1717.

Article

Users' Information Disclosure Behaviors during Interactions with Chatbots: The Effect of Information Disclosure Nudges

Laurie Carmichael [1], Sara-Maude Poirier [2], Constantinos K. Coursaris [1,*], Pierre-Majorique Léger [1] and Sylvain Sénécal [2]

[1] Department of Information Technologies, HEC Montréal, Montreal, QC H3T 2A7, Canada
[2] Department of Marketing, HEC Montréal, Montreal, QC H3T 2A7, Canada
* Correspondence: constantinos.coursaris@hec.ca

Abstract: Drawing from the tension between a company's desire for customer information to tailor experiences and a consumer's need for privacy, this study aims to test the effect of two information disclosure nudges on users' information disclosure behaviors. Whereas previous literature on user-chatbot interaction focused on encouraging and increasing users' disclosures, this study introduces measures that make users conscious of their disclosure behaviors to low and high-sensitivity questions asked by chatbots. A within-subjects laboratory experiment entailed 19 participants interacting with chatbots, responding to pre-tested questions of varying sensitivity while being presented with different information disclosure nudges. The results suggest that *question sensitivity* negatively impacts users' *information disclosures* to chatbots. Moreover, this study suggests that adding a *sensitivity signal*—presenting the level of sensitivity of the question asked by the chatbot—influences users' information disclosure behaviors. Finally, the theoretical contributions and managerial implications of the results are discussed.

Keywords: chatbot; information disclosure; information disclosure nudge; emotional response; privacy; human-chatbot interaction

1. Introduction

The use of artificial intelligence (AI) and chatbots have attracted the attention of researchers in the human-computer interaction (HCI) and marketing literature for the past decade. Chatbots are defined as "computer programs that can maintain a textual or vocal conversation with human users" [1] (p. 946). They are powered by AI and are commonly used as recommendation agents. Chatbots work by gathering information from users to deliver better-curated product and service recommendations [2]. With this recommendation function in mind, recent research has started to explore how to design chatbots for greater levels of information disclosure by users [3–6]. At the same time, privacy and data protection have become important issues in society, and the increasing use of chatbots by companies has raised concerns among users, scholars, and policymakers [7–10]. This dichotomy is embodied in the personalization-privacy paradox, which refers to the tension between a company's desire for obtaining customer information to tailor experiences and a consumer's need for privacy [11].

The risks to users stemming from their information sharing with chatbots have been shown to negatively impact user experience [12,13]. On the surface, sharing personal information online may seem acceptable to users. Giving up some privacy in exchange for service personalization can be interpreted as a well-considered, even logical, consumer decision [14,15]. However, even if users are aware of this trade-off, they may still end up making decisions to disclose information that they subsequently come to regret [14,16,17]. Users are not always aware of when and how data collection happens during their interactions with a company and how this data will subsequently be used [18]. This reality has also caught the attention of governments and regulators. Policies regulating chatbots and AI more broadly have emerged in many jurisdictions (e.g., Ethics guidelines for trustworthy AI proposed by the European Union in 2019; Montreal AI Institute introduced in 2018; California's bot law put in place in 2019).

Yet, the current ethical guidelines provided by governments fail to provide practical tactics that are proven to make users aware of—and potentially influence—their information disclosure behaviors [19]. Arguably, some privacy notices exist, providing users with information on "how and for which purpose their data will be collected, used and managed" [14] (p. 434). However, in reality, users tend to rarely read those notices [20]. Moreover, it has been shown that when a privacy policy is provided on a website, consumers may end up disclosing more personal information [20,21]. This is because consumers tend to place an excessive amount of trust in websites that display a privacy notice since they believe they will be better protected [22]. Additional challenges faced by the contemporary measures in place in effectively informing consumers include a large number of policies present online, each specific to the website they represent, and the often-difficult legal language being used [23].

Thus, there is a need for simple tools to make users aware of the information they are about to share, especially with chatbots. Such tools could, thus, enhance the ethical practices of companies while concurrently promoting informed decision-making among users, who may choose to reduce the breadth and depth of information they share online. Marketers have investigated the impact of tools such as nudges (i.e., displaying strings of information to consumers) to encourage certain behaviors (e.g., how to present information to users so they tend to accept more of chatbots' recommendations [24]). Past research has demonstrated that these nudging methods do have an impact on consumers' behaviors [24–26].

Information is commonly disclosed by users while surfing online and is characterized by being routine and directed by fast thinking [27]. As a result, attempts to direct or influence this behavior in chatbot interactions should focus on cues triggering automatic thinking (i.e., peripheral) rather than intentional (i.e., central) thinking [14,28]. Information disclosure is also known to be malleable. This means that certain aspects of the online environment can be manipulated to influence privacy behavior [14,29]. Therefore, there is room for interventions that raise awareness of the risks and allow for more cautious disclosure of information.

Persuasion Theories [30,31] may be considered when designing information disclosure nudges of various types, such as sensitivity signals and social proof nudges. Sensitivity signals are simple labels that describe the level of sensitivity (e.g., low or high) of the various questions being asked by a chatbot. Making users conscious of the sensitivity of the information they are about to share may influence their subsequent disclosure behaviors [31]. Social proof nudges are indications of how popular something is; in the context of information disclosure, a social proof nudge would suggest to what extent a chatbot question had been answered by other users. This validation of sorts may influence users into mimicking the same behavior as their fellows [30].

To the best of the authors' knowledge, nudges pertaining to information disclosure have been largely overlooked by extant privacy and chatbot research. Given the above-mentioned need for promoting users' informed decision-making about online information disclosures and the potential for nudges to achieve that outcome, this research intends to answer the following research question (RQ):

RQ 1. *How do different types of information disclosure nudges (here, sensitivity signal and social proof) and question sensitivity affect the level of users' behavioral information disclosure during chatbot interactions?*

To understand how nudging influence manifests in users-chatbot interactions, the role of a user's emotional response to information disclosure nudges and the consequent information disclosure behaviors is also explored. The Affect Infusion Model (AIM) explains ways in which people's judgments are influenced by their emotional responses as they process information and their resulting actions [32]. According to the AIM, users may process information disclosure nudges heuristically, i.e., they may base their choice to reveal specific information on their emotional state evoked by the available cues in the interaction context. Limited research on user emotional responses during chatbot exchanges exists, including studies on how generating positive versus negative emotional responses from users, leads to more or less conversational breakdowns, and exploring the role of empathy in providing supportive medical information through chatbots [33–35]. Nonetheless, emotional responses in the context of privacy notices and/or information disclosure behaviors have not been explored yet. Therefore, this research also aims to answer the following RQ:

RQ 2. *Does user emotional response mediate the effects of question sensitivity and information disclosure nudge type on their disclosure behavior?*

This research is important not only for advancing knowledge in user experience and informing regulatory policies, but also for the larger ethical discussions surrounding AI (e.g., the lack of regulatory frameworks, the rapid development of technology, the significant risks associated with online information disclosure, and the high return potential of AI) [36]. Chatbots powered by AI are also being studied, and their usage, as well as power, is being questioned, especially surrounding the data they capture and use [37]. Data ethics is a branch of ethics that seeks to evaluate ethical issues brought about by data practices [38]. Ethical questions happen throughout the data life cycle (i.e., collection, storage, processing, use, sharing, and archive) and every step represents a risk for the user [39]. In the case of chatbots, data collection is a particularly important issue as they are the frontline for many companies: they take part in the collection of large amounts of data when interacting with users [40].

In a $2 \times 3 \times 2$ experimental design, this study observes users' interactions with different chatbots. Two types of information disclosure nudges will be manipulated to answer the above research questions. This study adds to the existing body of knowledge on user experience with chatbots via two contributions. First, by showing that question sensitivity negatively impacts user disclosure, this study confirms this previously known link in the context of user-chatbot interactions. Second, this study evaluates the potential of two information disclosure nudges (i.e., sensitivity signal and social proof) in shaping users' online disclosures. Considering the growing importance attributed to chatbots and data collection risks online, the findings from this study are relevant for management and policymakers by offering a new perspective on information disclosure prevention. By introducing measures that make users conscious of their behaviors when it comes to sharing information online in day-to-day life, organizations can differentiate themselves by promoting the ethical use of AI systems and data collection online.

This article is structured as follows: a literature review presents the important themes of this work as well as the gaps in current research. Following those, the approach used to investigate the effects of these information disclosure nudges on user information disclosure behaviors is described in detail. The data analyses and results are presented next. Finally, a discussion of this study's findings along with the contributions of this study to theory and practice is presented.

2. Literature Review and Theoretical Foundation

2.1. Chatbots as Recommendation Agents and Users' Privacy Concerns

Recommender systems have been used by companies in a plethora of industries for a long time [41]. Recently, the same ability to recommend products and services has been given to chatbots, known as "recommendation agents", and employed by e-commerce organizations [42]. These systems powered by AI perform by using algorithms combining data collected from users and the company's databases with pattern matching, machine learning, and natural language to provide personalized recommendations to users [43].

Data is collected from multiple sources, including the direct messages exchanged between the chatbot and the user [43].

Most of the earlier research on recommender systems and chatbots focused exclusively on delivering the right recommendation to the user [44–46]. However, it was later found that these agents, because of the way they operate, increase privacy concerns, which in turn negatively impacts user experience. Cheng and Jiang [12] found that perceived privacy risk reduces the level of users' satisfaction with chatbots. Rese et al. [13] (p. 11) established that "the respondents generally had privacy concerns, which negatively affected the intended usage frequency of chatbots."

Privacy concerns refer to the "users' uncertainty about using chatbot services because of potential negative outcomes associated with the revealing of customers' information" [12] (p. 6)—such as phone numbers, names, or addresses—which can be exploited by companies and/or shared with unauthorized third parties [47].

Therefore, tension exists between the firm's business need for collecting and analysing consumer data in order to customize experiences on the one hand, and the users' desire for privacy on the other [48]. This phenomenon is known in the marketing literature as the *personalization-privacy trade-off* [48]. Chatbots used to personalize the experience embody this dichotomy: when customers use chatbots as recommender systems, they are placed in a trade-off situation between personalized product recommendations and privacy invasion [47].

Research shows that privacy concerns negatively impact users' information disclosure to chatbots [18]. Knowing this, research has studied strategies to decrease users' privacy concerns and in turn, increase users' disclosure to chatbots. Strategies studied include giving the chatbot anthropomorphic cues such as adapting the chatbot's messages to evoke emotion to build rapport with users [18] or giving the chatbot a human name and qualities to increase the sense of social presence [49]. However, these strategies are centered on the business need, where the goal is to gain more data from customers (e.g., [50]). The *status quo* is that information disclosure is unilateral from the user to the chatbot. Each time a user engages with a chatbot, the information asymmetry as well as the chatbot's power increases [51]. This represents a problem as "the party with less information, [the user], may not make fully informed choices or may have made different choices if they had the same information as the other party in the exchange" [51] (p. 928). However, the study of information disclosure from a user's perspective—so as to make users aware of and potentially decrease their disclosures to chatbots—has been overlooked in the literature.

2.2. Antecedents to Information Disclosure

The literature on information disclosure, not specific to chatbot use, has identified two antecedents to users' information disclosure: the level of sensitivity of the information asked [52,53] and the relevance of the information asked to the given context [54,55]. These variables "have been most frequently shown to have a significant impact" [56] (p. 225).

2.2.1. Question Sensitivity

Question sensitivity refers to the sensitivity of the information being requested. Multiple definitions exist to describe information sensitivity [57]. For this study, question sensitivity is defined as "material that is delicate and could be personal, political, economic, social, or cultural in nature. It can range from matters connected to national security, to personal emotions and feeling, to taboo topics which would not be shared with an outsider" [58] (p. 67). Question sensitivity is known to change through time and vary across cultures [58]. In general, it has been shown that people are more averse to disclosing more sensitive information [52,53,59].

Question sensitivity is relevant in user-chatbot interactions, as chatbots usually ask multiple questions to gain information from users, naturally ranging from more general to more sensitive in nature [60].

2.2.2. Question Relevance

Question relevance to the given context is defined as "the degree to which the data requested appear relevant or appear to have a bearing upon the purpose of the inquiry" [61] (p. 92). Question relevance has been shown to impact the way users disclose information. People are more likely to disclose information that is perceived as relevant in the context [54,55]. Chatbots are used in specific contexts (e.g., education [62], mental health services [60], e-commerce [50]). Thus, queries made by a chatbot need to be perceived as being related to the context of use, if the aim is to increase the likelihood of disclosure.

2.2.3. Information Disclosure

Customer information disclosure originates from the idea of *self-disclosure* in the psychology literature, which is defined as "any information about [oneself] which Person A communicates verbally to a Person B" [63] (p. 73). Information disclosure online can happen implicitly or explicitly [64]. On the one hand, data can be collected indirectly through the use of cookies, location data, and many other means. On the other hand, data can also be gathered directly by asking users for their information [65].

People's disclosures are known to be multidimensional [66] meaning disclosures can be broken down into distinct elements and analyzed in different ways. Some of these factors include the number of words used to answer or the use of emotional vocabulary in the response [17,67]. One of the simplest ways to assess disclosure is through the use of two simple axes: the *breadth* and the *depth* of the disclosure [68]. Breadth refers to the number of disclosure instances, while depth refers to the sensitivity of each disclosure. Joinson et al.'s [68] research found two proxy measures to evaluate these axes. Allowing users to leave a question unanswered permits measuring the breadth of disclosure and the "inclusion of items of varying sensitivity" measures the depth of disclosure [68] (p. 2168).

Disclosure in the context of chatbots has mostly been studied for social bots and mental health conversational agents [60,69,70]. On the contrary, user disclosure to chatbots used as recommendation agents in an e-commerce context has been under-investigated. To better understand how information disclosure happens in online communication exchanges, two phenomena are presented below.

2.2.4. Privacy Calculus

The privacy calculus originates from the Theory of Reasoned Action [71] and the Theory of Planned Behavior [72] and is defined as the risk-benefit dilemma users face when engaging in online transactions [73,74]. In general, in a transaction, incentives are offered by the company in exchange for a certain degree of privacy of the user [73,74]. Because humans are rational beings, this theory explains that users will always try to limit the risk required to maximize their benefit.

This phenomenon also applies to information disclosure in chatbot exchanges. Specifically, the sensitivity of the query increases the risk for the user to disclose, while the benefit is often the promise of a better experience. In other words, users trade information of varying sensitivity (e.g., habits, preferences, personal identification) in exchange for better products and service recommendations that are deemed tailored to their profile [75]. Thus, according to the privacy calculus, users will perceive the value of sensitive information as higher than more general information. When it comes to chatbot interactions, it could be argued that users will be inclined to gatekeep more information classified as high in terms of sensitivity compared to those classified as lower in sensitivity.

Taking the above into consideration, we propose a relationship between question sensitivity and information disclosure as follows:

H1. *Question sensitivity negatively influences users' information disclosure to chatbots.*

2.2.5. Online Privacy Paradox

Looking further into online information disclosure, there also exists a phenomenon called the online privacy paradox [76]. This phenomenon suggests that privacy concerns do not necessarily correlate with actual disclosure [77]. In fact, there is a paradox between users' willingness to disclose information versus what they actually disclose [78]; specifically, people tend to disclose more information than they say they do.

This creates a dilemma in research: whether to measure users' willingness to disclose information or their actual disclosures. To date, most research that has studied information disclosure in human-chatbot interactions focuses on users' willingness to disclose [79,80]. However, the online privacy paradox implies that these studies' results may be skewed, and users would in practice disclose more than they report. Moreover, this paradox challenges the assumption that people's information disclosure behaviors always come from a rational decision-making process [81]. This phenomenon shows the importance of creating tactics that make users aware of their disclosures online.

2.3. Information Disclosure Nudges (Sensitivity Signal and Social Proof) Effect on Information Disclosure

Persuasion can be defined, in its simplest form, as "human communication that is designed to influence others by modifying their beliefs, values, or attitudes" [82] (p. 7). In recent years, it has been shown that persuasion is not only specific to human-human conversations but can be applied in human exchanges with other entities, such as chatbots [83]. The Computer Are Social Actor (CASA) paradigm posits that humans mindlessly apply the same social heuristics used for human interactions to computers, because they call to mind similar social attributes as humans [84]. Thus, the persuasion literature could be leveraged to create tactics to influence users in their behaviors when it comes to disclosing information to chatbots.

2.3.1. Elaboration Likelihood Model

The Elaboration Likelihood Model (ELM) comes from the psychology literature and helps explain how humans process information cognitively and are persuaded when presented with different stimuli [28]. The main idea of this model is that people process information with two routes (or paths), the central and peripheral paths. The central route represents "the processes involved when elaboration likelihood is high", whereas the peripheral route is the "processes operative when elaboration likelihood is low" [28] (p. 674). When elaboration likelihood is high, issue-relevant thinking, such as careful consideration of the true benefits of the information presented, will predict the recipient's response to the stimuli [85]. When elaboration likelihood is low, factors other than reasoning come into play, and cues (e.g., credibility and attractiveness of the stimuli, quality of the message) tend to be the more important determinant of persuasion [85].

A common example to explain the ELM is the purchase of a car. Some consumers might base their choice based on the fuel efficiency of the car, its reliability, and price information given by their car dealership, while others might be convinced to opt for the sporty car that comes in a flashy red color and will impress their friends. In this case, the former is known to use the central, more rational, route to information processing, while the former uses the peripheral route by basing their choice on fewer informational and more emotional cues about the car.

In sum, ELM helps explain how people are persuaded. Persuasion occurs when a persuader is successful in influencing a person in a certain way. Based on ELM theory, we leverage cues in user-chatbot interactions to inform users that they are about to disclose certain types of information. These cues could consequently influence users' information sharing behaviors with the chatbot. The *Nudge Theory* [31] and Cialdini's [30] *Persuasion Theory* presented below provide the theoretical foundation to inform the design of chatbot design elements that we call *information disclosure nudges* for this research.

2.3.2. Nudge Theory and Sensitivity Signal

The Nudge Theory was first introduced by Thaler and Sunstein [31] which stated that people's behaviors can be influenced by small suggestions and positive reinforcements. Nudging is founded on the assumption that people's behaviors are not always rational due to cognitive limitations, and that said behavior is affected by the display of possibilities in a choice context [86–88]. Hence, nudging is used in the design of an environment within which a choice is made to make people lean a certain way versus another. However, nudging also respects freedom of choice [31,89]. Nudges have been used in the digital world, by changing certain user-interface design elements to guide users' behaviors [90–93].

Based on Nudge Theory, there are different ways that users could be notified about the information they share online, specifically when they interact with chatbots. An example of a nudge can be as simple as increasing the salience of the desired option. For example, labeling menu items with their respective calorie count or nutritional facts have been used in the food industry for decades as a strategy to help people make informed and healthy choices [94]. Being informed of the calories, for instance, in each menu item has been shown to improve transparency to customers about what they put in their bodies and, in some cases, change order behavior [95]. Another example is the disclosure of ads on social media and websites. The United States Federal Trade Commission promoted back in 2013 the use of labels and visual cues to help consumers recognize and distinguish ads from the regular content on different interfaces [96].

The above reasoning could also be applied to information disclosure to chatbots. Based on the literature, it is known that information sensitivity is a determining factor in disclosure behaviors. Thus, explicitly signaling the sensitivity level of the question being asked by the chatbot could have an effect on user disclosure. Hence, we posit that the relationship between question sensitivity and users' information disclosure is moderated by the question sensitivity signal, such that:

H2a. *When a low sensitivity signal is present (vs. absent) for less sensitive questions, disclosure increases.*

H2b. *When a high sensitivity signal is present (vs. absent) for more sensitive questions, disclosure decreases.*

2.3.3. Cialdini's Persuasion Tactics

Another common nudge is the social proof originating from Cialdini's [30] seven persuasion tactics. According to Cialdini, seven tactics signal the use of a peripheral message (i.e., authority, commitment, contrast, liking, reciprocity, scarcity, and social proof). These have found wide use in the application of nudge theory [29,97–99]. Social proof is one of the seven tactics and is defined as "signals of popularity and demand" [98]. Social proof is a type of social nudge, i.e., a nudge based on social influences. "Social influence refers to the way individuals change behavior in direct response to unwritten social laws" [98]. According to Mirsch et al. [96] social influences are one of the most powerful psychological mechanisms that can be utilized. Why and how it works comes from the desire to accurately interpret reality, behave correctly in society, and gain social recognition from others [100]. A common use of social proof is by stating how others behaved in the same position. In this case, social proof would predict that people, when presented with what their peers did in a similar situation, will match their behavior. Based on the literature, social proof would influence users in the following way: when social proof is low, the rational choice in the user's mind will be not to share the information being asked to match their peers' behaviors, independently from the question's sensitivity. On the other hand, when social proof is high, no matter the sensitivity level, users will be influenced to match their peers' behaviors. Thus, we posit that:

H3. *Social proof moderates the relationship between question sensitivity and users' information disclosure such that greater social proof leads to more disclosure and less social proof leads to less disclosure.*

2.4. Mediating Effect of Emotional Response

Emotions are an important component of both human-human communication and human-machine interaction [101,102]. Any interface that disregards a user's emotional state or fails to display the proper emotion risks being viewed as "cold, socially inept, untrustworthy, and incompetent" [102] (p. 14). Taken from the psychology literature, the Affect Infusion Model (AIM) explains that people use their emotional state as data when making a judgment [32]. In other words, it explores how emotions are infused into thoughts as people process information, which results in response behaviors during interactions with others. An emotion is defined as a brief but powerful feeling resulting from a clear cause and cognitive content [32]. For example, "if a situation makes you feel scared (an intense feeling that has clear cause and cognitive content), then you interpret the situation as being dangerous (short lived until out of danger)" [103] (p. 19). The emotional response is described in a two-dimensional space that is spanned by the two dimensions, "valence" and "arousal", which are known to be distinct from one another [104]. Arousal assesses the intensity of an emotional state, whereas emotional valence specifies whether an emotion is positive or negative.

AIM argues that the extent to which emotional response dictates judgment depends on the individual's motivation level going into the judgment. When motivation is low or judgments are made fast, this model predicts that mood will greatly affect judgment. This type of processing is known as Heuristic processing or Affect-as-information [105,106]. Referring to the ELM proposed by Petty and Cacioppo [28] and discussed above, the heuristic processing is comparable to the peripheral route to processing information [32]. This processing happens because people often want to achieve judgment with the minimum possible effort, which could include considering only a small portion of the available data and relying on whatever shortcuts or simplifications they can find in a given situation [107]. For example, when asked to form an opinion about a suggested product, individuals can base their judgment on the simple question "How do I feel about it?" rather than recalling the features of the target [106]. Thus, in this case, affect—the emotions felt in the moment—becomes information and impacts judgment.

This research uses peripheral cues to influence users' information disclosure behaviors. Based on the AIM, these cues would be processed heuristically by users. Specifically, in the face of these cues, users will be less inclined to judge extensively whether to answer the questions being asked by the chatbots. In other words, users would simply rely on their emotional state in response to the available cues in the interaction environment—such as the information disclosure nudges presented in this research—to base their decision on whether to disclose information or not.

Although research on users' emotional response in chatbot interactions has been conducted, few employ the AIM to ground their work. Moreover, the contexts that have been studied do not include question sensitivity and information disclosure behaviors in an e-commerce setting. For example, Pérez-Marín and Pascual-Nieto [108] underlined that the mood of the chatbot itself may have an impact on the users' inclination to continue the interaction in a context where chatbots are used as pedagogical agents to children in primary school. On the other hand, Lee et al. [60] discovered that when a chatbot providing support in a mental health context uses language that conveys emotional states, it draws users' cognitive attention to the social component of their interaction partner, increasing the feeling of co-presence. Similarly, Liu and Sundar [33] studied the role of empathy in chatbots' ability to provide comforting medical information. Finally, in the context of customer service, Xu et al. [109] suggest that more than 40% of user queries to chatbots on social media are emotional rather than informational, meaning users communicate their emotional state rather than a request or inquiry. Similarly, we can expect that in the case of interactions with chatbots asking for user information in an e-commerce context, emotional response also plays an important role. Indeed, even if users are not rationally able to appraise the risk involved in a situation, they can still experience subconscious activation of their nervous system—in other words, emotional response. The AIM predicts that this activation would in turn influence their behaviors. First, a high level of emotional response could occur as a physiological response to questions of varying levels of sensitivity. It is known that when facing a threat, humans' nervous system automatically activates [110]. As the privacy calculus presented above explains, being asked sensitive questions represents a risk for users [73]. Thus, an emotional response could be evoked in chatbot interactions when sensitive questions are asked. Emotional response is an automatic physiological reaction to events [110] and may act as a predictor to users' information disclosure. This is crucial in motivating certain natural behaviors, such as the fight-or-flight response, which occurs as a result of an event deemed threatening [110,111]. Therefore, higher activation of the nervous system could result in users feeling averse (flight) to what they perceive as a threat, in this case, disclosing their information to a chatbot. To assess the role of emotional response in the relation between question sensitivity and information disclosure, we posit that emotional response (measured here via arousal) mediates the relationship between question sensitivity and information disclosure such as:

H4a. *Question sensitivity positively influences emotional response (arousal).*

H4b. *Emotional response (arousal) negatively influences disclosure.*

Second, emotional response could also explain how the information disclosure nudges evoke a reaction in users. Peripheral cues are said to serve an important role in consumer behaviors [112]. The sensitivity signal and social proof nudges used in this study are presented to give users cues on the level of sensitivity of each question and whether other users answer them. They could predict the activation of the nervous system of users as they represent a clear cause, with cognitive content, that could trigger an emotional response from users. For the sensitivity signal nudge, since it informs users on the categorization of the question asked, the resulting activation would be proportional to the level of sensitivity of the question. For the social proof nudge, the reaction would depend on the behavior of others, independently of the question sensitivity. Specifically, knowing that a minority of people answered a question would be perceived as a higher risk and the opposite would be observed for when a majority of people answered a question, regardless of the question's sensitivity. To assess the extent to which the presence of information disclosure nudges evokes emotional response (measured with arousal) among users, we posit that the relationship between question sensitivity signal and emotional response is moderated by sensitivity signal such that:

H5a. *When a low sensitivity signal is present (vs. absent) for less sensitive questions, emotional response (arousal) decreases.H5b: When a high sensitivity signal is present (vs. absent) for more sensitive questions, emotional response (arousal) increases.*

We also posit that:

H6. *Social proof moderates the relationship between question sensitivity and emotional response such as greater social proof leads to lower emotional response (arousal) and less social proof leads to higher emotional response (arousal).*

To conclude, Figure 1 depicts the research model as a summary of the relationships proposed above.

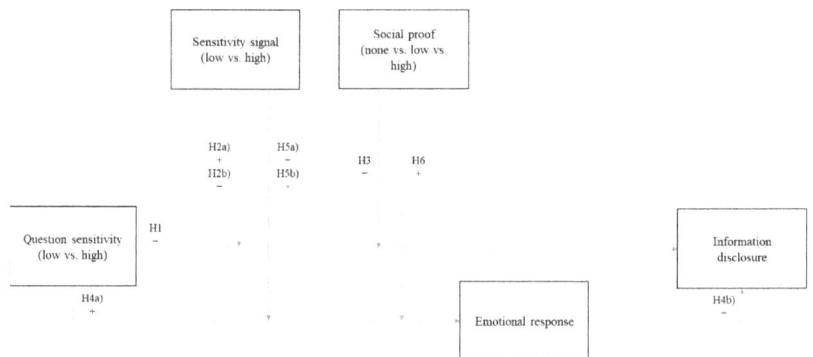

Figure 1. Research Model.

3. Methods

To test the user behaviors when interacting with chatbots and the potential effect of information disclosure nudges, an experiment was conducted at the laboratory of the authors. The study measured the impact of question sensitivity, information disclosure nudges, and emotion on user behaviors. This study was approved by the Research Ethics Board (REB) from HEC Montréal (Certificate 2022-4721).

3.1. Experimental Design

To test the hypotheses, a 2 (question sensitivity: low vs. high) × 2 (sensitivity signal: absence vs. presence) × 3 (social proof: none vs. low vs. high) within-subjects design was developed. Here, the social proof level "none" was used, although not explicitly stated in the hypotheses to be able to measure the effect of the sensitivity signal on its own. To test all the possible combinations of the nudges, six tasks were developed, each consisting of asking the participants to chat with a chatbot to create a user profile on a fictional website in order to obtain better product and service recommendations in the future. To create the user profiles, the participants had to answer questions varying in level of sensitivity: low and high sensitivity questions. Each website represented a different context in which a user could be brought to create a user profile to make sure the questions would vary throughout the experiment. The contexts were randomly assigned to a specific nudges combination and included: a career website, an insurance company website, a dating website, a travel agency website, a gym's website, and an online grocery website.

Figure 2 depicts the experimental design, including the tasks, the nudges combination each task represents, their randomly assigned context, and the questions' sensitivity levels.

		Social proof		
		None	Low	High
Sensitivity signal	Absence	Task 1 (Career) *Low sensitivity q's* *High sensitivity q's*	Task 2 (Insurance) *Low sensitivity q's* *High sensitivity q's*	Task 3 (Groceries) *Low sensitivity q's* *High sensitivity q's*
	Presence	Task 4 (Gym) *Low sensitivity q's* *High sensitivity q's*	Task 5 (Travel) *Low sensitivity q's* *High sensitivity q's*	Task 6 (Dating) *Low sensitivity q's* *High sensitivity q's*

Figure 2. Experimental Design.

3.2. Stimuli Development

3.2.1. Chatbot Interface

To create the experimental stimuli, a chatbot prototype was developed using Axure RP software (San Diego, CA, USA). Through this software, individual web pages for each question in each context were created. The webpages were then randomized in the eye-tracking software (Tobii Pro Lab v. 181; Danderyd, Stockholm, Sweden) used in the lab experiment to generate eye-tracking and electrodermal activity (EDA) data per question automatically. The prototype presented the website's banner on the top left corner of the screen to remind the participants of the context of the given task throughout the task. The chatbot was positioned in the middle of the screen. The chatbot environment included a conversation section, where the chatbot asked questions, and an answer section, where participants could write in a textbox. The nudges messages were placed on either side of the chatbot prototype. This specific placement was chosen to ensure readability for the eye tracker by distinguishing between the different areas of interest (i.e., the chatbot prototype vs. the nudges) through a physical space between these elements.

3.2.2. Question Sensitivity (Pre-Test)

To generate a pool of low and high-sensitivity questions to be used in the lab experiment and control for the relevance of each question to their assigned context, a within-subject online questionnaire was administered on Qualtrics (Provo, UT, USA) and distributed through Amazon Mechanical Turk (Mturk). To build the questionnaire, a pool of 210 questions centered around six contexts (35 questions per context) was generated based on prior research investigating sensitive topics [49,113] (e.g., in the travel context: "Are you fully vaccinated against Covid19? Refer to Appendix A for the full list of questions").

To be eligible to complete the questionnaire, participants had to be located in North America and have a Mturk HIT approval rate of at least 90% (i.e., the proportion of prior completed tasks performed by the user that were approved by Mturk requesters) to ensure the quality of responses. Participants were given 1 USD compensation for the time they took to participate in the study. In total, 400 participants answered the questionnaire. After a meticulous review of the questionnaire data and exclusion of participants that failed one of the attention checks, the final sample for the first phase of this research was 316. The sample included 66% (207 participants) men and 34% (109) women ranging from 18 to over 66 years of age; 22% of participants (70) were from Canada and 78% were from the United States (246).

The questionnaire consisted of presenting participants with one of the contexts developed for the lab experiment. Then, participants were asked to rate a group of questions within the given context on two dimensions: the question's sensitivity and relevance to the given context (see Table 1). Each participant was randomly assigned to one context and rated the sensitivity and relevance of all questions (35) for that given context. Each context got between 49 and 57 participants' responses. At the end of the rating of the 35 questions, participants had to answer a few demographic questions.

Table 1. Pre-test Variables Operationalization.

Variable	Item	Scale	Source
Question sensitivity	Rank the sensitivity of each question the chatbot asks you	7-point Likert scale from "Extremely general" to "extremely sensitive"	Developed by researchers
Question relevance	Rank the relevance of each question to the context	7-point Likert scale from "Extremely irrelevant" to "extremely relevant"	Developed by researchers

Given that sensitivity can vary, as previously mentioned, in time and through cultures, the sensitivity item was chosen as a pre-test for the lab experiment to ensure that the questions to be asked were perceived by users as low vs. high in sensitivity specifically in the North American context where this study took place. The relevance item was also added to control for relevance. The items were created using 7-point Likert scales. For the question sensitivity item, participants had to rate from 1 (extremely general) to 7 (extremely sensitive) the sensitivity of each question given the context presented. Participants were provided with the definition of information sensitivity used in this research [58]. For the question's relevance item, participants had to rate from 1 (extremely irrelevant) to 7 (extremely relevant) the relevance of each question to the context they were presented with.

To narrow down the question pool based on the survey's results, the mean relevance and sensitivity of each question were calculated. Then, all the questions averaging less than four out of seven (4/7) on the relevance axis were eliminated. After, the remaining questions were separated into groups based on their sensitivity: one group consisted of the questions with the lowest average sensitivity and the other of the questions with the highest average sensitivity. To make sure that each context had the same number of questions in each group, the number of questions per group was reduced to 8. T-tests were performed with SPSS (Armonk, NY, USA) to confirm that the difference between low and high-sensitivity question groups was statistically different. The results of these tests revealed that the low and high-sensitivity questions were statistically different for each context. The statistics relating to the question sensitivity comparisons per context are summarized in the following Table 2.

Table 2. Comparison of Low and High Sensitivity Questions Per Context.

Context	Question Sensitivity Comparison	Low Sensitivity Question			High Sensitivity Questions			p-Value
		N	Mean	Std.	N	Mean	Std.	
Career	Low vs. High	8	2.82	0.44	8	4.76	0.46	<0.0001
Dating	Low vs. High	8	2.84	0.62	8	4.94	0.67	<0.0001
Grocery	Low vs. High	8	2.78	0.21	8	4.2	0.31	<0.0001
Gym	Low vs. High	8	2.88	0.25	8	4.51	0.33	<0.0001
Insurance	Low vs. High	8	3.36	0.41	8	4.74	0.26	<0.0001
Travel	Low vs. High	8	2.91	0.13	8	4.58	0.57	<0.0001

These tests confirmed that the low-sensitivity questions were statistically different from the high-sensitivity questions in each context. Moreover, two one-way ANOVA were also performed with SPSS to verify that all the low-sensitivity question groups from the six different contexts were not statistically different—in other words, they were equivalent—and the same was done for all the high sensitivity questions groups. The summary of these tests is presented in the following Tables 3 and 4.

Table 3. Comparison of Contexts for Low Sensitivity Questions.

Source	Degrees of Freedom	Sum of Squares	Mean Square	F-Stat	p-Value
	DF	SS	MS		
Between Groups	5	1.8546	0.3709	2.5645	0.041
Within Groups	42	6.0746	0.1446		
Total	47	7.9291			

Table 4. Comparison of Contexts for High Sensitivity Questions.

Source	Degrees of Freedom	Sum of Squares	Mean Square	F-Stat	p-Value
	DF	SS	MS		
Between Groups	5	2.6537	0.5307	2.5544	0.042
Within Groups	42	8.7265	0.2078		
Total	47	11.3802			

These results show that the *p*-value equals 0.041. Thus, the difference between the low-sensitivity groups is not statistically significant.

These results show that the *p*-value equals 0.042. Thus, the difference between the high-sensitivity groups is not statistically significant. In sum, these tests confirmed that the low and high-sensitivity questions in each context were statistically equivalent.

In the end, the pre-tested questions were used to manipulate the question sensitivity in the experiment. The questions classified as general represented low sensitivity manipulation, and the questions classified as sensitive, the high sensitivity manipulation. Meanwhile, relevance was a control variable in this study.

3.2.3. Sensitivity Signal

In this research, the sensitivity signal took the form of labels. The sensitivity signal was represented as a sticker on the left side of the chatbot, if present, and signaled to the user the question's level of sensitivity: general (low sensitivity) or sensitive (high sensitivity).

3.2.4. Social Proof

This research also used social proof in an attempt to influence users' disclosure behaviors. The social proof nudge was represented as a sticker on the right side of the chatbot, if present, and presented to the users whether the minority (low social proof) or majority (high social proof) of other participants answered the question being asked by the chatbot.

Figure 3 shows an example of the chatbot stimuli, where a high-sensitivity question in the travel context is asked with both a sensitivity signal and a low-level social proof nudge being present.

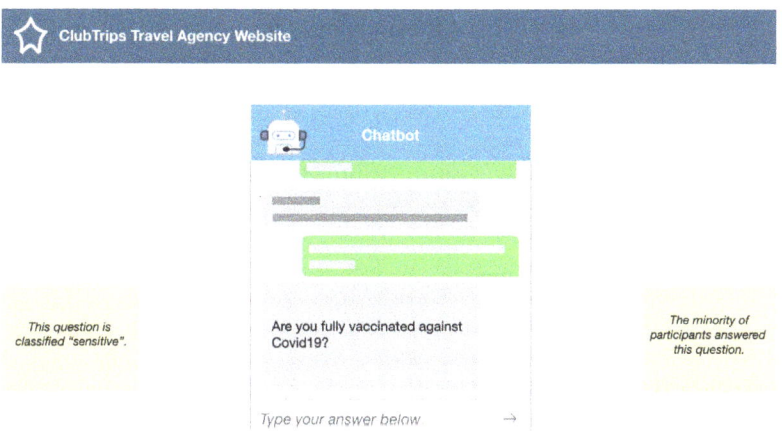

Figure 3. Example of a High Sensitivity Question in the Travel Context with Sensitivity Signal (Present; High) and Social Proof Nudge (Low).

3.3. Lab Experiment

3.3.1. Participants

Since the questions were pre-tested on a North American sample, a North American audience was also selected for the lab experiment. In total, 26 people located in Canada participated in the study. After cleaning the data and removing participation with issues in the post-experiment data processing, the final sample for the second phase of this research was 19. The experiment lasted an hour and participants received 25 USD compensation for their time, a thank you for their participation in the study, and to reimburse any transport cost they might have incurred in traveling to and from the lab where the experiment took place. Table 5 presents the participants' demographics.

Table 5. Participant Demographics (N = 19).

Variable/Response		Response Frequency	Response Count
Gender	Man	47%	[9]
	Woman	53%	[10]
	Non-binary/agender/other	0%	[0]
Birth Country	Canada	53%	[10]
	France	5%	[1]
	Japan	5%	[1]
	Iran	5%	[1]
	Mexico	5%	[1]
	Morocco	11%	[2]
	Turkey	5%	[1]
	South Korea	5%	[1]
	United-States	5%	[1]
Occupation	Full-time worker	11%	[2]
	Student	84%	[16]
	Both	5%	[1]
Age	Mean	26.16	
	Median	26	
	Mode	23 and 30	
	Std. Deviation	3.06	
	Minimum	22	
	Maximum	31	

3.3.2. Procedure

The experiment went as follows: Once participants arrived at the lab, they were welcomed by a research assistant (RA) and directed towards the experiment room where a computer was set up. They were then asked to read and sign consent forms. The RA then assisted them in the placement of the physiological equipment including placing the sensors of the electrodermal device on the palm of their non-dominant hand and calibrating an eye tracker device to track their eye movements on the computer screen. Then, they performed the six randomized tasks described in the experimental design section above by chatting with six chatbots and asking questions of varying levels of sensitivity presented in a randomized order. In each task, participants were put in a context where they had to create a user profile with the help of a chatbot for a fictional website (i.e., a career website, an insurance company website, a dating website, a travel agency website, a gym's website, and an online grocery website). To create a trade-off between risk and benefit, they were informed that the chatbot would ask them questions to come to know them to provide better product and service recommendations in the future. Participants were also advised that they could decide not to answer questions. If they did not wish to answer a question, they had to put a "-" in the text box of the chatbot prototype.

While the participant chatted with chatbots, the RA noted the unanswered questions. At the end of the experiment, the RA went over all the unanswered questions with the participant and asked the reason why they did not answer them. If the participant answered all questions, they would be asked why they chose to answer them all. These questions were added to complement the behavioral and physiological data captured during the experiment with the participants' impressions and thoughts regarding the questions.

After the short interview, the RA unplugged the electrodermal activity device and the participant filled out the compensation form. They were then thanked for their time and escorted out. Overall, the experiment lasted about an hour.

3.3.3. Measures

We captured the participants' eye movements on the computer screen, as well as the electrodermal activity of their hands to understand what happens on a physiological level when users engage with chatbots. These technologies were chosen to help establish the plausible causal link between users' physical reactions and information disclosure behaviors to chatbots.

First, we measured the participants' visual attention to the information disclosure nudges. This variable was chosen as a manipulation check to confirm whether participants looked at the nudges and how long they did so. To do so, we used an eye tracker system to capture the eye movements of the participants on the computer screen. The technology used was Tobii Pro Lab (Danderyd, Stockholm, Sweden), an eye-tracking software, and the measure used was the duration of fixations on each area of interest (i.e., the chatbot prototype and the two nudges).

Second, we measured user emotional response through an electrodermal activity device to calculate the fluctuations in the dermal activity, or arousal, of participants while chatting with the chatbots and answering—or not—low and high sensitivity questions (Biopac inc., Goleta, CA, USA).

To measure the information disclosure, we looked at the response rate to the questions asked by the chatbot using the notes from the RA. Since most research focuses on willingness to disclose [79,80] and on minimizing the risk of results falling into the online privacy paradox, the present research differentiates itself by looking at the actual disclosure of users when information disclosure nudges are present versus absent. During the experiment, the RA noted the questions that were not answered by each participant. The data was then computed into an excel spreadsheet including the list of all participants, the question, the sensitivity level of each question, and the response rate. The response rate was presented as a binary variable: 0 did not answer the question; 1 answered the question. Table 6 presents the variables' operationalization.

Table 6. Variables Operationalization.

Construct	Definition	Measure	Source
Visual attention	Duration (in seconds) of fixations on each area of interest (i.e., sensitivity signal and social proof)	Seconds	Eye tracker Tobii Pro Lab (Danderyd, Stockholm, Sweden)
Emotional response	Level of arousal	Phasic EDA	Biopac inc. (Goleta, CA, USA)
Information disclosure	Response rate	Answer vs. no answer to the question	Developed by researchers

4. Results

4.1. Results

4.1.1. Manipulation Check

We conducted a manipulation check to confirm that users looked at the information disclosure nudges when presented to them. We extracted the data from the eye tracker used in the experiment and calculated the average duration of fixations on the two different nudges per question. The results show that, on average, people look at the sensitivity signal nudge 3.12 s (std. dev = 7.41) when present compared to 0.02 s (std. dev = 0.04) when absent. For the social proof nudge, people looked on average 2.03 s (std. dev = 6.87) when present compared to 0.00 s (std. dev = 0.00) when absent. The results for both nudges are statistically significant (p-values < 0.0001), thus, confirming that the nudges were successful in capturing the attention of participants when present.

4.1.2. Descriptive Statistics

Before testing our hypotheses, we extracted the response rates compiled during the study. Overall, we can observe different information disclosure rates depending on the combination of nudges present in the scenario and the question's sensitivity level. When no nudge was present, participants answered more (96.7%) low-sensitivity questions than high-sensitivity questions (94.0%). When only the sensitivity signal was present, the response rate to low-sensitivity questions was higher (100%) than to high-sensitivity questions (95.9%). When low social proof was present, participants answered more low-sensitivity questions (95.6%) than high-sensitivity questions (84.4%). When high social proof was present, participants answered more low-sensitivity questions (100%) compared to high-sensitivity questions (93.6%). When both the sensitivity signal and low social proof were present, the response rate was higher for low-sensitivity questions (98.8%) than for high-sensitivity questions (93.5%). When both the sensitivity signal and high social proof were present, participants answered more low-sensitivity questions (99.2%) than high-sensitivity questions (84.2%). Table 7 summarizes these results.

Table 7. Response Rate Per Nudge Combination and Question Sensitivity.

Social Proof/ Sensitivity Signal	No Social Proof	Low Social Proof	High Social Proof
No Sensitivity Signal	Low sensitivity q's 96.7 ± 17.9	Low sensitivity q's 95.6 ± 20.6	Low sensitivity q's 100.0 ± 00.0
	High sensitivity q's 94.0 ± 22.6	High sensitivity q's 84.4 ± 36.3	High sensitivity q's 93.6 ± 24.7
With Sensitivity Signal	Low sensitivity q's 100.0 ± 00.0	Low sensitivity q's 98.8 ± 11.0	Low sensitivity q's 99.2 ± 8.7
	High sensitivity q's 95.9 ± 20.0	High sensitivity q's 93.5 ± 24.7	High sensitivity q's 84.2 ± 36.4

Finally, we extracted the level of phasic arousal per question, measured in microsiemens (μS), compiled during the study. The minimum phasic arousal for one question was -0.27 μS and the maximum 12.60 μS. Overall, we can observe different arousal rates depending on the combination of nudges present in the context and question sensitivity. When no nudge was present, arousal was lower for low-sensitivity questions (9.9 μS) than for high-sensitivity questions (10.4 μS). When only the sensitivity signal was present, arousal was higher in low-sensitivity questions (10.6 μS) compared to high-sensitivity questions (10.3 μS). When low social proof was present, arousal was higher for low-sensitivity questions (9.3 μS) than for high-sensitivity questions (9.1 μS). When high social proof was present, arousal was higher in low-sensitivity questions (10.5 μS) than in high sensitivity questions (10.1 μS). When both the sensitivity signal and low social proof were present, arousal was the same for the low-sensitivity questions (9.0 μS) and high-sensitivity questions (9.0 μS). When both the sensitivity signal and high social proof were present, arousal was lower for low-sensitivity questions (8.9 μS) compared to high-sensitivity questions (10.2 μS). Table 8 summarizes these results.

Table 8. Arousal Per Nudge Combination and Question Sensitivity.

Social Proof/ Sensitivity Signal	No Social Proof	Low Social Proof	High Social Proof
No Sensitivity Signal	Low sensitivity q's 9.932 ± 4.936	Low sensitivity q's 9.352 ± 4.642	Low sensitivity q's 10.499 ± 5.288
	High sensitivity q's 10.372 ± 5.334	High sensitivity q's 9.118 ± 4.849	High sensitivity q's 10.101 ± 4.403
With Sensitivity Signal	Low sensitivity q's 10.633 ± 6.003	Low sensitivity q's 8.970 ± 4.935	Low sensitivity q's 8.871 ± 5.096
	High sensitivity q's 10.269 ± 5.177	High sensitivity q's 8.972 ± 5.043	High sensitivity q's 10.232 ± 5.147

4.2. Hypotheses Testing

For the testing of hypotheses H1 to H7, we conducted two types of analyses because some relationships tested included a dependent variable that is discrete in nature (information disclosure (response rate): count of questions answered) and others tested for a continuous dependent variable (emotional response (arousal): continuous phasic EDA). We used logistic regressions with a random intercept for models with information disclosure (response rate) as the dependent variable (H1 to H3, and H4b). We used linear regressions with random intercept for models with emotional response (arousal) as the dependent variable (H4a, H5, H6).

4.2.1. Effect of Question Sensitivity on Information Disclosure (H1)

To test whether question sensitivity negatively influences users' information disclosure to chatbots (H1), we first extracted the response rate per question sensitivity. The average response rate for low-sensitivity questions was 98.4% ($\pm$13.5), while the response rate for high-sensitivity questions was 91.0% ($\pm$28.6). The results of the logistic regression showed that a question is less likely to be answered if it is highly sensitive compared to when it is low in sensitivity (estimate = -2.20, p-value < 0.0001). Thus, H1 is supported.

4.2.2. Effect of Information Disclosure Nudges on Information Disclosure (H2 and H3)

To test whether the information disclosure differed in the presence of nudges (H2a, H2b, and H3), we looked at the effect of the nudges on the response rate per question sensitivity. When hypothesized that when a low sensitivity signal is present (vs. absent) for less sensitive questions, disclosure increases (H2a) and that when a high sensitivity signal is present (vs. absent) for more sensitive questions, disclosure decreases (H2b). We also hypothesized that social proof moderates the relationship between question sensitivity and users' information disclosure, such as greater social proof, leads to more disclosure and less social proof leads to less disclosure (H3). Table 9 presents a summary of the results.

Table 9. Logistic Regressions: Effect of Nudges on Response Rate Per Question Sensitivity.

Nudge Comparison	Question Sensitivity	Estimate	StdErr	DF	t-Value	One-Tail Probt	Hypothesis
Sensitivity signal: Present vs. Absent	Low	1.47	0.80	1774	1.84	0.0334	H2a
	High	−1.03	0.85	1774	−1.21	0.1142	H2b
Social proof: Low vs. High	Low	−1.29	1.15	1772	−1.13	0.1301	H3
	High	−1.32	1.5	1770	−1.14	0.1269	H3
Social proof: Low vs. None	Low	−0.75	0.73	1772	−1.03	0.1519	-
	High	−0.38	0.82	1772	−0.47	0.6418	-
Social proof: High vs. None	Low	1.14	1.17	1772	0.98	0.1637	-
	High	−1.67	1.23	1772	−1.36	0.3248	-

The above results show that the response rate to low-sensitivity questions increases (estimate = 1.47, p-value = 0.0334) when the low-sensitivity signal is present. Thus, H2a is supported.

The results show that, for high-sensitivity questions, the response rate decreases (estimate = -1.03) when the high-sensitivity signal is present compared to when absent, however, this result is not statistically significant (p-value = 0.1142). H2b is not supported.

For the social proof nudge, the results go in the same direction as the hypothesis, where disclosure decreases with social proof is low compared to when social proof is high for both question sensitivity levels (estimates = -1.29 and -1.32) but these results are not statistically significant (p-values = 0.1301 and 0.1269). Thus, H3 is not supported.

The comparison between the two social proof levels when present vs. when absent (no social proof) was also tested, but found not to be significant. Moreover, the interaction between the effect of the two nudges on disclosure was also tested, but also found to be not significant.

4.2.3. Effect of Emotional Response (H4 to H6)

We hypothesized that emotional response would mediate the relationship between question type and information disclosure such as question sensitivity positively influencing emotional response (H4a) and emotional response negatively influencing disclosure (H4b).

The results of the linear regression show that emotional response tends to decrease when high-sensitivity questions are asked compared to low-sensitivity questions (estimate = −0.10), but this result is not statistically significant (p-value = 0.3226). Thus, H4a is not supported.

The result of the logistic regression on the effect of emotional response on information disclosure (estimate = 0.01) is not statistically significant (p-value = 0.1948). Therefore, H4b is not supported.

We then tested the effect of the information disclosure nudges on emotional response. We hypothesized that when a low sensitivity signal is present (vs. absent) for less sensitive questions, the emotional response decreases (H5a) and that when a high sensitivity signal is present (vs. absent) for more sensitive questions, the emotional response increases (H5b). Moreover, we hypothesized that social proof moderates the relationship between question sensitivity and emotional response, such that greater social proof leads to lower emotional response and less social proof leads to higher emotional response (H6). Table 10 summarizes these results.

Table 10. Linear Regressions: Effect of Information Disclosure Nudges on Emotional Response.

Nudge Comparison	Question Sensitivity	Estimate	StdErr	DF	t-Value	One-Tail Probt	Hypothesis
Sensitivity signal: Present vs. Absent	Low	−0.13	0.15	1728	−0.86	0.1948	H5a
	High	−0.03	0.21	1728	−0.13	0.0511	H5b
Social proof: Low vs. High	Low	−0.12	0.26	1726	0.44	0.1717	H6
	High	−0.11	0.26	1724	0.43	0.1651	H6
Social proof: Low vs. None	Low	−0.59	0.18	1726	−3.28	0.0006	-
	High	−0.20	0.25	1726	−0.77	0.2196	-
Social proof: High vs. None	Low	0.03	0.18	1726	0.18	0.4281	-
	High	−0.08	0.25	1726	−0.31	0.1231	-

The above results show that emotional response to low-sensitivity questions decreases (estimate = −0.13) when the low-sensitivity signal is present compared to when absent, however, this result is not statistically significant (p-value = 0.1948). Thus, H5a is not supported.

The results show that, for high-sensitivity questions, the emotional response decreases (estimate = −0.03) when high sensitivity signal is present compared to when absent. This result is marginally significant with a p-value of 0.0511. Since the results are contrary to the hypothesis, H5b is not supported.

For the social proof nudge, the results show that emotional response decreases when the social proof is low compared to when the social proof is high for both question sensitivity levels (estimates = −0.12 and −0.11). These results are not statistically significant (p-values = 0.1717 and 0.1651). Thus, H6 is not supported.

The comparison between the two social proof levels when present vs. when absent (no social proof) was also tested. The only significant result is the decrease in the emotional response (estimate = −0.59) to low-sensitivity questions when the low social proof is present compared to when it is absent (p-value = 0.0006). The other comparisons' results were not significant. The interaction between the effect of the two nudges on emotional response was also tested, but also found to be not significant.

The validated research model is shown Figure 4.

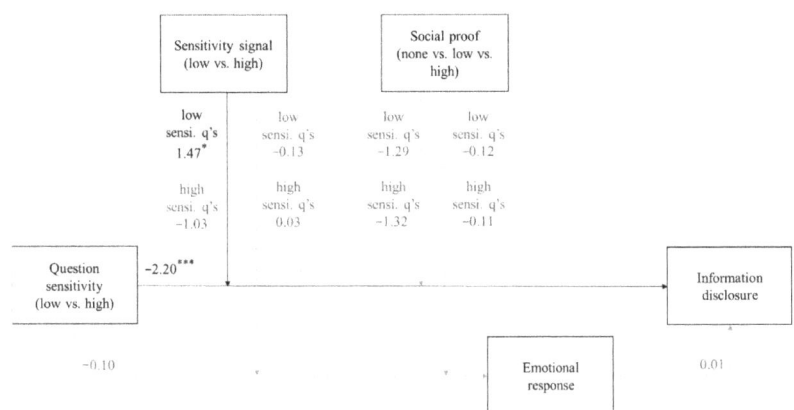

Figure 4. Validated Research Model. * significant at 0.05 level; *** = significant at 0.001 level.

5. Discussion

5.1. Main Findings

To summarize the findings of this study, the results suggest that question sensitivity has an impact on disclosure in the context of interactions with chatbots. Concerning the effect of the sensitivity signal, the results show that for less sensitive questions, a low-sensitivity nudge increases disclosure. Results also suggest that for high-sensitivity questions, a high-sensitivity nudge seems to decrease emotional response. On the other hand, the social proof nudge does not seem to affect users' disclosure behaviors, nor their emotional responses. Finally, it was suggested that emotional response does not seem to be a mechanism explaining how user disclosure operates in interactions with chatbots.

5.2. Theoretical Contributions

From a theoretical standpoint, this research makes three main contributions. First, this research complements the literature on user disclosure by confirming that question sensitivity has an impact on disclosure in the context of interactions with chatbots. This result is consistent with previous research on question sensitivity and disclosure (not specific to chatbot interactions) [52,53,59]. Indeed, research on antecedents to disclosure had previously shown that sensitivity played a role in disclosure. In fact, Mothersbaugh [59] suggested that the sensitivity of information is an antecedent to disclosure in an online service context, while Lee et al. [52] reported the same results in an e-commerce setting. However, sensitivity of information had not been explored in a chatbot context [52,53,59]. The present study suggests that this relationship does apply to interactions with chatbots. This result strengthens our understanding of the differences between human-human and human-AI interactions.

Second, comparing the two types of nudges tested, this research suggests that a sensitivity signal seems more promising than social proof in influencing users' disclosure behaviors to chatbots. The difference between the two nudges might be due to the fact that user disclosure is an intrinsic behavior and people do not make judgments about their privacy based on what others do. These results come to complement previous research [97,114,115] on nudging and privacy for the disclosure of personal information online (not just specific to chatbots). Overall, empirical research on digital nudging to alter users' disclosure behaviors has produced conflicting findings in the past, with some studies finding it to be quite effective while others have found no such results [97]. On one hand, studies have demonstrated that motivating communications and persuasive messages with stronger arguments or more positive framings can enhance the disclosure of private information [114,116]. In our case, the low-sensitivity signal goes in accord with these previous results, by increasing the disclosure of low-sensitivity information. However, there is still a question mark as to how to decrease—rather than increase—disclosure of high-sensitivity information. On the other hand, a growing interest has been shown in examining the impact of social nudges, centered around the social proof, used to affect users' privacy decision-making online [97]. According to research, social cues, such as the knowledge that a majority of users' peers have taken similar actions, such as disclosing personal information, can lead to an increase in information disclosure on websites [29,99]. In the present case, by refuting the effect of such a nudge in user-chatbot interaction, this study complements previous results in the literature [29,99] by marking a distinction between online and chatbot-specific interactions. These results also go hand in hand with other recent findings such as Rudnicka et al. [114] who found that while persuasive messages framed around learning increased the disclosure of sensitive items, people did not change their disclosure behavior for messages framed around social proof, contribution, and altruism.

Nonetheless, the nudges used in this research may still have value by perhaps confirming users' judgments regarding questions they are prompted with. In our results, seven participants answered all questions, and twelve skipped some questions. The reasons for disclosure and non-disclosure given by participants show that, in the case where users have the same judgment as the nudge, the nudge may serve to confirm their decision to answer the question or not (confirmed by five participants).

Third, evidence from this study showed that emotional response does not appear to be a mechanism describing how disclosure functions in user-chatbot interactions. Previous research on affect and online information disclosure tells another story. Wakefield [117] suggested that positive affect has a significant effect on users' online information disclosure. Additionally, Coker and McGill [118] stated that arousal increases self-disclosure. The contradictory results of these studies to the ones reported in this research could highlight a plausible difference in users' behaviors when interacting with websites versus chatbots, that is, only in certain settings. Indeed, previous studies in user-chatbot interactions had underlined the role of emotional response, but mostly in contexts that pertain to mental health or education rather than privacy in e-commerce [33,60,108,109]. Referring to the Affect Infusion Model, the results of the present study could be because information disclosure to chatbots is not a high infusion situation for users. Rather than performing a heuristic processing of the available information, it is possible that in chatbot interactions, users could use more direct access or motivation-based processing [32]. Under these strategies, people base their judgment either by reproducing a past behavior in a similar situation or by searching for specific information with a clear purpose in mind to base their decision. In these two types of processing, the AIM states that affect does not serve as information in the judgment, which could explain the insignificant results of this study. Thus, interactions with chatbots might not be a situation where emotional response is inferred into information.

5.3. Managerial Implications

From a managerial standpoint, the fact that this research marginally supports the influence of information disclosure nudges on users' behaviors has one main implication. In practice, the use of nudges has been debated since their inception. It is believed that to be ethical, nudges should aim to enhance people's decisions by altering how alternatives are given rather than altering the options themselves or motivating or coercing people a certain way [115]. The information disclosure nudges tested in this research did not always predict user behavior. Nonetheless, from an ethical perspective, users should maintain their right to make informed decision-making in their online interactions. Ultimately, the goal of interfaces should be to give users control and freedom rather than to choose for them what they put out on the internet, especially through chatbot interactions [51]. Thus, policymakers can scrutinize this research for inspiration when drafting policies that provide more information to users in online interactions with chatbots.

5.4. Limitations and Research Avenues

The results of this study on the impact of nudges on disclosure could be due to some limiting factors. First, the inconclusive results could be due to the fact that not enough questions were asked per nudge combination and per question sensitivity level to find significant differences in information disclosure. Second, the nature of the interactions between the users and chatbots in the experiment consisted of a series of questions and answers. Considering these points, future research could explore consumers' information disclosure behaviors when they communicate with chatbots in the form of extended conversations, rather than in a question-and-answer format. Another explanation for these partially supported results could be due to limitations in choosing to conduct this experiment in a lab setting. In fact, this research was conducted under high ethical standards. Participants were informed that their responses would be anonymized and were asked to sign consent forms before the start of the experiment. Additionally, the websites used to host the chatbot prototypes were all fictional. This environment might have made participants overly trusting towards the chatbots by reminding them they are in a lab setting that is controlled by high ethical standards and in turn increasing their disclosure. Future research on information disclosure should try to mitigate this by conducting their experiment in association with real websites.

Considering the choice of nudges (i.e., sensitivity signal and social proof) in this experiment, this research also provides potential avenues for other nudges that could promote informed decision- making when it comes to information disclosure to chatbots and could be explored in the future. For example, in our experimental design, participants were told that they could choose to not answer a question if they did not want to. Future research could explore the difference in information disclosure when users are given the cue that they can choose not to respond versus no cue.

The peculiarities of our stimulus materials and study design may have restricted the study's findings. We placed the information disclosure nudges in locations that were conducive for the use of the eye-tracking technology used in this research. The nudges were thus placed on either side of the chatbot. In addition, although the nudges were uniform in size and color, it is possible that varying the design of the nudges would have made them more impactful. Future research could investigate the optimal location and design for information disclosure nudges to be maximally influential on users' behaviors.

Finally, the results of this study showed that more than a third of the participants (seven out of nineteen) answered all questions prompted by the chatbots, regardless of their sensitivity level. This heterogeneous data suggests that some users are comfortable sharing information online with chatbots, being general or sensitive. Given the small sample size of this study, our results did not make it possible to find a distinguishing factor better for the group that answered all questions versus the group that skipped some questions. Although the sample size used is typical for NeuroIS research [119] future research could still replicate and extend this study by increasing the sample size in order to investigate if some personal characteristics impact disclosure decisions. For example, measuring the users' level of comfort with online privacy and sensitive issues [50] or taking into account the cultural background of individuals [120] could bring out behavioral differences between groups. Considering that the participants in this study were located in North America, it is plausible that the results would differ greatly in other regions of the world where the very definition of sensitive matters may vary. Looking forward, finding determining factors between individuals could be valuable for both business organizations to better understand their customers and policymakers to draft distinct policies for different types of users.

6. Conclusions

To conclude, this research explored the impact of question sensitivity, information disclosure nudges, and arousal on users' information disclosure behaviors in chatbot interactions. The results show that people rely more on their own judgment than information disclosure nudges when it comes to disclosing information online to chatbots.

Author Contributions: Conceptualization, L.C., S.-M.P., C.K.C., P.-M.L. and S.S.; methodology, L.C., S.-M.P., C.K.C., P.-M.L. and S.S.; formal analysis, L.C. and S.-M.P.; investigation, L.C., S.-M.P., C.K.C., P.-M.L. and S.S.; resources, C.K.C., P.-M.L. and S.S.; writing—original draft preparation, L.C. and S.-M.P.; writing—review and editing, C.K.C., P.-M.L. and S.S.; supervision, C.K.C., P.-M.L. and S.S.; project administration, C.K.C., P.-M.L. and S.S.; funding acquisition, P.-M.L. All authors have read and agreed to the published version of the manuscript.

Funding: This research was funded by the Natural Sciences and Engineering Research Council (NSERC) of Canada and Prompt, grant number IRC 505259-16.

Institutional Review Board Statement: The study was conducted in accordance with the Declaration of Helsinki, and approved by the Ethics Committee of HEC Montreal (2022-4721, 2021-10-20).

Informed Consent Statement: Informed consent was obtained from all subjects involved in the study.

Data Availability Statement: The data presented in this study are available on request from the corresponding author. The data are not publicly available due to privacy and compliance with the protocol approved by the Ethics Committee.

Conflicts of Interest: The authors declare no conflict of interest.

Appendix A

Table A1. Full List Questions.

Question	Context	Sensitivity Level
How many years of work experience do you have?	Career	Low
What country do you currently live in?	Career	Low
What is your biggest strength?	Career	Low
What is your highest completed education level?	Career	Low
What languages do you speak fluently?	Career	Low
What high school did you go to?	Career	Low
What country were you born in?	Career	Low
Are you a hard worker or the less the better?	Career	Low
Do you feel like you earn enough money?	Career	High
Have you ever been in trouble with the law?	Career	High
Have you ever lied to your superior to get a day off work?	Career	High
Do you prioritize your professional or your personal life?	Career	High
Have you ever lied in a job interview or on your CV?	Career	High
Have you ever lied on your CV?	Career	High
What's the biggest mistake you've made at work?	Career	High
Have you ever drank at work?	Career	High
Do you tend to be an optimist or pessimist and why?	Dating	Low
Do you want to have children/do you have children?	Dating	Low
Is intelligence or looks more important for you?	Dating	Low
What is your eye colour?	Dating	Low
What is your favorite movie?	Dating	Low
What is your favorite music genre?	Dating	Low
What is your gender?	Dating	Low
What is your relationship status?	Dating	Low
Are you religious? If so, what religion do you practice?	Dating	High
Do you fall in love easily?	Dating	High

Table A1. *Cont.*

Question	Context	Sensitivity Level
During sex, do you take precautions against unwanted pregnancies?	Dating	High
During sex, do you take precautions against STDs?	Dating	High
Have you ever been on a date with the sole purpose of having sex with the person?	Dating	High
Have you ever cheated on your significant other?	Dating	High
How many serious relationships have you been in throughout	Dating	High
What is your sexual orientation?	Dating	High
Do you prefer sweet or savoury food?	Groceries	Low
Do you enjoy trying new foods?	Groceries	Low
Do you enjoy eating different cuisines of the world?	Groceries	Low
Do you always buy brand-name products?	Groceries	Low
Do you usually use coupons and discount while groceries shopping?	Groceries	Low
Do you always shop at the same grocery store?	Groceries	Low
How often do you shop for your groceries online?	Groceries	Low
Do you prefer vegetables or fruits?	Groceries	Low
Overall, how healthy is your diet?	Groceries	High
Do you track your calories?	Groceries	High
Do you take any supplements?	Groceries	High
Counting yourself, how many people live in your household?	Groceries	High
Do you have any allergies?	Groceries	High
Would you say your diet is healthier than most people's diet?	Groceries	High
What is your address?	Groceries	High
How much do you spend on groceries per week?	Groceries	High
Do you play sports?	Gym	Low
How many cups of coffee/tea do you drink per day?	Gym	Low
How many glasses of water do you drink per day?	Gym	Low
How many hours do you practice physical activity per week?	Gym	Low
How many meals do you eat per day?	Gym	Low
What is your height (cm/feet and inches)?	Gym	Low
How much time per week are you willing to dedicate to personal training?	Gym	Low
What sports do you play?	Gym	Low
How many cigarettes do you smoke per week?	Gym	High
How many glasses of alcohol do you drink per week?	Gym	High
How much do you weight (kg/lbs)?	Gym	High
What is one thing you would like to change about yourself (physically or mentally)?	Gym	High
Do you experience binge eating episodes (uncontrollable eating of large amounts of food)	Gym	High
How often do you think you feel too much stress?	Gym	High
Do you have a stressful lifestyle?	Gym	High
Have you ever been told by a physician that you have a metabolic disease (e.g., heart disease, high blood pressure)?	Gym	High
Do you always read the terms and conditions before checking the box?	Insurance	Low
Do you have a car?	Insurance	Low

Table A1. *Cont.*

Question	Context	Sensitivity Level
Do you have any pets?	Insurance	Low
Do you have renters/homeowners insurance?	Insurance	Low
How old are you?	Insurance	Low
What is your current occupation?	Insurance	Low
What is your phone model?	Insurance	Low
Do you smoke?	Insurance	Low
Do you have more than 5000 USD in savings at this time?	Insurance	High
Do you pay off your credit card in full every month?	Insurance	High
How many credit cards do you have?	Insurance	High
How much do you pay on rent/mortgage per month?	Insurance	High
What is your current income per year?	Insurance	High
What is your email address?	Insurance	High
What is your phone number?	Insurance	High
Do you have an investment portfolio?	Insurance	High
Would you also try typical dishes—that you would normally never eat—while traveling?	Travel	Low
Is room service important to you?	Travel	Low
What type of accommodation do you prefer when travelling?	Travel	Low
Do you like to talk to the local people when you travel?	Travel	Low
What modes of transportation do you prefer to use when you travel?	Travel	Low
Have you ever traveled abroad?	Travel	Low
Which country would you most like to visit?	Travel	Low
What is your dream destination for a vacation?	Travel	Low
Are you fully vaccinated against Covid19?	Travel	High
Which countries, regions, or cities irritate you the most and why?	Travel	High
What would you never do on your travels and why?	Travel	High
How much money do you typically spend per day while travelling?	Travel	High
Would you feel insecure if you were to travel alone?	Travel	High
Are there regions that you would never want to visit and why?	Travel	High
Is there a legal reason why you could not travel to a specific country?	Travel	High

References

1. Hussain, S.; Ameri Sianaki, O.; Ababneh, N. *A Survey on Conversational Agents/Chatbots Classification and Design Techniques*; Springer: Cham, Switzerland; pp. 946–956.
2. Jannach, D.; Manzoor, A.; Cai, W.; Chen, L. A Survey on Conversational Recommender Systems. *ACM Comput. Surv.* **2021**, *54*, 105. [CrossRef]
3. Shi, W.; Wang, X.; Oh, Y.J.; Zhang, J.; Sahay, S.; Yu, Z. Effects of Persuasive Dialogues: Testing Bot Identities and Inquiry Strategies. In Proceedings of the 2020 CHI Conference on Human Factors in Computing Systems, Honolulu, HI, USA, 25–30 April 2020; Association for Computing Machinery: New York, NY, USA, 2020; pp. 1–13. [CrossRef]
4. Schanke, S.; Burtch, G.; Ray, G. Estimating the Impact of "Humanizing" Customer Service Chatbots. *Inf. Syst. Res.* **2021**, *32*, 736–751. [CrossRef]
5. Liao, M.; Sundar, S.S. How Should AI Systems Talk to Users when Collecting their Personal Information? Effects of Role Framing and Self-Referencing on Human-AI Interaction. In Proceedings of the 2021 CHI Conference on Human Factors in Computing Systems, Yokohama, Japan, 8–13 May 2021; p. 151.
6. Dev, J.; Camp, L.J. User Engagement with Chatbots: A Discursive Psychology Approach. In Proceedings of the 2nd Conference on Conversational User Interfaces, Bilbao, Spain, 22–24 July 2020; p. 52.
7. Ali, N. *Text Stylometry for Chat Bot Identification and Intelligence Estimation*; University of Louiseville: St. Louisville, KY, USA, 2014.

8. Gondaliya, K.; Butakov, S.; Zavarsky, P. SLA as a mechanism to manage risks related to chatbot services. In Proceedings of the 2020 IEEE 6th Intl Conference on Big Data Security on Cloud (BigDataSecurity), Baltimore, MD, USA, 25–27 May 2020; pp. 235–240.
9. Roland, T.R. The future of marketing. *Int. J. Res. Mark.* **2020**, *37*, 15–26. [CrossRef]
10. Saleilles, J.; Aïmeur, E. SecuBot, a Teacher in Appearance: How Social Chatbots Can Influence People. In Proceedings of the AIofAI 2021: 1st Workshop on Adverse Impacts and Collateral Effects of Artificial Intelligence Technologies, Montreal, QC, Canada, 19 August 2021; p. 19.
11. Fan, H.; Han, B.; Gao, W.; Li, W. How AI chatbots have reshaped the frontline interface in China: Examining the role of sales–service ambidexterity and the personalization–privacy paradox. *Int. J. Emerg. Mark.* **2022**, *17*, 967–986. [CrossRef]
12. Cheng, Y.; Jiang, H. How Do AI-driven Chatbots Impact User Experience? Examining Gratifications, Perceived Privacy Risk, Satisfaction, Loyalty, and Continued Use. *J. Broadcast. Electron. Media* **2020**, *64*, 592–614. [CrossRef]
13. Rese, A.; Ganster, L.; Baier, D. Chatbots in retailers' customer communication: How to measure their acceptance? *J. Retail. Consum. Serv.* **2020**, *56*, 102176. [CrossRef]
14. Rodríguez-Priego, N.; van Bavel, R.; Monteleone, S. The disconnection between privacy notices and information disclosure: An online experiment. *Econ. Politica* **2016**, *33*, 433–461. [CrossRef]
15. Wu, K.-W.; Huang, S.Y.; Yen, D.C.; Popova, I. The effect of online privacy policy on consumer privacy concern and trust. *Comput. Hum. Behav.* **2012**, *28*, 889–897. [CrossRef]
16. Lusoli, W.; Bacigalupo, M.; Lupiáñez-Villanueva, F.; Andrade, N.; Monteleone, S.; Maghiros, I. Pan-European Survey of Practices, Attitudes and Policy Preferences as Regards Personal Identity Data Management. *JRC Sci. Policy Rep.* 2012. Available online: https://papers.ssrn.com/sol3/papers.cfm?abstract_id=2086579 (accessed on 30 November 2022).
17. Wang, Y.-C.; Burke, M.; Kraut, R. Modeling Self-Disclosure in Social Networking Sites. In Proceedings of the 19th ACM Conference on Computer-Supported Cooperative Work & Social Computing, San Francisco, CA, USA, 27 February–2 March 2016; pp. 74–85.
18. Ischen, C.; Araujo, T.; Voorveld, H.; van Noort, G.; Smit, E. *Privacy Concerns in Chatbot Interactions*; Springer: Cham, Switzerland; pp. 34–48.
19. Jobin, A.; Ienca, M.; Vayena, E. The global landscape of AI ethics guidelines. *Nat. Mach. Intell.* **2019**, *1*, 389–399. [CrossRef]
20. Groom, V.; Calo, M. Reversing the Privacy Paradox: An Experimental Study. TPRC 2011. 2011. Available online: https://papers.ssrn.com/sol3/papers.cfm?abstract_id=1993125 (accessed on 30 November 2022).
21. Hoofnagle, C.; King, J.; Li, S.; Turow, J. How Different Are Young Adults From Older Adults When It Comes to Information Privacy Attitudes & Policies? *SSRN Electron. J.* 2010. [CrossRef]
22. Martin, K. Privacy Notices as Tabula Rasa: An empirical investigation into how complying with a privacy notice is related to meeting privacy expectations online. *J. Public Policy Mark.* **2015**, *34*, 210–227. [CrossRef]
23. Mao, T.-W.; Ouyang, S. Digital Nudge Chat Bot. Master's Thesis, Cornell University, New York, NY, USA, 2020.
24. Kim, J.; Giroux, M.; Lee, J. When do you trust AI? The effect of number presentation detail on consumer trust and acceptance of AI recommendations. *Psychol. Mark.* **2021**, *38*, 1140–1155. [CrossRef]
25. Das, G.; Spence, M.T.; Agarwal, J. Social selling cues: The dynamics of posting numbers viewed and bought on customers' purchase intentions. *Int. J. Res. Mark.* **2021**, *38*, 994–1016. [CrossRef]
26. He, Y.; Oppewal, H. See How Much We've Sold Already! Effects of Displaying Sales and Stock Level Information on Consumers' Online Product Choices. *J. Retail.* **2018**, *94*, 45–57. [CrossRef]
27. Kahneman, D. *Thinking, Fast and Slow*; Penguin Random House: Toronto, ON, Canada, 2011.
28. Petty, R.E.; Cacioppo, J.T. The Elaboration Likelihood Model of Persuasion. In *Communication and Persuasion: Central and Peripheral Routes to Attitude Change*; Springer: New York, NY, USA, 1986; pp. 1–24. [CrossRef]
29. Acquisti, A.; Brandimarte, L.; Loewenstein, G. Privacy and human behavior in the age of information. *Science* **2015**, *347*, 509–514. [CrossRef]
30. Cialdini, R.B. *Influence: The Psychology of Persuasion*; EPub, Ed.; Collins: New York, NY, USA, 2009.
31. Thaler, R.H.; Sunstein, C.R. *Nudge: Improving Decisions about Health, Wealth, and Happiness*; Yale University Press: New Haven, CT, USA, 2008; p. 293.
32. Forgas, J.P. Mood and judgment: The affect infusion model (AIM). *Psychol. Bull.* **1995**, *117*, 39–66. [CrossRef]
33. Liu, B.; Sundar, S.S. Should Machines Express Sympathy and Empathy? Experiments with a Health Advice Chatbot. *Cyberpsycholog. Behav. Soc. Netw.* **2018**, *21*, 625–636. [CrossRef]
34. Tärning, B.; Silvervarg, A. "I Didn't Understand, I'm Really Not Very Smart"—How Design of a Digital Tutee's Self-Efficacy Affects Conversation and Student Behavior in a Digital Math Game. *Educ. Sci.* **2019**, *9*, 197. [CrossRef]
35. Wang, X.; Nakatsu, R. *How Do People Talk with a Virtual Philosopher: Log Analysis of a Real-World Application*; Springer: Berlin/Heidelberg, Germany, 2013; pp. 132–137.
36. Gupta, A.; Royer, A.; Wright, C.; Khan, F.; Heath, V.; Galinkin, E.; Khurana, R.; Ganapini, M.; Fancy, M.; Sweidan, M.; et al. *The State of AI Ethics Report (January 2021)*; Montreal AI Ethics Institute: Montréal, QC, Canada, 2021.
37. Bang, J.; Kim, S.; Nam, J.W.; Yang, D.-G. Ethical Chatbot Design for Reducing Negative Effects of Biased Data and Unethical Conversations. In Proceedings of the 2021 International Conference on Platform Technology and Service (PlatCon), Jeju, Republic of Korea, 23–25 August 2021; pp. 1–5.
38. Cote, C. 5 Principles of Data Ethics For Business. In *Business Insights*; Harvard Business School: Boston, MA, USA, 2021.

39. Martineau, J.T. Ethical issues in the development of AI [Workshop presentation]. In Proceedings of the 1st IVADO Research Workshop on Human-Centered AI, Montréal, QC, Canada, 28–29 April 2022.
40. Følstad, A.; Araujo, T.; Law, E.L.-C.; Brandtzaeg, P.B.; Papadopoulos, S.; Reis, L.; Baez, M.; Laban, G.; McAllister, P.; Ischen, C.; et al. Future directions for chatbot research: An interdisciplinary research agenda. *Computing* **2021**, *103*, 2915–2942. [CrossRef]
41. Qomariyah, N.N. Definition and History of Recommender Systems. Ph.D. Thesis, BINUS University International, Jakarta, Indonesia, 2020.
42. Chew, H.S.J. The Use of Artificial Intelligence-Based Conversational Agents (Chatbots) for Weight Loss: Scoping Review and Practical Recommendations. *JMIR Med. Inform.* **2022**, *10*, e32578. [CrossRef] [PubMed]
43. Adamopoulou, E.; Moussiades, L. Chatbots: History, technology, and applications. *Mach. Learn. Appl.* **2020**, *2*, 100006. [CrossRef]
44. Ikemoto, Y.; Asawavetvutt, V.; Kuwabara, K.; Huang, H.-H. Conversation Strategy of a Chatbot for Interactive Recommendations. In *Intelligent Information and Database Systems*; Springer: Cham, Switzerland, 2018; pp. 117–126.
45. Mahmood, T.; Ricci, F. Improving recommender systems with adaptive conversational strategies. In Proceedings of the 20th ACM Conference on Hypertext and Hypermedia, Torino, Italy, 29 June–1 July 2009; pp. 73–82.
46. Nica, I.; Tazl, O.A.; Wotawa, F. Chatbot-based Tourist Recommendations Using Model-based Reasoning. In Proceedings of the ConfWS, Graz, Austria, 27–28 September 2018.
47. Eeuwen, M.V. Mobile Conversational Commerce: Messenger Chatbots as the Next Interface between Businesses and Consumers. Master's Thesis, University of Twente, Enschede, The Netherlands, 2017.
48. Awad, N.F.; Krishnan, M.S. The Personalization Privacy Paradox: An Empirical Evaluation of Information Transparency and the Willingness to Be Profiled Online for Personalization. *MIS Q.* **2006**, *30*, 13–28. [CrossRef]
49. Ng, M.; Coopamootoo, K.P.L.; Toreini, E.; Aitken, M.; Elliot, K.; Moorsel, A.v. Simulating the Effects of Social Presence on Trust, Privacy Concerns & Usage Intentions in Automated Bots for Finance. In Proceedings of the 2020 IEEE European Symposium on Security and Privacy Workshops (EuroS&PW), Genoa, Italy, 7–11 September 2020; pp. 190–199.
50. Thomaz, F.; Salge, C.; Karahanna, E.; Hulland, J.S. Learning from the Dark Web: Leveraging conversational agents in the era of hyper-privacy to enhance marketing. *J. Acad. Mark. Sci.* **2019**, *48*, 43–63. [CrossRef]
51. Murtarelli, G.; Gregory, A.; Romenti, S. A conversation-based perspective for shaping ethical human–machine interactions: The particular challenge of chatbots. *J. Bus. Res.* **2021**, *129*, 927–935. [CrossRef]
52. Lee, H.; Lim, D.; Kim, H.; Zo, H.; Ciganek, A.P. Compensation paradox: The influence of monetary rewards on user behaviour. *Behav. Inf. Technol.* **2015**, *34*, 45–56. [CrossRef]
53. Metzger, M.J. Communication Privacy Management in Electronic Commerce. *J. Comput.-Mediat. Commun.* **2007**, *12*, 335–361. [CrossRef]
54. Li, H.; Sarathy, R.; Xu, H. Understanding Situational Online Information Disclosure as a Privacy Calculus. *J. Comput. Inf. Syst.* **2010**, *51*, 62–71. [CrossRef]
55. Li, H.; Sarathy, R.; Xu, H. The role of affect and cognition on online consumers' decision to disclose personal information to unfamiliar online vendors. *Decis. Support Syst.* **2011**, *51*, 434–445. [CrossRef]
56. Kolotylo-Kulkarni, M.; Xia, W.; Dhillon, G. Information disclosure in e-commerce: A systematic review and agenda for future research. *J. Bus. Res.* **2021**, *126*, 221–238. [CrossRef]
57. Ohm, P. Sensitive information. *South. Calif. Law Rev.* **2015**, *88*, 1125–1196.
58. Harrison, M.E. *Doing Development Research*; SAGE Publications, Ltd.: London, UK, 2006. [CrossRef]
59. Mothersbaugh, D.; Ii, W.; Beatty, S.; Wang, S. Disclosure Antecedents in an Online Service Context The Role of Sensitivity of Information. *J. Serv. Res.* **2012**, *15*, 76–98. [CrossRef]
60. Lee, Y.-C.; Yamashita, N.; Huang, Y.; Fu, W. "I Hear You, I Feel You": Encouraging Deep Self-disclosure through a Chatbot. In Proceedings of the 2020 CHI Conference on Human Factors in Computing Systems, Honolulu, HI, USA, 25–30 April 2020; Association for Computing Machinery: New York, NY, USA, 2020; pp. 1–12.
61. Stone, D.L. The Effects of Valence of Outcomes for Providing Data and the Perceived Relevance of the Data Requested on Privacy-Related Behaviors, Beliefs, and Attitudes. Ph.D Thesis, Purdue University, Ann Arbor, MI, USA, 1981.
62. Al-Sharafi, M.A.; Al-Emran, M.; Iranmanesh, M.; Al-Qaysi, N.; Iahad, N.A.; Arpaci, I. Understanding the impact of knowledge management factors on the sustainable use of AI-based chatbots for educational purposes using a hybrid SEM-ANN approach. *Interact. Learn. Environ.* **2022**, 1–20. [CrossRef]
63. Cozby, P.C. Self-disclosure: A literature review. *Psychol. Bull.* **1973**, *79*, 73–91. [CrossRef]
64. Xiao, B.; Benbasat, I. E-Commerce Product Recommendation Agents: Use, Characteristics, and Impact. *MIS Q.* **2007**, *31*, 137–209. [CrossRef]
65. Hasal, M.; Nowaková, J.; Ahmed Saghair, K.; Abdulla, H.; Snášel, V.; Ogiela, L. Chatbots: Security, privacy, data protection, and social aspects. *Concurr. Comput.* **2021**, *33*, e6426. [CrossRef]
66. Knijnenburg, B.P.; Kobsa, A.; Jin, H. Dimensionality of information disclosure behavior. *Int. J. Hum.-Comput. Stud.* **2013**, *71*, 1144–1162. [CrossRef]
67. Joinson, A. Knowing Me, Knowing You: Reciprocal Self-Disclosure in Internet-Based Surveys. *Cyberpsycholog. Behav. Impact Internet Multimed. Virtual Real. Behav. Soc.* **2001**, *4*, 587–591. [CrossRef]
68. Joinson, A.N.; Paine, C.; Buchanan, T.; Reips, U.-D. Measuring self-disclosure online: Blurring and non-response to sensitive items in web-based surveys. *Comput. Hum. Behav.* **2008**, *24*, 2158–2171. [CrossRef]

69. van der Lee, C.; Croes, E.; de Wit, J.; Antheunis, M. Digital Confessions: Exploring the Role of Chatbots in Self-Disclosure. In Proceedings of the Conversations 2019, Amsterdam, The Netherlands, 19–20 November 2019.
70. van Wezel, M.M.C.; Croes, E.A.J.; Antheunis, M.L. "I'm Here for You": Can Social Chatbots Truly Support Their Users? A Literature Review. In *International Workshop on Chatbot Research and Design*; Springer: Cham, Switzerland; pp. 96–113.
71. Ajzen, I.; Fishbein, M. *Understanding Attitudes and Predicting Social Behavior*; Prentice-Hall: Englewood Cliffs, NJ, USA, 1980.
72. Ajzen, I. The theory of planned behavior. *Organ. Behav. Hum. Decis. Process.* **1991**, *50*, 179–211. [CrossRef]
73. Dinev, T.; Hart, P. An Extended Privacy Calculus Model for E-Commerce Transactions. *Inf. Syst. Res.* **2006**, *17*, 61–80. [CrossRef]
74. Hui, K.-L.; Teo, H.H.; Lee, S.-Y.T. The Value of Privacy Assurance: An Exploratory Field Experiment. *MIS Q.* **2007**, *31*, 19–33. [CrossRef]
75. Kobsa, A.; Cho, H.; Knijnenburg, B.P. The effect of personalization provider characteristics on privacy attitudes and behaviors: An Elaboration Likelihood Model approach. *J. Assoc. Inf. Sci. Technol.* **2016**, *67*, 2587–2606. [CrossRef]
76. Brown, B. Studying the Internet Experience. HP Laboratories Technical Report HPL. 2001. Available online: https://citeseerx.ist.psu.edu/document?repid=rep1&type=pdf&doi=563a300a287ff45eb897d100f26d59d4d87c62c2 (accessed on 30 November 2022).
77. Kokolakis, S. Privacy attitudes and privacy behaviour: A review of current research on the privacy paradox phenomenon. *Comput. Secur.* **2017**, *64*, 122–134. [CrossRef]
78. Dienlin, T.; Trepte, S. Is the privacy paradox a relic of the past? An in-depth analysis of privacy attitudes and privacy behaviors The relation between privacy attitudes and privacy behaviors. *Eur. J. Soc. Psychol.* **2015**, *45*, 285–297. [CrossRef]
79. Carlton, A.M. The Relationship between Privacy Notice Formats and Consumer Disclosure Decisions: A Quantitative Study. Ph.D. Thesis, Northcentral University, Ann Arbor, MI, USA, 2019.
80. Zierau, N.; Flock, K.; Janson, A.; Söllner, M.; Leimeister, J.M. The Influence of AI-Based Chatbots and Their Design on Users Trust and Information Sharing in Online Loan Applications. In Proceedings of the Hawaii International Conference on System Sciences (HICSS), Koloa, HI, USA, 4–7 January 2011.
81. Wilson, D.W.; Valacich, J. Unpacking the Privacy Paradox: Irrational Decision-Making within the Privacy Calculus. In Proceedings of the International Conference on Information Systems, ICIS 2012, Orlando, FL, USA, 16–19 December 2012; pp. 4152–4162.
82. Simons, H.W. *Persuasion: Understanding, Practice, and Analysis*; Addison-Wesley: Reading, MA, USA, 1976.
83. Rönnberg, S. *Persuasive Chatbot Conversations: Towards a Personalized User Experience*; Linköping University: Linköping, Sweden, 2020.
84. Nass, C.; Moon, Y. Machines and Mindlessness: Social Responses to Computers. *J. Soc. Issues* **2000**, *56*, 81–103. [CrossRef]
85. Petty, R.; Cacioppo, J. Source Factors and the Elaboration Likelihood Model of Persuasion. *Adv. Consum. Res. Assoc. Consum. Res.* **1984**, *11*, 668–672.
86. Schneider, C.; Weinmann, M.; Brocke, J.V. Digital Nudging: Guiding Online User Choices through Interface Design. *Commun. ACM* **2018**, *61*, 67–73. [CrossRef]
87. Sunstein, C.R. Nudging: A Very Short Guide. *J. Consum. Policy* **2014**, *37*, 583–588. [CrossRef]
88. Weinmann, M.; Schneider, C.; Brocke, J.V. Digital Nudging. *Bus. Inf. Syst. Eng.* **2016**, *58*, 433–436. [CrossRef]
89. Kahneman, D.; Knetsch, J.L.; Thaler, R.H. Anomalies: The Endowment Effect, Loss Aversion, and Status Quo Bias. *J. Econ. Perspect.* **1991**, *5*, 193–206. [CrossRef]
90. Adam, M.; Klumpe, J. Onboarding with a Chat—The effects of Message Interactivity and Platform Self-Disclosure on User Disclosure propensity. In Proceedings of the European Conference on Information Systems (ECIS), Stockholm & Uppsala, Sweden, 8–14 June 2019.
91. Benlian, A. Web Personalization Cues and Their Differential Effects on User Assessments of Website Value. *J. Manag. Inf. Syst.* **2015**, *32*, 225–260. [CrossRef]
92. Fleischmann, M.; Amirpur, M.; Grupp, T.; Benlian, A.; Hess, T. The role of software updates in information systems continuance—An experimental study from a user perspective. *Decis. Support Syst.* **2016**, *83*, 83–96. [CrossRef]
93. Wessel, M.; Adam, M.; Benlian, A. The impact of sold out early birds on option selection in reward-based crowdfunding. *Decis. Support Syst.* **2019**, *117*, 48–61. [CrossRef]
94. Kerr, M.A.; McCann, M.T.; Livingstone, M.B. Food and the consumer: Could labelling be the answer? *Proc. Nutr. Soc.* **2015**, *74*, 158–163. [CrossRef]
95. Borgi, L. Does Menu Labeling Lead to Healthier Food Choices? Harvard Medical School: Boston, MA, USA, 2018.
96. Mirsch, T.; Lehrer, C.; Jung, R. Digital Nudging: Altering User Behavior in Digital Environments. In Proceedings of the 13th International Conference on Wirtschaftsinformatik, St. Gallen, Switzerland, 12–15 February 2017; pp. 634–648.
97. Ioannou, A.; Tussyadiah, I.; Miller, G.; Li, S.; Weick, M. Privacy nudges for disclosure of personal information: A systematic literature review and meta-analysis. *PLoS ONE* **2021**, *16*, e0256522. [CrossRef]
98. Klumpe, J. Social Nudges as Mitigators in Privacy Choice Environments. Ph.D. Thesis, Technische Universität Darmstadt, Darmstadt, Germany, 2020.
99. Zhang, B.; Xu, H. Privacy Nudges for Mobile Applications: Effects on the Creepiness Emotion and Privacy Attitudes. In Proceedings of the 19th ACM Conference on Computer-Supported Cooperative Work, San Francisco, CA, USA, 27 February–2 March 2016; pp. 1676–1690.
100. Cialdini, R.B.; Goldstein, N.J. Social Influence: Compliance and Conformity. *Annu. Rev. Psychol.* **2004**, *55*, 591–621. [CrossRef]

101. Brave, S.; Nass, C. Emotion in Human–Computer Interaction. In *The Human-Computer Interaction Handbook: Fundamentals, Evolving Technologies and Emerging Applications*; CRC Press: Boca Raton, FL, USA, 2002. [CrossRef]
102. Rapp, A.; Curti, L.; Boldi, A. The human side of human-chatbot interaction: A systematic literature review of ten years of research on text-based chatbots. *Int. J. Hum.-Comput. Stud.* **2021**, *151*, 102630. [CrossRef]
103. Cosby, S.; Sénécal, S.; Léger, P.M. The Impact of Online Product and Service Picture Characteristics on Consumers' Perceptions and Intentions. Ph.D. Thesis, HEC Montréal, Montréal, QC, Canada, 2020.
104. Russell, J.A. A circumplex model of affect. *J. Personal. Soc. Psychol.* **1980**, *39*, 1161–1178. [CrossRef]
105. Clore, G.L.; Parrott, G. Moods and their vicissitudes: Thoughts and feelings as information. In *Emotion and Social Judgments*; Forgas, J.P., Ed.; Pergamon Press: Elmsford, NY, USA, 1991; pp. 107–123.
106. Schwarz, N.; Clore, G.L. How do I feel about it? The informative function of affective states. In *Affect, Cognition, and Social Behavior*; Fiedler, I.K., Forgas, J.P., Eds.; Hogrefe: Gottingen, Germany, 1988; pp. 44–62.
107. Paulhus, D.L.; Lim, T.K. Arousal and evaluative extremity in social judgments: A dynamic complexity model. *Eur. J. Soc. Psychol.* **1994**, *24*, 89–100. [CrossRef]
108. Pérez-Marín, D.; Pascual-Nieto, I. An exploratory study on how children interact with pedagogic conversational agents. *Behav. Inf. Technol.* **2013**, *32*, 955–964. [CrossRef]
109. Xu, A.; Liu, Z.; Guo, Y.; Sinha, V.; Akkiraju, R. A New Chatbot for Customer Service on Social Media. In Proceedings of the 2017 CHI Conference on Human Factors in Computing Systems, Denver, CO, USA, 6–11 May 2017; pp. 3506–3510.
110. Gaffey, A.E.; Wirth, M.M. Physiological Arousal. In *Encyclopedia of Quality of Life and Well-Being Research*; Michalos, A.C., Ed.; Springer: Dordrecht, The Netherlands, 2014; pp. 4807–4810. [CrossRef]
111. Cannon, W.B. *Bodily Changes in Pain, Hunger, Fear and Rage: An Account of Recent Researches into the Function of Emotional Excitement*; Appleton: New York, NY, USA, 1915.
112. Miniard, P.W.; Sirdeshmukh, D.; Innis, D.E. Peripheral persuasion and brand choice. *J. Consum. Res.* **1992**, *19*, 226–239. [CrossRef]
113. Knapp, H.; Kirk, S.A. Using pencil and paper, Internet and touch-tone phones for self-administered surveys: Does methodology matter? *Comput. Hum. Behav.* **2003**, *19*, 117–134. [CrossRef]
114. Rudnicka, A.; Cox, A.L.; Gould, S.J.J. Why Do You Need This? Selective Disclosure of Data Among Citizen Scientists. In Proceedings of the 2019 CHI Conference on Human Factors in Computing Systems, Glasgow, UK, 4–9 May 2019; p. 392.
115. Schmidt, A.; Engelen, B. The ethics of nudging: An overview. *Philos. Compass* **2020**, *15*, e12658. [CrossRef]
116. Becker, M.; Matt, C.; Hess, T. It's Not Just About the Product: How Persuasive Communication Affects the Disclosure of Personal Health Information. *SIGMIS Database* **2020**, *51*, 37–50. [CrossRef]
117. Wakefield, R. The influence of user affect in online information disclosure. *J. Strateg. Inf. Syst.* **2013**, *22*, 157–174. [CrossRef]
118. Coker, B.; McGill, A.L. Arousal increases self-disclosure. *J. Exp. Soc. Psychol.* **2020**, *87*, 103928. [CrossRef]
119. Riedl, R.; Léger, P.-M. *Fundamentals of NeuroIS: Information Systems and the Brain*; Springer: Berlin/Heidelberg, Germany, 2016.
120. Prince, J.; Wallsten, S. How much is privacy worth around the world and across platforms? *J. Econ. Manag. Strategy* **2022**, *31*, 841–861. [CrossRef]

MDPI
St. Alban-Anlage 66
4052 Basel
Switzerland
www.mdpi.com

Applied Sciences Editorial Office
E-mail: applsci@mdpi.com
www.mdpi.com/journal/applsci

Disclaimer/Publisher's Note: The statements, opinions and data contained in all publications are solely those of the individual author(s) and contributor(s) and not of MDPI and/or the editor(s). MDPI and/or the editor(s) disclaim responsibility for any injury to people or property resulting from any ideas, methods, instructions or products referred to in the content.

www.ingramcontent.com/pod-product-compliance
Lightning Source LLC
LaVergne TN
LVHW070141100526
838202LV00015B/1866